BIOSALINITY IN ACTION:
BIOPRODUCTION WITH SALINE WATER

Developments in Plant and Soil Sciences

Volume 17

1. J. Monteith and C. Webb, eds., Soil Water and Nitrogen in Mediterranean-type Environments. 1981. ISBN 90-247-2406-6
2. J.C. Brogan, ed., Nitrogen Losses and Surface Run-off from Landspreading of Manures. 1981. ISBN 90-247-2471-6
3. J.D. Bewley, ed., Nitrogen and Carbon Metabolism. 1981. ISBN 90-247-2472-4
4. R. Brouwer, I. Gašparíková, J. Kolek and B.C. Loughman, eds., Structure and Function of Plant Roots. 1981. ISBN 90-247-2510-0
5. Y.R. Dommergues and H.G. Diem, eds., Microbiology of Tropical Soils and Plant Productivity. 1982. ISBN 90-247-2624-7
6. G.P. Robertson, R. Herrera and T. Rosswall, eds., Nitrogen Cycling in Ecosystems of Latin America and the Caribbean. 1982. ISBN 90-247-2719-7
7. D. Atkinson et al., eds., Tree Root Systems and their Mycorrhizas. 1983. ISBN 90-247-2821-5
8. M.R. Sarić and B.C. Loughman, eds., Genetic Aspects of Plant Nutrition. 1983. ISBN 90-247-2822-3
9. J.R. Freney and J.R. Simpson, eds., Gaseous Loss of Nitrogen from Plant-Soil Systems. 1983. ISBN 90-247-2820-7
10. United Nations Economic Commission for Europe. Efficient Use of Fertilizers in Agriculture. 1983. ISBN 90-247-2866-5
11. J. Tinsley and J.F. Darbyshire, eds., Biological Processes and Soil Fertility. 1984. ISBN 90-247-2902-5
12. A.D.L. Akkermans, D. Baker, K. Huss-Danell and J.D. Tjepkema, eds., *Frankia* Symbioses. 1984. ISBN 90-247-2967-X
13. W.S. Silver and E.C. Schröder, eds., Practical Application of Azolla for Rice Production. 1984. ISBN 90-247-3068-6
14. P.G.L. Vlek, ed., Micronutrients in Tropical Food Crop Production. 1985. ISBN 90-247-3085-6
15. T.P. Hignett, ed., Fertilizer Manual. 1985. ISBN 90-247-3122-4
16. D. Vaughan and R.E. Malcolm, eds., Soil Organic Matter and Biological Activity. 1985. ISBN 90-247-3154-2
17. D. Pasternak and A. San Pietro, eds., *Biosalinity in Action: Bioproduction with Saline Water* 1985. ISBN 90-247-3159-3.
18. M. Lalonde, C. Camiré and J.O. Dawson, eds., *Frankia* and Actinorhizal Plants. 1985. ISBN 90-247-3214-X

Biosalinity in Action: Bioproduction with Saline Water

Edited by

D. PASTERNAK
The Institutes for Applied Research
Ben Gurion University of the Negev
Beer-Sheva
Israel

and

A. SAN PIETRO
Department of Biology
Indiana University
Bloomington, Indiana
U.S.A.

First published as *Plant and Soil*, Vol. 89 (1985)

1985 **MARTINUS NIJHOFF PUBLISHERS**
a member of the KLUWER ACADEMIC PUBLISHERS GROUP
DORDRECHT / BOSTON / LANCASTER

Proceedings of the Third International Workshop on Biosaline Research, Beer-Sheva, Israel, 19–23 March, 1984

Distributors

for the United States and Canada: Kluwer Academic Publishers, 190 Old Derby Street, Hingham, MA 02043, USA
for the UK and Ireland: Kluwer Academic Publishers, MTP Press Limited, Falcon House, Queen Square, Lancaster LA1 1RN, UK
for all other countries: Kluwer Academic Publishers Group, Distribution Center, P.O. Box 322, 3300 AH Dordrecht, The Netherlands

Library of Congress Catalogue Card Number: 85-15366

Library of Congress Cataloging-in-Publication Data

International Workshop on Biosaline Research (3rd : 1984 : Beersheba, Israel)
Biosalinity in action.

(Developments in plants and soil sciences ; v. 17)
1. Biosaline resources--Congresses. 2. Salt-tolerant crops--Congresses. 3. Algae culture--Congresses. 4. Mariculture--Congresses. 5. Halophytes--Congresses. I. Pasternak, Dov, 1940- . II. San Pietro, Anthony Gordan, 1922- . III. Title. IV. Series.
S620.I57 1984 631.5'8 85-21623

ISBN-13: 978-94-010-8759-9 e-ISBN-13: 978-94-009-5111-25
DOI: 10.1007/978-94-009-5111-2

Contents

Section 4: Seawater agriculture

Section 5: Mariculture

Preface

Historically, scientists and laymen have regarded salinity as a hazardous, detrimental phenomenon. This negative view was a principal reason for the lack of agricultural development of most arid and semi-arid zones of the world where the major sources of water for biological production are saline.

The late Hugo Boyko was probably the first scientist in recent times to challenge this commonly held, pessimistic view of salinity. His research in Israel indicated that many plants can be irrigated with saline water, even at seawater strength, if they are in sandy soil – a technique that could open much barren land to agriculture. This new, even radical, approach to salinity was clearly enunciated in the book he edited and most appropriately entitled 'Salinity and Aridity: New Approaches to Old Problems' (1966).

A decade later, three members of the United States National Science Foundation (NSF), Lewis Mayfield, James Aller and Oskar Zaborsky, formulated the 'Biosaline Concept'; namely, that *poor soils, high solar insolation and saline water, which prevail in arid lands, should be viewed as useful resources rather than as disadvantages,* and that these resources can be used for *non-traditional production of food, fuels and chemicals.*

The First International Workshop on Biosaline Research was convened at Kiawah Island, South Carolina, in 1977 by A. San Pietro. This workshop, which was supported by the NSF, focussed primarily on an assessment of the state of the science in Biosaline Research. It permitted a bold exchange on a wide scientific front by many researchers, but little information on actual applications could yet be reported. The proceedings of this workshop were published in 1979 under the title 'The Biosaline Concept: An Approach to the Utilization of Underexploited Resources,' edited by A. Hollaender *et al.*

The Second International Workshop on Biosaline Research was convened in La Paz, Mexico, in 1980, again with primary support from the NSF but also with partial support and the full approval of the Mexican Council for Science and Technology (CONACYT). Mexico was chosen specifically as the workshop site because it is a country that could benefit greatly from the application of Biosaline Research. The proceedings of the Second Workshop were edited by A. San Pietro in a volume entitled 'Biosaline Research: A Look to the Future' (1982).

The Third International Workshop on Biosaline Research was convened in Beer-Sheva, Israel, in March, 1984. This workshop was supported by the Israeli National Council for Research and Development and hosted by Ben-Gurion University of the Negev. The site of the Third Workshop was ideal for two reasons: (1) the Middle East is a region where application of the Biosaline Concept can reap tremendous benefits; and, more importantly, (2) Israel is already far advanced in Biosaline Research and can serve as a proven model for other arid-zone countries. While the first two workshops addressed mainly the science of Biosaline Research, the Third Workshop was dedicated to the practical application of the Biosaline Concept.

This volume is the proceedings of the Third Workshop and is divided into five sections. The first presents the current state of knowledge on the mechanisms of salt tolerance in algae and terrestrial plants, at all levels of cellular organization. It is encouraging to note the overall agreement among the contributors to this section on the basic mechanisms of salt tolerance, which testifies to the major advances made in recent years towards an understanding of these processes. The relatively small sacrifice in energy required of halotolerant algae and plants to counteract the low osmotic potential of their growth media provides a good explanation for the high productivity of these organisms under seawater salinity, as demonstrated in subsequent sections.

The second section provides up-to-date information on production and utilization of algae as sources of both foods and valuable chemicals. The production of algae in controlled and managed aquaculture is an infant industry. One cannot but admire the rapid scientific and technological progress that has taken place and support the authors in their imaginative speculation on the potential of this new field.

Crop production by irrigation with brackish water (water of relatively low salinity, *ca* 500–5000 ppm dissolved salts) is the subject of the third section. This is not a new concept and is now practiced in many parts of the world. To date, much of the scientific work in this field has been dedicated to the management of salt and water in the soil to minimize salt stress at the root level. Alternative approaches, such as breeding and selection for salt-tolerant crops, have never been attempted in great depth. We therefore decided to devote a relatively large part of this section to the genetic approach. Other aspects reviewed in the third section are water management, crop management, crop nutrition, and the effects of salinity on the quality of agricultural products.

The fourth section presents research currently being pursued at three centers, two in the United States and one in Israel, on irrigation of halophytes with seawater. All three groups report that the yields of organic biomass from seawater-irrigated halophytes are comparable to those obtained with conventional crops irrigated with nonsaline water. Fodder might become the first economical halophytic crop. However, the salt in the fodder may have detrimental effects on animals eating it. This implies that more work should be directed to yet another area of Biosaline Research – the responses of animals to high salt concentrations in the diet.

The intensive production of fish with seawater is the subject of the final section of this volume. The workshop organizers intended to present the excellent work that is being carried out on the mass propagation and culture of plankton organisms in several nations and the original methods developed in Israel for pond culture of marine fish along desert coasts. Circumstances beyond our control prevented the presentation of these studies. We are fortunate, however, to be able to include the comprehensive review by Professor T. Watanabe on seawater fish production in Japan, which is by far the leading country in the world today in research and implementation of this important area of Biosalinity.

We hope that the publication of this volume will accelerate the advance of Biosaline Research. We would like to see the data presented here being used not only by scientists but also by planners and decision-making authorities as a valuable aid with which to promote the development of the world's arid zones.

Beer-Sheva, Israel
Bloomington, USA
November 1984

DOV PASTERNAK
ANTHONY SAN PIETRO

To Hugo Boyko, who started it all

Organizing Committee

Shoshana (Malis) Arad, The Boyko Institute for Agriculture and Applied Biology, The Institutes for Applied Research, Ben-Gurion University of the Negev, Beer-Sheva

Ami Ben-Amotz, Israel Oceanographic and Limnological Research, Tel-Shikmona, Haifa

Shulamit Cahana, National Council for Research and Development, Jerusalem

Yona Chen, The Faculty of Agriculture, Hebrew University of Jerusalem, Rehovot

Baruch Eyal, National Council for Research and Development, Jerusalem

Rami Keren, The Institute of Soils and Water, Agricultural Research Organization, The Volcani Center, Bet-Dagan

Dov Pasternak, The Boyko Institute for Agriculture and Applied Biology, The Institutes for Applied Research, Ben-Gurion University of the Negev, Beer-Sheva

Anthony San Pietro (honorary member), Department of Biology, Indiana University, Bloomington, Indiana, USA

Miriam Waldman, National Council for Research and Development, Jerusalem

Acknowledgements

The Third International Workshop on Biosaline Research was supported by the National Council for Research and Development of the State of Israel and hosted by Ben-Gurion University of the Negev.

The editors and authors would like to express their gratitude to Dr. Marjorie A. Tiefert for her competent, professional aid in editing and preparing this book for publication.

List of contributors

Numbers in parentheses indicate the page on which the author's contribution begins.

Michel Adda (117), The Institutes for Applied Research, Ben-Gurion University of the Negev, P.O. Box 1025, Beer-Sheva 84110, Israel

Shoshana (Malis) Arad (117), The Institutes for Applied Research, Ben-Gurion University of the Negev, P.O. Box 1025, Beer-Sheva 84110, Israel

James Alon Aronson (337), The Institutes for Applied Research, Ben-Gurion University of the Negev, P.O. Box 1025, Beer-Sheva 84110, Israel

Gozal Ben-Hayyim (243), Institute of Horticulture, Agricultural Research Organization, The Volcani Center, Bet-Dagan 50250, Israel

Roger W. Benjamin (337), The Agricultural Research Organization, Gilat Station, Negev Mobile Post, Israel

Ephraim Cohen (117), The Institutes for Applied Research, Ben-Gurion University of the Negev, P.O. Box 1025, Beer-Sheva 84110, Israel

Avihai Danon (337), The Institutes for Applied Research, Ben-Gurion University of the Negev, P.O. Box 1025, Beer-Sheva 84110, Israel

Emanuel Epstein (183), Department of Land, Air and Water Resources, University of California, Davis, CA 95616, USA

Amos Feigin (285), Institute of Soils and Water, Agricultural Research Organization, The Volcani Center, P.O. Box 6, Bet-Dagan 50250, Israel

T. J. Flowers (41), School of Biological Sciences, University of Sussex, Falmer, Brighton, BN1 9QG Sussex, UK

Rachel Gabbay (107), Department of Agricultural Botany, The Hebrew University of Jerusalem, Rehovot 76100, Israel

Joseph Gale (57), The Institute for Desert Research, Ben-Gurion University of the Negev, Sede Boker 84990, Israel

John L. Gallagher (323), College of Marine Studies, University of Delaware, Lewes, DE 19958, USA

E. P. Glenn (311), Environmental Research Laboratory, University of Arizona, Tucson International Airport, Tucson, AZ 85706, USA

John Gorham (15), Centre for Arid Zone Studies, School for Agriculture, Forestry and Allied Sciences, University College of North Wales, Bangor, Gwynedd, Wales LL57 2UW, UK

Johan A. Hellebust (69), Department of Botany, University of Toronto, Toronto, Ontario M5S 1A1, Canada

H. R. Lerner (3), Department of Botany, The Hebrew University of Jerusalem, Jerusalem 91904, Israel

Yaakov Lipkin (159), Department of Botany, The George S. Wise Faculty of Life Sciences, Tel-Aviv University, Tel-Aviv 69978, Israel

Eugene V. Maas (273), United States Department of Agriculture, Agricultural Research Service, U.S. Salinity Laboratory, 4500 Glenwood Dr., Riverside, CA 92501, USA

E. McDonnell (15), Department of Biochemistry and Soil Science, University College of North Wales, Bangor, Gwynedd, Wales LL57 2UW, UK

J. McLachlan (137), Atlantic Research Laboratory, National Research Council of Canada, 1411 Oxford St., Halifax, N.S., B3H 3Z1, Canada

A. Meiri (253), Institute of Soils and Water, Agricultural Research Organization, The Volcani Center, P.O. Box 6, Bet-Dagan 50250, Israel

Yosef Mizrahi (301), The Institutes for Applied Research, Ben-Gurion University of the Negev, P.O. Box 1025, Beer-Sheva 84110, Israel

D. J. Moore (97), Department of Biological Sciences, The University of Dundee, Dundee DD1 4HN, Scotland, UK

James W. O'Leary (311), Environmental Research Laboratory, University of Arizona, Tucson International Airport, Tucson, AZ 85705, USA

Leslie G. Paleg (83), Department of Plant Physiology, Waite Agricultural Research Institute, The University of Adelaide, Glen Osmond, South Australia 5064, Australia

Dov Pasternak (ix, 301, 337), The Institutes for Applied Research, Ben-Gurion University of the Negev, P.O. Box 1025, Beer-Sheva 84110, Israel

Z. Plaut (253), Institute of Soils and Water, Agricultural Research Organization, The Volcani Center, Bet-Dagan 50250, Israel

R. H. Reed (97), Department of Biological Sciences, The University of Dundee, Dundee DD1 4HN, Scotland, UK

D. L. Richardson (97), Department of Biological Sciences, The University of Dundee, Dundee DD1 4HN, Scotland, UK

Amos Richmond (129), The Institute for Desert Research, Ben-Gurion University of the Negev, Sede Boker 84990, Israel

Anthony San Pietro (ix), Department of Biology, Indiana University, Bloomington, IN 47405, USA

Michael C. Shannon (227), United States Department of Agriculture, Agricultural Research Service, U.S. Salinity Laboratory, 4500 Glenwood Dr., Riverside, CA 92501, USA

P. Spiegel-Roy (243), Institute of Horticulture, Agricultural Research Organization, The Volcani Center, Bet-Dagan 50250, Israel

R. Starr (83), Department of Plant Physiology, Waite Agricultural Research Institute, University of Adelaide, Adelaide, South Australia 5064, Australia

G. R. Stewart (83), Department of Botany, Birbeck College, University of London, Malet St., London WC1E 7HX, UK

W. D. P. Stewart (97), A.F.R.C. Research Group on Cyanobacteria, The University of Dundee, Dundee DD1 4HN, Scotland, UK

Moshe Tal (199), Department of Biology, Ben-Gurion University of the Negev, P.O. Box 653, Beer-Sheva 84105, Israel

Elisha Tel-Or (107), Department of Agricultural Botany, The Hebrew University of Jerusalem, Rehovot 76100, Israel

Avigad Vonshak (129), The Institute for Desert Research, Ben-Gurion University of the Negev, Sede Boker 84990, Israel

S. R. C. Warr (97), Department of Biological Sciences, The University of Dundee, Dundee DD1 4HN, Scotland, UK

Takeshi Watanabe (351), Laboratory of Fish Nutrition, Department of Aquaculture, Tokyo University of Fisheries, Konan 4, Minato-ku, Tokyo 108, Japan

M. C. Watson (311), Environmental Research Laboratory, University of Arizona, Tucson International Airport, Tucson, AZ 85706, USA

R. G. Wyn Jones (15), Department of Biochemistry and Soil Science, University College of North Wales, Bangor, Gwynedd, Wales LL57 2UW, UK

M. Zeroni (57), The Institute for Desert Research, Ben-Gurion University of the Negev, Sede Boker 84990, Israel

Section 1

Mechanisms of salt tolerance in algae and terrestrial plants

Plant and Soil 89, 3–14 (1985). Ms. BR 01

Adaptation to salinity at the plant cell level

H. R. LERNER
Department of Botany, The Hebrew University of Jerusalem, Jerusalem, 91904, Israel

Key words Adaptation to stress Salinity Suspension culture of higher plant cells

Summary Various mechanisms of adaptation of plant cells to salinity are reviewed: (1) protection of enzymes and maintenance of turgor by organic solutes; (2) prevention of ion toxicity by compartmentation; and (3) energization of solute transport by the proton pump. All these mechanisms seem to play a role in adaptation. The particular advantages of using salt-adapted cells in suspension culture to identify mechanisms of adaptation are pointed out.

Introduction

It is logical to imagine that plants growing in saline environments exploit various strategies at the whole plant as well as at the cell level that allow them to overcome the salinity stress, which has two components, ion toxicity and osmotic stress. Our understanding of these mechanisms of survival would be extremely valuable in the development of salt-resistant crops. With this in view, during the past 30 years great efforts have been invested in studying the physiological differences between salt-resistant and salt-sensitive plants (for recent reviews see [5, 8, 12, 18, 22, 27, 29]).

At the plant cell level a few strategies can be imagined for survival in a saline environment. Salt may be taken up by the cell and used as a major osmoticum, and ion toxicity could be prevented by compartmentation or by some other mechanism of protecting enzyme and ribosomal activities. Another strategy would be pumping out of the cell most of the salt that diffuses in; in this case osmotic stress would be relieved by accumulation of organic solutes to maintain turgor. It is most probable that in most cases cells utilize both of these strategies to various extents.

We have been studying different aspects of adaptation of plant cells to salinity to gain some insight into the mechanisms functioning at the cellular level. These studies have dealt with (1) the effect of the osmoregulating solute glycerol on the conformation of dihydroxyacetone kinase (DHAK); (2) the compartmentation of Na^+ and K^+ in cells in

Abbreviations; DHA, dihydroxyacetone; DHAK, dihydroxyacetone kinase; $\Delta\psi$, difference of potential across a membrane; ΔpH, difference of pH across a membrane; SDS, sodium dodecyl sulfate; O.P., osmotic pressure; ABA, abscisic acid and FCCP, p-trifluoromethoxyphenylhydrazone.

suspension culture; (3) a comparison of the plasmalemma proton pump from the halophyte *Atriplex* with similar pumps from glycophytes; and (4) adaptation of cells in suspension culture to salinity.

Effect of an osmoregulating solute on enzyme conformation

During a search for the metabolic pathway of the osmoregulating solute in Dunaliella it became evident that DHAK is responsible for the exit reaction from the glycerol pool [13, 16]. This enzyme was stable in the crude centrifuged extract but rapidly lost activity when purified more than three fold. Attempts to stabilize DHAK in the crude preparation against thermal inactivation at 50°C showed that 2*M* glycerol provided protection (Fig. 1). In the presence of 2*M* glycerol the enzyme was sufficiently stable to be purified 134 fold by conventional methods.

Citri[2] has reviewed that evidence showing that changes in an enzyme's conformation modify its stability. A suggestion that glycerol may induce a conformational change in DHAK is provided by the change in the kinetics of the reaction – in water, DHAK shows substrate (DHA) inhibition, which disappears in the presence of 2*M* glycerol (Fig. 2).

As Paleg has proposed (this volume, [19, 21]), it seems that osmoregulating solutes affect the conformation of enzymes, thereby

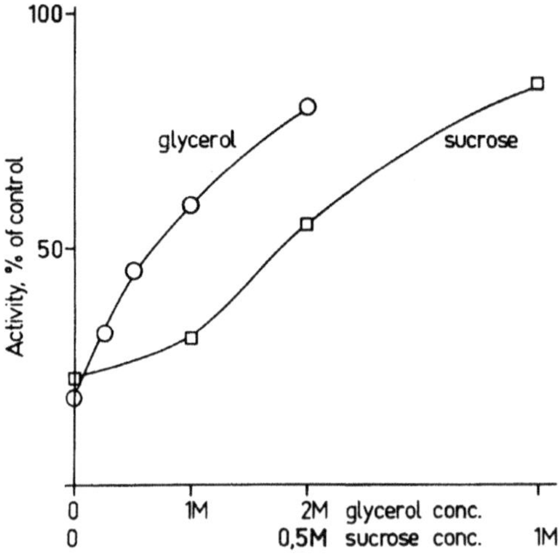

Fig. 1. Glycerol and sucrose protection of dihydroxyacetone kinase (DHAK) against thermal inactivation. Crude DHAK was incubated in the presence or absence of glycerol or sucrose for 60 min at 50°C and then assayed [13].

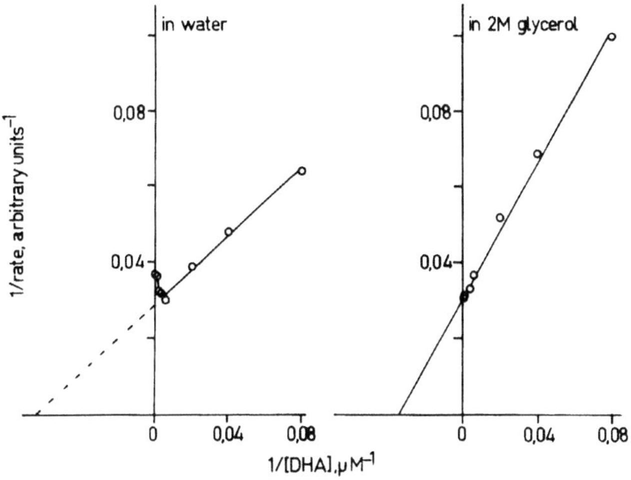

Fig. 2. Double-reciprocal plots of dihydroxyacetone phosphorylation catalyzed by crude DHAK in the absence and presence of 2 *M* glycerol. Both experiments were performed under identical conditions [13].

stabilizing their active conformation, and in this way protect enzymes against conformational perturbations caused by mineral ions. The mechanism of this stabilization has not been demonstrated; one hypothesis is that the osmoregulating solutes affect the state of the water in the cell, which would modify protein-water interactions [28]. It seems, in conclusion, that osmoregulating solutes not only maintain turgor and cytosol volume but also protect enzymes.

Study of solute compartmentation

To measure solute compartmentation in plant cells we have developed a method for isolation of the low-molecular-weight solutes present in the cytosol[15]. Nicotiana cells from suspension culture are incubated with an appropriate amount of poly-L-lysine (140 residues long). This treatment selectively induces pore formation in the plasmalemma while not affecting the membranes of the organelles. The solutes remaining in the cells represent mainly those present in the vacuole. After quantitative determination of the solutes leaked to the medium and those remaining in the cells it is possible to calculate the approximate concentrations in the cytosol and vacuole if the relative volumes of these two compartments are known. In our calculations we assumed that the cytoplasm represents about 5 percent and the vacuole about 80 percent of the total cell volume. Addition of 85 m*M* NaCl to the growth

Table 1. Compartmentation of K^+ and Na^+ in *Nicotiana tabacum* cells that were not preadapted to salinity

Experiment 1		Experiment 2	
Growth medium at inoculation time			
K^+	25 m*M*	25 m*M*	
Na^+	0.5 m*M*	85 m*M*	
π	140 mOsm	300 mOsm	
O.P. of medium after 7 days of growth			
π	40 mOsm	200 mOsm	
	K^+	K^+	Na^+
Cytosol:	224 m*M*	128 m*M*	227 m*M*
Vacuole:	106 m*M*	25 m*M*	105 m*M*

Selectivity coefficient for K^+/Na^+ cytosol/vacuole:

$$K_c^+ : Na_c^+ / K_v^+ : Na_v^+ = 2.4$$

Selectivity coefficient for K^+/Na^+ cytosol/medium:

$$K_c^+ : Na_c^+ / K_o^+ : Na_o^+ = 1.9$$

In experiment 1, cells of *Nicotiana tabacum* cv Xanthi were grown in Gamborg Medium B5. In experiment 2, the cells were grown in the same medium with 85 m*M* NaCl added. Cells were collected 7 days after inoculation, in the late exponential growth phase. Cytoplasmic K^+ and Na^+ were extracted by incubating cells for 20 min with poly-L-lysine: salt-grown cells with 350 μg poly-L-lysine/ml packed cells, non-salt-grown cells with 550 μg poly-L-lysine/ml packed cells. Subscripts: c, cytosol; v, vacuole; o, outside of cell. No values are given for cytosol or vacuole Na^+ in Expt. 1 because these values were very low and within the noise of the method.

medium causes a decrease of 43% in K_c (cytosol) while monovalent cations ($K^+ + Na^+$) in the cytosol increase from 224 m*M* to 355 m*M* (Table 1). This is a $K^+ + Na^+$ increase of 58% in the cytosol for a 340% increase in the medium, while in the vacuole there is only a 23% increase. The selectivity for K^+/Na^+ for cytosol/vacuole compartmentation is 2.4 while the selectivity for K^+/Na^+ for uptake of K^+ in the cytosol from the medium is 1.9 (Table 1). These results indicate a limited selectivity for K^+ accumulation in the cytosol. However, they do not indicate a preferential accumulation of inorganic ions in the vacuole. We know that cell suspensions of Nicotiana can be adapted to salinity and that this adaptation is a slow process requiring growth in a saline environment for many generations (see section on Adaptation of cells in suspension culture to salinity). The cells used in this experiment were grown for only about 4 generations in 85 m*M* NaCl and may be showing the lack of preferential compartmentation of inorganic ions in cells not adapted to a saline environment.

Atriplex plasmalemma proton pump

In plant cells the transport of many solutes across membranes is believed to be energized by a proton gradient ($\Delta\tilde{\mu}H^+ = \Delta\psi + 58.8\,\Delta\,pH$) which is formed by pumping protons out of the cytoplasm, making this compartment more negative ($\Delta\psi$) and more basic (Δ pH) than the free space or the vacuole. The $\Delta\psi$ and Δ pH are potential energies, where the electrical gradient can provide energy for cation uptake while the $\Delta\tilde{\mu}H^+$ can drive solute uptake by solute/proton cotransport and solute efflux by solute/proton antiport. Indications for the presence of a proton pump in membrane vesicles can be obtained by observing the setting up of a $\Delta\psi$ or Δ pH across the membrane or, more indirectly, by measuring ATPase activity.

Compartmentation of mineral ions away from the cytosol in cells of plants accumulating large amounts of salt should give rise to a large mineral ion gradient across both the plasmalemma and tonoplast. It may therefore be expected that these cells have mechanisms for pumping the mineral ions from the cytosol outwards to the free space and inwards to the vacuole. Na^+ could be driven out of the cytoplasm by H^+/Na^+ antiport at the plasmalemma and at the tonoplast, using the H^+ gradient as an energy source [17].

A Na^+/K^+ exchange mechanism has also been proposed ([10] and references cited therein). However, in this case the source of energy for driving the exchange in a particular direction is not evident, since the $[K^+]_{in}/[K^+]_{out}$ ratio has frequently been reported to be close to that predicted for passive distribution according to the Nernst equilibrium. Therefore, an additional source of energy would be required to drive Na^+/K^+ exchange.

It should be noted that, while studying the effect of NaCl on Photosystem II in isolated thylakoids, Critchley[3] did not find any evidence for NaCl compartmentation away from the chloroplasts in the various halophytes studied; that is, PS II showed maximum activity between 200 and 500 m*M* NaCl (higher concentrations were not tested).

To better understand the physiology of the halophylic cell, membrane function was studied with membrane vesicles isolated from halophytes and compared to those from glycophytes. In preliminary studies we compared the effects of salt on root membrane ATPases prepared from the salt-sensitive pea with those from the halophyte *Atriplex nummularia*. This study shows that increasing NaCl concentrations inhibit the MgATPase activity while at the same time stimulating the monovalent cation-activated portion of the activity. The inhibition outweighs the stimulation at far lower NaCl concentrations for the pea than for Atriplex (see Figs. 1–4 in [14]).

We purified Atriplex plasmalemma fractions by centrifugation of the root membrane preparation on a sucrose gradient. The ATPase observed in these fractions is in fact an ATPase rather than a non-specific phosphatase, because the preparation shows practically no other phosphatase activity, except for considerable hydrolysis of pyrophosphate (see Fig. 5 in [14]). Stimulation by KCl is found only for the MgATPase activity (see Table 1 in [14]) and not for the pyrophosphatase activity.

Incubation of the Atriplex root plasmalemma preparation with the lipophylic radioactive anion thiocyanate (SCN^-) together with ATP, followed by filtration on a Millipore filter, showed SCN^- accumulation, indicating the formation of an electrical potential gradient ($\Delta\psi$) across the membrane (positive inside). This ATP-dependent $\Delta\psi$ is dissipated by FCCP, showing that it depends on a proton gradient. The preparation therefore contains H^+-tight inverted plasmalemma vesicles with functional proton pumps that can be fueled from the outside by ATP (Fig. 3). The pump shows a broad peak of activity between pH 6 and 7 which drops at higher pH values, and its apparent K_m for ATP is about 0.3 m*M*. So far we have not detected any essential difference between the Atriplex plasmalemma proton pump and the proton pumps isolated from glycophytes in other laboratories.

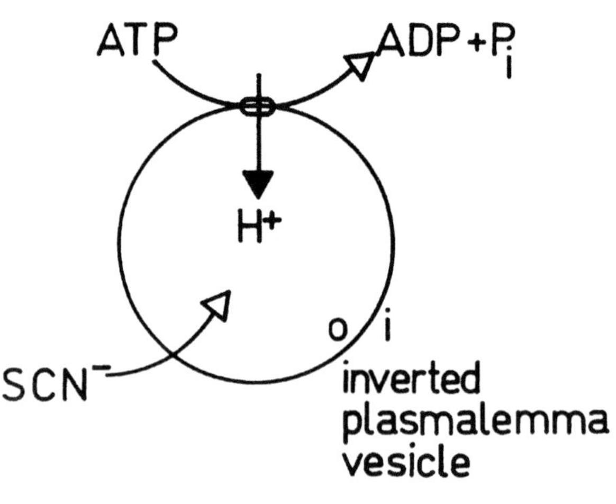

Fig. 3. Diagram representing an inverted plasmalemma vesicle showing proton transport coupled to ATPase activity. The ATP-dependent proton gradient (positive inside) induces accumulation of the lipid-soluble thiocyanate (SCN^-) anion. Outer side of plasmalemma = o; cytoplasmic side of plasmalemma = i.

Adaptation of cells in suspension culture to salinity

Salt-sensitive cells in suspension culture can in some cases be adapted to grow at high levels of salinity and therefore can serve as useful models for studying mechanisms of resistance to salinity at the cell level[9]. This approach has the particular advantage of providing two closely related cell lines which can be compared and which presumably differ mainly in their resistance toward salinity. Furthermore, in cell suspension culture the level of salinity to which the cells are exposed can be accurately and conveniently controlled.

We have adapted a salt-sensitive Nicotiana cell line to grow at high levels of salinity[26]. When the salt-sensitive cells are transferred to media containing increasing concentrations of NaCl the lag phase of the growth curve increases with increased NaCl concentration. Above a certain concentration, about 200 mM NaCl, the cells die soon after transfer. The cells were adapted to salinity by growing them in NaCl concentrations increasing by increments of 35 mM for about 10 generations at each NaCl level. Today these adapted cells are growing in 500 mM NaCl. They can be brought down stepwise to zero NaCl concentration, grown for a few generations without NaCl and then in one step transferred back to 500 mM NaCl, where they grow perfectly well without showing any increase in the lag phase. If the cells are grown for long periods at zero NaCl concentration (20 to 60 generations), the lag phase progressively increases when they are transferred back to 500 mM NaCl.

Figure 4 shows the intracellular K^+ and Na^+ contents of salt-adapted

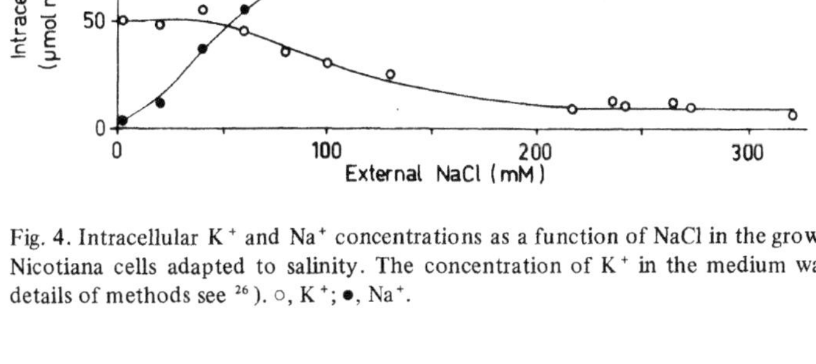

Fig. 4. Intracellular K^+ and Na^+ concentrations as a function of NaCl in the growth medium of Nicotiana cells adapted to salinity. The concentration of K^+ in the medium was 20 mM (for details of methods see [26]). ○, K^+; ●, Na^+.

cells growing in media containing 20 m*M* K^+ and various levels of NaCl. At the lower NaCl levels increasing media salinity causes an increase in intracellular Na^+ together with a decrease in intracellular K^+. Above about 200 m*M* external NaCl, the intracellular K^+ remains constant while the intracellular Na^+ increases more gradually. Especially above 200 m*M* NaCl in the medium, there seems to be a mechanism limiting K^+ loss and Na^+ accumulation in these cells.

The intracellular proline content of salt-adapted cells increases exponentially with increasing medium NaCl (Fig. 5). Proline accumulation is reversible: when the NaCl concentration is lowered, the internal proline pool decreases slowly and reaches its new steady-state level within 24 hrs (see [26], Fig. 4). In contrast to the salt-adapted cells, the salt-sensitive cells do not accumulate proline (see [26], Fig. 2). Since proline does not accumulate in the salt-sensitive cells, which suffer from salinity, but does accumulate in salt-adapted cells, which do not suffer from salinity, proline accumulation is most probably not due to stress damage as claimed by Stewart and Hanson [24].

Since proline accumulation is a frequent response both to various stress conditions (salinity, water stress, low-temperature stress, mineral

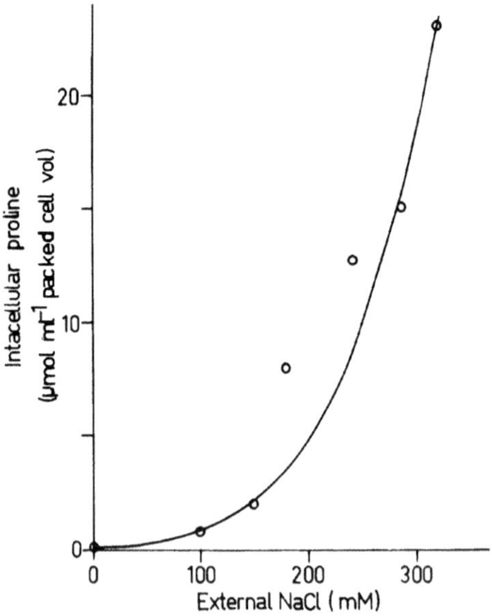

Fig. 5. Intracellular proline concentration as a function of NaCl in growth medium of Nicotiana cells adapted to salinity. A plot of log proline concentration versus external NaCl was linear. For details of methods see [26].

deficiency) and to abscisic acid, Göring suggests that there is some general signal for proline formation [6, 7], perhaps a decrease in cytoplasmic pH. This hypothesis has the advantages of explaining why proline accumulates under a wide variety of circumstances and of proposing an experimental approach to verify it, *i.e.* artificial acidification of the cytoplasm.

The transformation of salt-sensitive cells into salt-resistant cells is most probably due to an epigenetic transformation, because it occurs much too rapidly to be the result of a mutation. This implies that the genetic information for growth in a saline environment is present in the salt-sensitive cells but is not expressed. This in itself is not surprising, since in eukaryotic cells only a small fraction of the genetic information is expressed at any given time. During differentiation the expression of some genes is turned on while the expression of others is turned off. We can look at the transformation of salt-sensitive cells into salt-resistant cells as a phenomenon similar to differentiation. Characteristic of this differentiation is that it is triggered by salt itself and that it is a stepwise phenomenon. We still do not know whether the expression of salt-resistance at the cell level is a particular property of cells in suspension culture only or whether it also can be expressed in cells in the whole plant.

As part of the adaptation to salinity, we should expect a change in protein synthesis due to the change in gene expression. Singh and coworkers [23] did find new or enhanced protein bands in SDS gels after electrophoresis when comparing extracts from salt-adapted and salt-sensitive tobacco cells. The protein band at 26 kd was particularly intense in extracts from salt-adapted cells but was practically non-existent in non-adapted cells (Fig. 6). Interestingly, incubation of non-adapted cells with $10^{-5} M$ abscisic acid caused the appearance of the 26 kd band as well as enhancement of salt resistance[24]. Furthermore, adaptation of these cells to water stress, by growth in the presence of polyethylene glycol, induced the same changes in protein synthesis as adaptation to saline stress (Fig. 6.).

While we do not know the mechanism by which salt affects the genome, it is interesting to speculate that salt itself could modulate gene expression by its effect on translation. It is known that salt selectively inhibits protein synthesis in animal cells: the synthesis of some proteins is inhibited while that of others may be enhanced due to a reduction in competition for available ribosomes [11, 20]. Naturally, we can imagine other mechanisms. Salt stress could induce a change in phytohormone synthesis, *e.g.* an enhancement of ABA formation. ABA is known to modulate gene expression [4], which could lead to salt

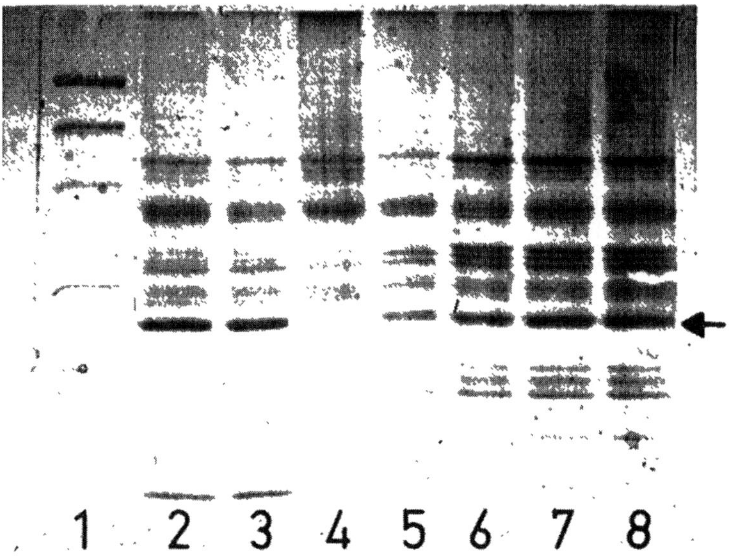

Fig. 6. SDS-polyacrylamide gel electrophoresis of extracts of Nicotiana cells. Lane 1 – MW standards; lane 2 – cells grown in 25% polyethylene glycol; lane 3 – cells grown in 30% polyethylene glycol; lane 4 – unadapted cells; lane 5 – cells grown in 10 g/l NaCl; lane 6 – cells grown in 16 g/l NaCl; lane 7 – cells grown in 20 g/l NaCl; lane 8 – cells grown in 25 g/l NaCl. Each lane was loaded with an equal amount of protein. Arrow indicates 26 kd protein. For details of methods see Singh *et al* [23, 24].

resistance. However, so far neither of these proposed mechanisms explains why adaptation to salt by cells in suspension culture occurs only incrementally.

Conclusion

In conclusion, in plant cells growing under saline conditions there is an accumulation of a *controlled* amount of salt as an osmoticum (which is different for different cells). That is, as the salt concentration in the cell increases, mechanisms, most probably ion pumps, that limit its accumulation are activated. With increasing salinity in the environment, organic solutes also accumulate, providing part of the osmoticum required for turgor maintenance as well as protecting enzymes against ion toxicity. It has often been stated that mineral ions are mainly compartmented in the vacuole; however, this may not be as general as thought at present and requires further documentation.

Finally, the conversion of plant cells in suspension culture from salt-sensitive to salt-resistant is a promising approach. It has already yielded information on the unexpressed genetic potential of certain cell lines and may in the near future reveal some of the mechanisms responsible for salt resistance at the cell level.

Acknowledgements The studies performed in the Department of Botany of The Hebrew University are the result of collaboration with A Poljakoff-Mayber and L Reinhold, and are taken from theses of our students: Y Braun, R Guy, M Hasidim, M Reuveni and A Watad. These studies were supported in part by a grant from the United States-Israel Binational Science Foundation (BSF), and in part by a grant from the United States-Israel Binational Agricultural Research and Development Fund (BARD).

I am very grateful to R A Bressan for unpublished data provided for this review, for providing Figure 6, and for stimulating discussions.

References

1 Chaleff R S 1981 Genetics of Higher Plants. Applications of Cell Culture, Cambridge University Press, London.

2 Citri N 1973 Conformational adaptability in enzymes. *In* Advances in Enzymology, vol. 37, Ed. A Meister. pp 397–648. Interscience Pub., New York.

3 Critchley C 1983 Further studies on the role of chloride in photosynthetic O_2 evolution in higher plants. Biochim. Biophys. Acta 724, 1–5.

4 Dure III L, Galav G, Chlan C and Pyle J 1983 Developmentally regulated gene sets in cotton embryogenesis. *In* Plant Molecular Biology: Proceedings of the UCLA Symposia on Molecular and Cellular Biology, new series, vol. 12. Ed. R B Goldberg. pp 331–342. Alan R. Liss Inc., New York.

5 Flowers T J, Troke P F and Yeo A R 1977 The mechanism of salt tolerance in halophytes. Annu. Rev. Plant Physiol. 28, 89–121.

6 Göring H 1981 Plant responses to effects of adverse physical and chemical environmental factors. *In* Genetic Responses and Plant Breeding for Resistance to Diseases, Pests and Abiotic Environmental Conditions. Eds. D D Brezhnev and V F Dorofeef. pp 152–164. Vavilov All-Union Scientific Research Institute of Plant Industry, Leningrad.

7 Göring H and Dui Huy Thien 1979 Influence of nutrient deficiency on proline accumulation in the cytoplasm of *Zea mays* L. seedlings. Biochem. Physiol. Pflanzen 174, 9–16.

8 Greenway H and Munns R 1980 Mechanism of salt tolerance in nonhalophytes. Annu. Rev. Plant Physiol. 31, 149–190.

9 Hasegawa, P M, Bressan R A and Handa A K 1980 Growth characteristics of NaCl-selected and nonselected cells of *Nicotiana tabacum* L. Plant Cell Physiol. 21, 1347–1355.

10 Jeschke W D, Stelter W, Reising B and Behl R 1983 Vacuolar Na/K exchange, its occurrence in root cells of *Hordeum, Atriplex* and *Zea* and its significance for K/Na discrimination in roots. J. Exp. Bot. 34, 964–979.

11 Kaempfer R 1984 Regulation of eukaryotic translation. *In* Comprehensive Virology, vol. 19. Eds. H Fraenkel-Conrat and R R Wagner. Plenum, New York (*In press*).

12 Kramer D 1983 The possible role of transfer cells in the adaptation of plants to salinity. Physiol. Plant. 58, 549–555.

13 Lerner H R and Avron M 1977 Dihydroxyacetone kinase activity in *Dunaliella parva*. Plant Physiol. 59, 15–17.

14 Lerner H R, Reinhold L, Guy R, Braun Y, Hasidim M and Poljakoff-Mayber A 1983 Salt activation and inhibition of membrane ATPase from roots of the halophyte *Atriplex nummularia*. Plant Cell Environ. 6, 501–506.

15 Lerner H R and Reuveni M 1982 Induction of pore formation selectively in the plasmalemma of plant cells by poly-L-lysine treatment: a method for the direct measurement of cytosol solutes in plant cells. *In* Plasmalemma and Tonoplast: Their Functions in the

Plant Cell. Eds. D Marmé, E Marrè and R Hertel. pp 49–52. Elsevier Biomedical Press, Amsterdam.

16 Lerner H R, Sussman I and Avron M 1980 Characterization and partial purification of dihydroxyacetone kinase in *Dunaliella salina*. Biochim. Biophys. Acta 615, 1–9.

17 Marrè E 1979 Fusicoccin: a tool in plant physiology. Annu. Rev. Plant Physiol. 30, 273–288.

18 Munns R, Greenway H and Kirst G D 1983 Halotolerant eukaryotes. *In* Encyclopedia of Plant Physiology, New Series, vol 12C. Eds. O L Lange, P S Nobel, C B Osmond and H Ziegler. pp 59–135. Springer-Verlag, Berlin.

19 Nash D, Paleg L G and Wiskich J T 1982 Effect of proline, betaine and some other solutes on the heat stability of mitochondrial enzymes. Aust. J. Plant Physiol. 9, 47–57.

20 Nuss D L and Koch G 1976 Variation in the relative synthesis of immunoglobulin G and non-immunoglobulin G proteins in cultured MPC11 cells with changes in the overall rate of polypeptide chain initiation and elongation. J. Mol. Biol. 102, 601–612.

21 Paleg L G, Douglas T J, van Dall A and Keech D B 1981 Proline, betaine and other organic solutes protect enzymes against heat inactivation. Aust. J. Plant Physiol. 8, 107–114.

22 Pitman M G and Luttge V 1983 The ionic environment and plant ionic relations. *In* Encyclopedia of Plant Physiology, New Series, vol 12C. Eds. O L Lange, P S Nobel, C B Osmond and H Ziegler. pp 5–34. Springer-Verlag, Berlin.

23 Singh N K, Handa A K, Hasegawa P M and Bressan R A 1984 Expression of salt-tolerance-specific proteins in cultured tobacco cells. 35th Annual Meeting of the Tissue Culture Association, Houston, Texas. June 3–7, 1984.

24 Singh N K, La Rosa P C, Bressan R A, Hasegawa P M and Handa A K 1984 Protein synthesis in response to NaCl adaptation and abscisic acid. 1984 UCLA Symposia.

25 Stewart C R and Hanson A D 1980 Proline accumulation as a response to water stress. *In* Adaptation of Plants to Water and High Temperature Stress. Eds. N C Turner and P J Kramer. pp 173–189. J. Wiley, New York.

26 Watad A A, Reinhold L and Lerner H R 1983 Comparison between a stable NaCl-selected *Nicotiana* cell line and the wild type. Plant Physiol. 73, 624–629.

27 Wyn Jones R G and Gorham J 1983 Osmoregulation. *In* Encyclopedia of Plant Physiology, New Series, vol 12C. Eds. O L Lange, P S Nobel, C B Osmond and H Ziegler. pp 35–58. Springer-Verlag, Berlin.

28 Wyn Jones R G and Pollard A 1983 Proteins, enzymes and inorganic ions. *In* Encyclopedia of Plant Physiology, New Series, vol 15. Eds. A Lauchli and R L Bieleski. pp 528–562. Springer-Verlag, Berlin.

29 Yeo A R 1983 Salinity resistance: physiologies and prices. Physiol. Plant. 58, 214–222.

Plant and Soil 89, 15–40 (1985). Ms. BR 02

Some mechanisms of salt tolerance in crop plants

J. GORHAM, R. G. WYN JONES and E. McDONNELL
Department of Biochemistry and Soil Science, University College of North Wales, Bangor, Gwynedd, Wales, LL57 2UW, UK

Key words Elytrigia Epicuticular waxes Halophytes Leymus Potassium Roots Salt tolerance Shoots Sodium Transpiration Triticum Water use efficiency

Summary In the first part of this review the main features of salt tolerance in higher plants are discussed. The hypothesis of intracellular compartmentation of solutes is used as a basis for models of tolerance mechanisms operating in roots and in leaves. Consideration is given to the implications of the various mechanisms for the yield potential of salt-tolerant crop plants.

Some work on the more salt-tolerant members of the Triticeae is then described. The perennial species *Elytrigia juncea* and *Leymus sabulosus* survive prolonged exposure to 250 mol m^{-3} NaCl, whereas the annual Triticum species are severely affected at only 100 mol m^{-3} NaCl. In the perennial species the tissue ion levels are controlled within narrow limits. In contrast, the more susceptible wheats accumulate far more sodium and chloride than is needed for osmotic adjustment, and the effects of salt stress increase with time of exposure.

Two different types of salt tolerance are exhibited in plants capable of growing at high salinities. In succulent Chenopodiaceae, for example, osmotic adjustment is achieved mainly by accumulation of high levels of sodium and chloride in the shoots, accompanied by synthesis of substantial amounts of the compatible solute glycinebetaine. This combination of mechanisms allows high growth rates, in terms of both fresh and dry weight. At the opposite end of the spectrum of salt tolerance responses are the halophytic grasses, which strictly limit the influx of salts into the shoots, but suffer from very much reduced growth rates under saline conditions. Another variation is shown in those species that possess salt glands. The development and exploitation of crop plants for use on saline soils is discussed in relation to the implications of these various mechanisms.

Introduction

Although a clear distinction is often made between halophytes and glycophytes, there is in reality a continuous spectrum of tolerance ranging from the most sensitive species, which are severely affected by as little as 1/10th seawater (*ca.* 50 mol m^{-3} NaCl), to those species able to complete their life cycle in full strength seawater (*ca.* 500 mol m^{-3} NaCl). There are, moreover, physiological differences in the way tolerance is achieved in species of similar gross sensitivity to salt. The halophytic Chenopodiaceae, which accumulate high levels of sodium salts in their tissues, may be contrasted with the salt-tolerant grasses, in which tolerance is mainly associated with exclusion of sodium salts from the shoots. In this chapter we will consider principally the mechanisms of tolerance that exist in crop plants. We will concentrate especially on the tribe Triticeae within the Gramineae (Poaceae), which includes some of the more tolerant of the conventional crop plants and the major grain crops wheat, barley, rye and triticale. To assess the significance of various tolerance mechanisms we will inevitably be forced into comparisons with more tolerant 'halophytic' species. An understanding of the mechanisms of tolerance in the Gramineae is of more than academic interest because of the dependence of many of the world's inhabitants on these crops.

Within the Triticeae are several species whose natural habitat implies a large degree of salt tolerance. Since many of these can be sexually hybridized with hexaploid and tetraploid wheats, the potential exists for the transfer of resistance characters from these 'aliens' to commercial varieties. While empirical field testing may eventually allow selection for the transfer of useful salt-tolerance characters, much would be gained by identifying the salt-tolerance mechanisms existing in the wild 'alien' species and determining their heritability. The development of simple selection criteria and an appreciation of the possible agronomic consequences of these characters is also required.

It must be emphasized that even the apparently straightforward comparison of tolerance in the wheats and alien wheatgrasses is complicated by variations in growth rate, growth habit and life history, by changes in tolerance with development, and by the dependence of tolerance on relative humidity, temperature, ionic balance, *etc*.

It is generally accepted that three major hazards are associated with saline habitats. These may be described as a) water stress arising from the more negative water potential (elevated osmotic pressure) of the rooting medium, b) specific ion toxicity usually associated with either excessive chloride or sodium intake, and c) nutrient ion imbalance when the excess of sodium or chloride leads to a diminished uptake of potassium, nitrate or phosphate or to impaired internal distribution of one or another of these ions. There are instances in the literature where the toxicity of salts has been ascribed to each of these hazards.

Mechanisms of tolerance

Intracellular compartmentation – the 'central dogma'

Physiological and biochemical studies in the last decade have produced a coherent and widely accepted hypothesis regarding the principal features of tolerance at the cellular level. Three interrelated propositions have been advanced:

a) Under saline conditions the large quantities of salts (mainly, but not exclusively, sodium chloride) that are absorbed into the leaves and contribute massively to osmotic adjustment are accumulated mainly in the vacuole when tissue concentrations exceed about 200 mol m^{-3} (*i.e.* osmotic pressure greater than 400 mOsmol kg^{-1}, 0.9 MPa).

b) The concentration of inorganic ions in the cytoplasm (especially of meristematic cells) is held in the range 100 to 200 mol m^{-3}, and the cytoplasm shows a strong selectivity for potassium over sodium, magnesium over calcium, and phosphate over chloride or nitrate.

c) Under hyperosmotic conditions (cell osmotic pressures greater than 0.9 MPa), the maintainance of osmotic (water potential) equilibrium across the tonoplast requires the accumulation in the cytoplasm of nontoxic organic solutes, which are usually referred to as 'compatible solutes'[7]. The chemical nature of the compatible solutes varies from one taxonomic group to another, but most are derivatives of polyols or nitrogen dipoles (Table 1).

These propositions, which will be referred to as the intracellular model of solute compartmentation, are summarized in the idealized diagram of a plant cell shown in Figure 1. The first two concepts may be stated in slightly different terms, namely that the ion selectivity and concentrations in the cytoplasm are highly conserved, while a wide range of solutes may be accumulated to varying extents in the vacuole, largely dependent on their external availability. Thus, cytoplasmic selectivity contrasts with vacuolar promiscuity.

It is neither possible nor useful in the space available to review the evidence that supports these propositions. This topic has been adequately discussed in several recent reviews[18, 28, 54, 56, 80, 81, 82, 84]. It is not clear, however, how large an effect factors such as changes in isoenzyme patterns, increased enzyme synthesis or increases in substrate levels have on the operation of this model, especially in older tissues.

While there is much circumstantial evidence supporting these concepts, obtaining precise quantitative evidence for the model presents many technical and interpretational problems. For example, it is difficult to obtain definitive evidence for the degree of potassium/sodium discrimination in the cytoplasm of mature mesophyll as opposed to meristematic cells in *Suaeda maritima*[22, 31]. Although this model is consistent with the biochemical data and allows the interpretation of much of the physiological and anatomical evidence, it is also apparent that we know little at the cellular level about the detailed operation of mechanisms for solute compartmentation. Nor do we understand how the biosynthesis and degradation of the various compatible solutes in higher plants is controlled, despite progress in the elucidation of several biosynthetic pathways[30].

Mechanisms operating in roots

It is clear that a wide range of physiological, anatomical and biochemical features are involved in the control of solute and water

Table 1. Taxonomic distribution of possible compatible solutes

Solute	Structure	Distribution
D-sorbitol	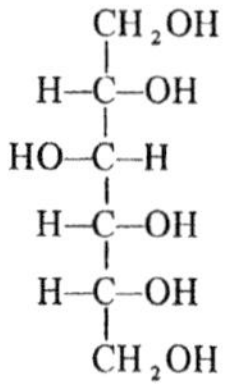	Plantaginaceae[2] Rosaceae[76]
D-mannitol	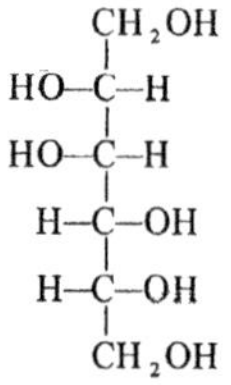	Combretaceae[60] Myrsinaceae[60] Rubiaceae[60]
D-pinitol	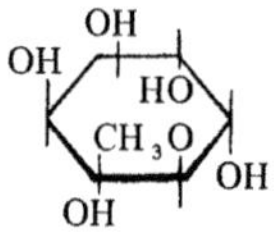	Leguminoseae[19, 27] Rhizophoraceae[60] Caryophyllaceae[24]
L-quebrachitol	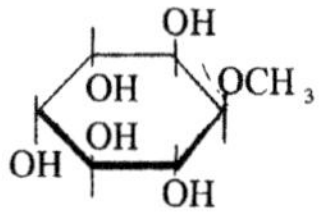	Euphorbiaceae[60]
Glycinebetaine	$(CH_3)_3N^+CH_2COO^-$	Chenopodiaceae[78, 79] Amaranthaceae Asteraceae Solanaceae Gramineae Avicenniaceae
B-alaninebetaine	$(CH_3)_3N^+CH_2CH_2COO^-$	Plumbaginaceae[79]
Proline	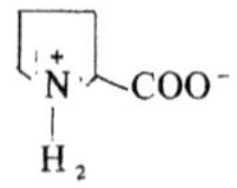	Juncaginaceae[71] Asteraceae Gramineae
Prolinebetaine (Stachydrine)	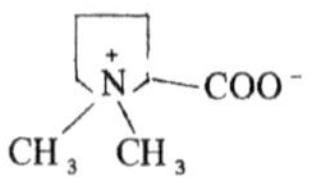	Labiateae[79] Capparidaceae Leguminoseae
3-dimethylsulfonio -propionate	$(CH_3)_2S^+CH_2CH_2COO^-$	Asteraceae[23] Gramineae[44]

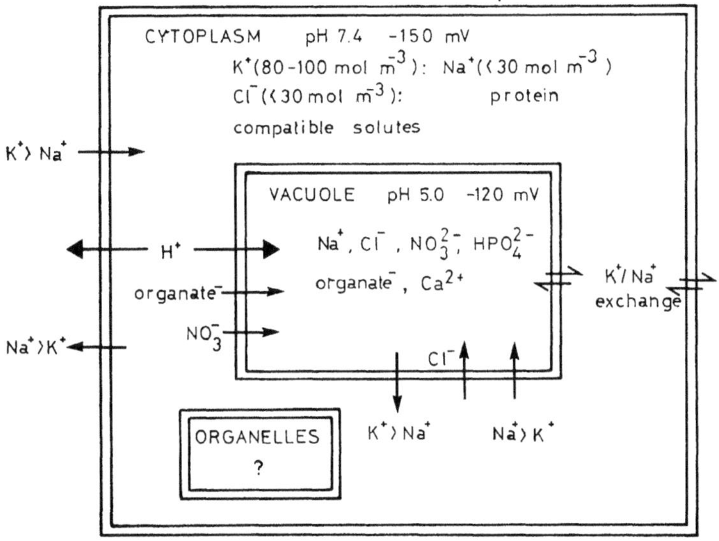

Fig. 1. Idealized model of a plant cell showing compartmentation of osmotica. Active electrogenic pumps are represented by thick arrows. No energetic implications are intended by the thin arrows representing other selective transport processes. The importance of these different processes will change with time and with changing external conditions.

balance and distribution on a whole-plant and tissue basis. These features will influence the water potential and solute concentrations to which individual cells throughout the plant are exposed. In this section we attempt to summarize some of the salt tolerance mechanisms operating in roots to control the influx of salts.

As a basis for describing these features, Figure 2 incorporates diagramatic representations of several sites at which salt-tolerance mechanisms could operate. Solute distribution will ultimately be controlled by the active and passive transport properties of the plasmalemma and tonoplast, and by any 'bypass' or 'leak' mechanisms operating *via* the apoplast. Such processes are not static, and steady-state concentrations represent a dynamic equilibrium to which several biochemical and electrochemical mechanisms contribute, together with diffusive and mass flow phenomena. Furthermore, exchange of monovalent cations has been observed in vacuoles, both as the root tissue matures and with changing external ion concentrations (at least in low-salt conditions)[35, 36, 37, 39]. At the next level of organization, the precise sites of selective transport through the root are determined by extracellular permeability barriers. The most important of these are the suberized cell walls of the endodermis, which force solutes moving

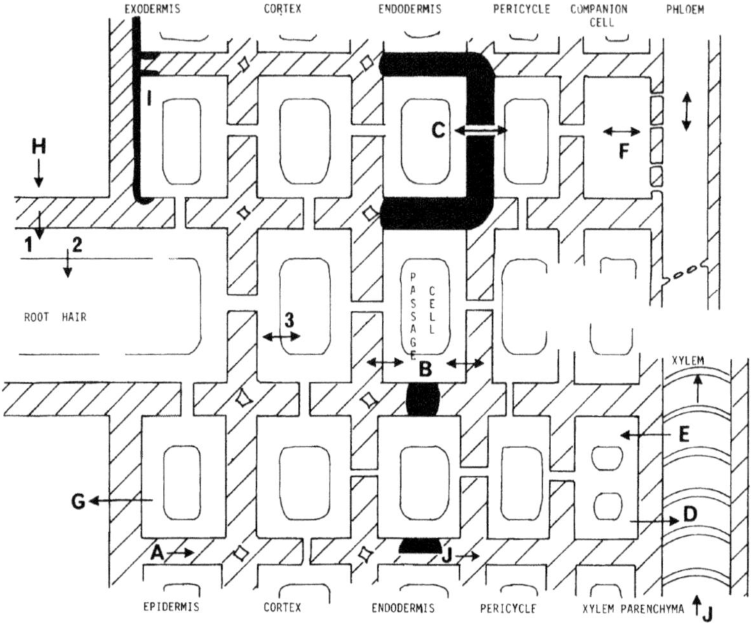

Fig. 2. Mechanisms of salt tolerance operating in roots. This diagramatic representation shows the following features:

1. Selective ion transport across the plasmalemma; 2. Selective ion transport across the tonoplast and 3. Exchange phenomena.

A. Apoplastic movement of solutes in the cortex.

B. Enforced passage through membranes/plasmodesmata at the endodermis. Routes to the endodermis are through the symplast or the apoplast or through a series of plasmalemmas and tonoplasts.

C. Enforced passage through plasmodesmata in secondary and tertiary endodermis (monocots).

D. Selective, active xylem loading.

E. Selective, active xylem unloading. Transfer cells.

F. Selective ion movement in the phloem. Supply of assimilates from the shoots.

G. Export from the roots.

H. Effect of mass flow of water on concentrations of solutes at the root surface and in the apoplast.

I. Suberization of the epidermal cells to form an exodermis (monocots). In dicots the function of the endo/exodermis is performed by the periderm, and the endodermis may be lost.

J. Bypass flow into the xylem.

from the cortex to the stele to pass through the plasmalemma and into the symplasm. There are unresolved debates about the extent to which water and ion transport proceeds through passage cells or plasmodesmata, and about the relative contributions of apoplastic, symplastic and transmembrane (vacuolar) pathways of water and ion movement across roots[77].

It is perhaps worth emphasizing the changes in transport characteristics that occur as the root ages[50]. In particular, the radial permeability of roots decreases as the suberization of the endodermis increases and the frequency of passage cells decreases. In some cases a second endodermis is formed by suberization of the innermost cortical cell walls as in *Puccinellia peisonis*[68]. The permeability of the epidermal layers also decreases with the loss of root hairs and the formation of a suberinized exodermis (in monocots) or a periderm (in dicots, in which the endodermis may be lost).

In the stele, active transport processes are involved in loading solutes into the xylem stream. This offers another site for selectivity. Conversely, active transport out of the xylem via transfer cells[43] may further regulate the proportions and concentrations of different ions supplied to the shoot. Such a mechanism has recently been described in maize mesocotyls, where sodium is transported through the endodermis of the mesocotyl from the xylem stream in the stele to the cortex[40, 41]. Obviously, the relatively small and limited volume of the mesocotyl limits the capacity of this system to deal with the influx of large quantities of sodium. In some of the most salt-sensitive crop plants sodium (and chloride) may be retransported from sites of abstraction from the xylem and exported from the roots. This has been reported for *Phaseolus vulgaris*[34, 46] and squash[11].

The rate of water movement through roots (as determined by transpiration) has two potentially important effects on ion uptake, apart from any direct interaction with transmembrane flow rates. The first of these is the local concentration of excluded ions in the region of the root surface and in the apoplast of the cortex caused by the mass flow of external solution into the root[64], which effectively increases the salt/water stress experienced by the plant[84]. Secondly, the bypass flow of solutes into the xylem via the young, incompletely suberinized endodermis and/or through damaged roots may be affected[50, 54, 84], thereby increasing the proportion of relatively non-selective uptake of sodium and divalent metal ions. Transport of calcium into the stele is thought to be mainly apoplastic[10] and transpiration-dependent. The effect of transpiration rate on ion levels in the xylem[77] requires further investigation in relation to the movement of solutes into the shoots of salt-stressed plants.

Although the mechanisms of salt tolerance operating in roots are here described separately from those in leaves, the underlying principles, embodied in the intracellular model of solute compartmentation described earlier, are the same in both tissues. At the whole-plant level, salinity often induces changes in the shoot/root ratio (see Table 4 below) with consequences for assimilate partitioning, water abstraction from the soil, *etc*. Assimilate partitioning might also be expected to be influenced by the extra respiratory demands of active solute transport, particularly in roots[20].

Mechanisms operating in leaves

The pathways of water and solute movement and the mechanisms of response to salt stress within leaves (Fig. 3) are equally complex and

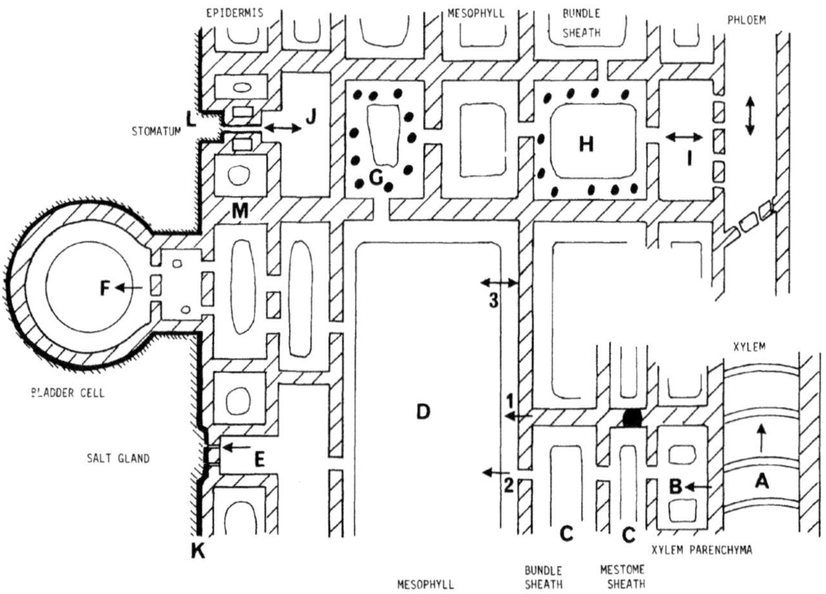

Fig. 3. Location and consequences of salt-tolerance mechanisms operating in leaves. The features represented in the diagram are: 1. Selective transport across the plasmalemma; 2. Selective transport across the tonoplast and 3. Exchange phenomena.
A. Composition of xylem sap; variation with mass flow of water and with 'upstream' abstraction processes.
B. Selective xylem unloading.
C. The bundle sheath and suberinized mestome of *Puccinellia peisonis*.
D. Succulence associated with Na^+ and Cl^- accumulation in vacuoles.
E. Salt glands.
F. Salt hairs and bladders.
G. Compartmentation in chloroplasts and effects on photosynthetic potential.
H. C-4 photosynthesis.
I. Selective movement of solutes in phloem. Supply to apical regions and recovery of nutrients from older leaves.
J. Effects on stomatal and 'mesophyll' resistances to gas exchange.
K. Effects on cuticular water loss, epicuticular waxes.
L. Effects on stomatal water loss. Xerophytic morphology. Effect of leaf temperature.
M. Salt load in the leaf apoplast.

probably vary more widely from species to species. Little is known about how the composition of xylem sap changes on its journey from the roots to the leaves and along the length of leaves. Furthermore, variations in the rates of water and solute movement can be expected to result in fluctuations in the concentrations of solutes in both the xylem and the leaf apoplast. Once again, transport across the plasmalemma and tonoplast is ultimately responsible for solute distribution at both the intracellular level and higher levels of organization. Movement into the symplasm may be enforced by suberization of the cell walls surrounding the vascular bundles as in the mestome sheath of *Puccinellia peisonis*[69]. In this species sodium accumulates initially, and to the greatest extent, in the vacuoles of the bundle sheath.

The capacity of the mesophyll vacuoles to accumulate sodium and chloride may be vastly increased by increasing the cell size and vacuole:cytoplasm ratio. This is seen at its most successful in the succulent Chenopodiaceae such as *Suaeda maritima*[54, 56], but also occurs to a lesser extent in many species[47]. To be effective in 'salt-including' halophytes, this mechanism must be accompanied by high growth rates. An alternative solution to the problem of salt accumulation in the leaf is its export to the outside of the cuticle via salt-exporting glands. These are found in a restricted number of species from a variety of monocot and dicot families including Gramineae (Spartina, Leptochloa (Diplachne)), Plumbaginaeae (Limonium) and Avicenniaceae (Avicennia)[32, 49]. Another means of disposing of excess salt is *via* bladder cells. These function as repositories for excess salt in young leaves of Atriplex species and release their contents to the environment by rupturing as the leaf ages[38, 56, 72].

Relatively few species possess effective mechanisms for excreting salt from the leaves or for sequestering very large quantities of salts in enlarged vacuoles. Moreover, mechanisms must exist in other species for the intracellular compartmentation of ions that accumulate in their shoots (see discussion of Fig. 1). Evidence on the effects of salinity on chloroplast ion contents and photosynthesis is conflicting. Recent studies with spinach have, however, shown that chloroplast ion contents in this species are lower than overall leaf values, and that potential photosynthetic capacity is only slightly impaired by salt stress[42, 62]. Reduced photosynthesis in stressed plants results from a combination of factors, including increased stomatal and mesophyll resistances and decreases in total leaf area[4, 42, 58, 62].

Phloem transport is important in relation to salinity at two stages of leaf growth, for the supply of solutes to young, expanding tissues and for the retranslocation of useful metabolites (including potassium) from senescing leaves to younger tissues. Most evidence suggests that sodium and chloride are relatively immobile in the phloem of halophytes[83] (but see also refs. [53, 54]). Export of sodium and chloride was found to be greatest in the most sensitive of a range of crop species[46], suggesting that poor control of symplastic ion levels might be a factor in both phenomena.

While the effects of salinity-induced increases in stomatal resistance on photosynthesis may be complex[4], reduction of transpiration rates is a characteristic response to salinity. Many halophytes show other, morphological, features such as trichomes (Halimione, Elytrigia) and succulence (Salicornia) associated with reduced transpiration and drought tolerance, and it should be remembered that salt stress also implies a simultaneous water stress. The problems of the water relations associated with drought and salinity have been extensively reviewed recently [6, 8, 17, 55, 59, 73, 74, 75] and will not be restated here. It is often not appreciated that non-stomatal water loss through the epidermis (cuticular transpiration) may be both substantial and also affected by salt stress as a result of increases in cuticular thickness and epicuticular wax deposition[16, 26, 29]. In species with salt glands or in those subjected to periodic inundation with saline water or salt spray the cuticle also functions as a barrier to the entry of salt into the leaf[1, 63]. Conversely, in *Suaeda maritima*, the cuticle may prevent rainwater from leaching salt from the cell walls, a process that could lead to the generation of excessive turgor pressures in the leaf cells (A. D. Tomos, personal communication). While the cuticle does not normally form a complete barrier to the movement of ions and water, there is almost no information concerning how cuticular permeability varies in plants in relation to salt stress[65]. Increased pubescence and epicuticular wax deposition may have an equally important role in reflecting incoming radiation, and hence in preventing excessive temperature rises in leaves in which transpirational cooling is reduced by increased stomatal resistance[73]. The interactions between stress, transpiration and photosynthesis have been considered in several recent reviews[12, 57, 66].

Mechanisms in meristems and pollen

Meristems are one of the few parts of higher plants that contain large proportions of cytoplasm, thus constituting a tissue in which the predictions of the intracellular model of ion compartmentation can be tested. Apical regions generally have higher K/Na ratios than more vacuolated tissues, and often lower overall inorganic salt levels[27, 37, 39, 72]. Vacuolation occurs very rapidly, and the region in which cytoplasm occupies the major volume of the cell may be no more than a few hundred μm^3, especially in shoots. In the region of expanding cells adjacent to the meristem the relative importance of phloem and xylem as sources of solutes for turgor generation is largely unknown. Recent theories on the causes of growth reduction in salt-stressed plants emphasize the possibility that phloem transport rates might limit solute delivery to expanding tissues[28, 53, 54, 84]. Xylem vessels may be seen in the apical regions of *Suaeda maritima* and certainly pass through the intercalary meristems of monocots. If, and to what extent, they are functional in supplying solutes for expansive growth is not known. Similarly, the rates of mass flow of water through such regions must be taken into account, although they will be low in comparison with the rates of water movement through leaves[52].

Table 2. Glycinebetaine content (mmol kg^{-1} fr. wt.) of leaves and pollen

Species	Glycinebetaine
Leymus giganteus	
Leaf	30
Pollen	142
Spinacia oleracea	
Leaf	2
Pollen	85

For details of methods, see references 24 and 25.

Another tissue which is highly cytoplasmic and which might be expected to have a high concentration of compatible or protective solutes is pollen. As can be seen from Table 2, pollen samples from both the grass *Leymus giganteus* and the chenopod *Spinacia oleracea* grown in nutrient solution alone have considerably higher glycinebetaine contents than the leaves of these species. In other species high levels of proline are found in pollen[5]. The effects of salinity on the compatible solute composition of pollen are being investigated.

Integrated model

We have emphasized the ultimate dependence of salt tolerance mechanisms on transport processes operating at the cellular level (Fig. 1), and have outlined how these mechanisms control solute levels passing through the roots (Fig. 2) and into, within and out of leaves (Fig. 3). It is more difficult to integrate these models into a description of processes operating at the whole-plant level, especially with respect to growth and changes in morphology (*e.g.* flowering). Figure 4 attempts

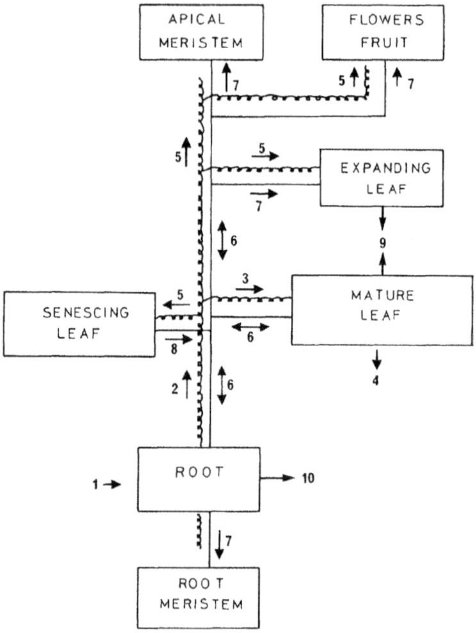

Fig. 4. Transport processes operating at the whole-plant level. 1–4. Main pathway of water movement through the plant. 5. Xylem supply to other organs. 6. Selective solute redistribution via the phloem. 7. Phloem supply to developing, less vacuolated tissues and regions of low transpiration. 8. Selective recovery of nutrients from senescing leaves. 9. Salt excretion from leaves. 10. Salt excretion from roots.

to bring together the main features of solute distribution operating within the plant as a whole.

Salt tolerance in the Triticeae

Progress in understanding the detailed mechanisms of salt tolerance operating in roots or leaves has necessarily been dependent to some extent on the use of specialized tissues (such as low-salt cereal roots) which are readily amenable to experimentation. It is also important to understand the responses of the plant as a whole to salinity. While this is principally a phenomenological approach, the aim must be to relate the observed changes in growth, solute accumulation and water relations to the underlying cellular and physiological mechanisms responsible for such changes. In the case of the Triticeae the desire for empirical knowledge is reinforced by the pressures to provide grain, *etc.*, in many developing countries. Such empirical studies should not be entirely divorced from a physiological and biochemical understanding of the processes involved.

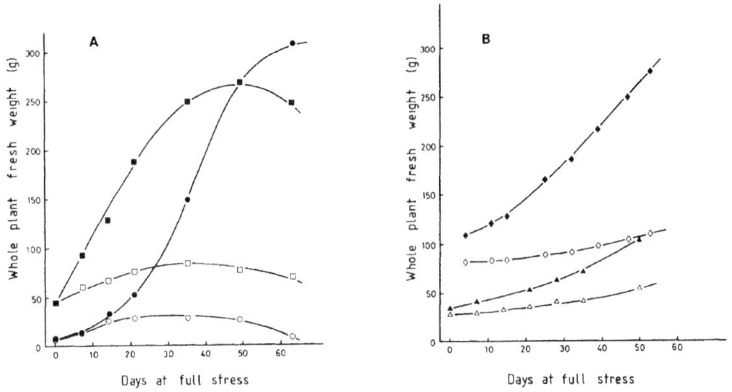

Fig. 5 (A) Growth of *Triticum aestivum* cv. Ciano 79 (■, □) and *T. dicoccoides* (●, ○) without (closed symbols) and with (open symbols) 100 mol m^{-3} NaCl added to dilute nutrient solution. (B) Growth of *Leymus sabulosus* (♦, ◇) and *Elytrigia juncea* (▲, △) without (closed symbols) and with (open symbols) 250 mol m^{-3} NaCl added to dilute nutrient solution.

We do not intend to review here the extensive literature on salt-tolerance mechanisms in the annual cereals, particularly barley (for references see [13, 30, 53, 54, 82, 84]). Instead we will discuss the mechanisms of salt tolerance operating in the salt-tolerant perennial grasses of the tribe Triticeae, and how these mechanisms may be used in the breeding of more salt-tolerant crops.

The first problem that we encounter in this approach is the definition of the terms 'tolerance' and 'sensitivity'. Figure 5A shows the response of a hexaploid bread wheat (Ciano 79) and the tetraploid wild wheat *Triticum dicoccoides* to a salinity level of 100 mol m^{-3} NaCl imposed starting at 21 days after germination. The bread wheat flowered earlier than the wild wheat and was already growing at its maximum rate when the salt stress was applied. In contrast, *T. dicoccoides* had not reached the phase of maximum absolute growth rate and the stressed plants died before they could flower. Had the stress been applied at the same stage in the growth cycle in each case the results might have been different. The yield of grain from the salt-stressed Ciano 79 was considerably lower than that from the controls (Table 3). Several factors contributed to the decrease in yield, but mainly the number of tillers produced, seed set and seed weight. Ideally, salt tolerance should be assessed for plants completing the whole of their life cycle under salt stress, although it is not always possible to incorporate this ideal into experimental designs (nor is it necessarily analogous to field conditions). The differences displayed in Fig. 5A could equally apply to two spring wheat varieties differing in time to

Table 3. Yield data for *Triticum aestivum* cv. Ciano 79

Parameter	Control	NaCl (100 mol m^{-3})	Sig. level (%)
Tillers/plant	19.6 + 1.8	6.6 ± 1.2	0.1
Spikelets/spike	18.1 + 0.3	15.7 + 0.6	0.2
Seeds/spike	62.0 + 2.2	34.8 + 1.9	0.1
Seed weight/spike (g)	1.9 + 0.1	0.6 + 0.1	0.1
Seed weight (mg)	30.0 + 1.6	16.1 + 0.2	0.2
Yield/plant (g)	37.2 + 2.6	4.0 + 1.4	0.1

flowering. Much greater problems arise when comparisons are made between spring and winter wheats or between annual and perennial species.

It is well established that maintaining a low Na:Ca ratio (of about 20:1) improves the salt tolerance of a range of plants compared with increasing NaCl alone[54]. The extent of this effect varies somewhat between different cultivars of wheat, as can be seen from Table 4. Root growth in particular was improved considerably by adding calcium chloride (5 mol m^{-3}) with sodium chloride (100 mol m^{-3}) in accession number 10150 of *T. durum*, but not in accession number 10139, compared with 100 mol m^{-3} sodium chloride alone. In *T. aestivum* cv. Ciano 79 sodium uptake was much lower in the presence of added calcium, whereas the chloride level was only slightly reduced. Table 4 also shows the differences in K/Na ratio, osmotic adjustment and overall ion levels between the hexaploid *T. aestivum* and the tetrapolid *T. durum*. Consistent effects of ploidy in wheats on such characters as ion uptake, cell size, photosynthetic performance and salt tolerance have been observed[3, 9, 26, 33]. Table 4 also illustrates the effect of salinity on shoot/root ratios, with opposing consequences for assimilate partitioning and for water uptake. The latter will be particularly important for crops growing in soils that are both saline and subjected to drought, and the benefits may substantially outweigh the disadvantage of greater heterotrophic root growth.

That some perennial Triticeae are more tolerant of salt than is wheat is clearly established[14,15,25,51,67] and can be seen from Fig. 5B. Both *Leymus sabulosus* and *Elytrigia juncea* survived and continued to grow when exposed to 250 mol m^{-3} NaCl for more than 50 days. Their tolerance is associated with an ability to tightly control osmotic adjustment (Table 5) by strictly regulating the influx of sodium and chloride. Osmotic adjustment was particularly tightly controlled in *E. juncea*, and the change in sap osmotic pressure closely followed the change in external osmotic pressure[26]. Similarly, the accumulation of inorganic ions, which were responsible for most of the osmotic adjustment, was

Table 4. The effects of calcium on the response of three wheat varieties to sodium chloride

	10034 *T. aestivum* Ciano 79			10150 *T. durum*			10139 *T. durum*		
Parameter	Control	NaCl[a]	$CaCl_2$[b] + NaCl	Control	NaCl	$CaCl_2$ + NaCl	Control	NaCl	$CaCl_2$ + NaCl
Whole plant fr. wr. (g)	17.6	7.4	11.3	25.2	5.9	12.3	15.7	7.7	9.2
% of control	100	42.2	64.4	100	23.6	48.9	100	49.3	58.6
Root fr. wt. (g)	5.1	3.3	3.6	6.8	2.3	4.4	3.9	3.6	3.3
% of control	100	64.3	69.6	100	34.5	65.3	100	91.6	84.3
Shoot fr. wt. (g)	11.4	4.2	6.3	17.6	3.5	7.1	9.6	4.3	4.6
% of control	100	36.4	59.6	100	20.2	40.2	100	44.8	48.4
Shoot/root	2.2	1.3	1.9	2.6	1.5	1.6	2.5	1.2	1.4
Sap O.P. (mOsmol kg^{-1})	439	573	571	475	699	680	450	658	654
Increase above control		134	132		213	205		208	204
Potassium (mol m^{-3})	180.8	132.5	151.7	208.3	90.8	92.5	214.2	80	105.0
Sodium (mol m^{-3})	3.0	74.7	44.0	3.7	228.3	196.7	5.0	193.3	170.0
Chloride (mol m^{-3})	25.7	156.3	150.7	33.3	191.7		24.7	134.7	156.3
K/Na ratio	60.3	1.8	3.4	56.3	0.4	0.5	42.8	0.4	0.6
(K + Na) × 1.8	330.8	372.9	352.3	381.6	574.4	520.6	394.6	491.9	495.0
Glycinebetaine (mol m^{-3})	1.1	5.0	4.5	1.9	10.2	5.6	2.8	8.5	8.2

[a] 100 mol m^{-3}; [b] 5 mol m^{-3}

Table 5. Osmotic adjustment (mOsmol kg^{-1}) in leaves of *Leymus sabulosus, Elytrigia juncea, Triticum aestivum* cv. Ciano 79 and *T. dicoccoides*

Parameter	*Leymus sabulosus*		*Elytrigia juncea*		*Triticum aestivum*		*Triticum dicoccoides*	
External NaCl (mol m^{-3})	0	250	0	250	0	100	0	100
Days at full stress		54		66		21		21
Measured sap osmotic pressure	502	1003	468	911	443	883	392	1184
Sap osmotic pressure corrected for sucrose hydrolysis	497	990	458	862				
Contribution of Na + K salts (% of total)	410 (82%)	810 (82%)	311 (68%)	650 (75%)	245 (55%)	637 (72%)	283 (72%)	851 (72%)
Sap O.P. – external O.P. (% of control)	487	530 (110%)	448	402 (98%)	433	693 (160%)	382	994 (260%)

For details of methods, see ref. [25]

Table 6. Inorganic ion contents of *Elytrigia juncea* and *Leymus sabulosus* expressed on fresh and dry weight bases

	Elytrigia juncea		*Leymus sabulosus*	
	Control	NaCl ($250\ mol\ m^{-3}$)	Control	NaCl ($250\ mol\ m^{-3}$)
Shoot				
Fresh weight/dry weight	4.28	3.93	5.41	4.12
Solutes ($mol\ m^{-3}\ H_2O$)				
Potassium	171	208	216	196
Sodium	2	153	12	254
Chloride	33	259	31	330
Solutes ($mmol\ kg^{-1}$ dry wt)				
Potassium	560	610	950	608
Sodium	6	448	53	787
Chloride	110	758	136	1023
Na/K	0.01	0.74	0.05	1.30
Root				
Fresh weight/dry weight	14.84	12.84	9.31	8.00
Solutes ($mol\ m^{-3}\ H_2O$)				
Potassium	18	19	59	28
Sodium	2	102	9	122
Chloride	0	79	0	64
Solutes ($mmol\ kg^{-1}$ dry wt)				
Potassium	249	223	490	196
Sodium	21	1208	75	854
Chloride	0	933	0	448
Na/K	0.11	5.37	0.15	4.36

initially rapid during the incremental imposition of salt stress, but thereafter was strictly limited. Note also that potassium concentrations in the leaves were increased by salt treatment, whether expressed on a dry or fresh weight basis (Table 6). Thus salt tolerance in these grasses implies strict control of tissue salt levels, not high exclusion of salt under all circumstances. The wheats displayed poor control of tissue salt loads, and the effects of stress increased with time. Conversely, a less tolerant species of *Elytrigia* (syn. *Agropyron*) has been reported to exhibit greater salt exclusion than the more tolerant *E. pontica* (syn. *Agropyron elongatum*)[15].

Problems of interpretation arise when ionic analyses are expressed only on a dry weight basis[45] (Table 6). Salinity frequently results in changes in the fresh weight/dry weight ratio such that the actual ion concentrations in the sap increase even when there is no change (or even a decrease) on a dry weight basis. Further complications arise when root and shoot data are compared, since hydrated roots contain a considerable free space component, particularly in the cortex. It is evident, however, from the Na/K ratios that sodium accumulation is greater in the roots than in the shoots of both *Leymus sabulosus* and *Elytrigia juncea* (Table 6). The potassium contents of the roots of *Leymus sabulosus* decreased considerably, whereas in *E. juncea* a decrease in potassium was only evident when expressed on a dry weight basis. Considerable differences have also been observed between different Triticum species in their ability to compartment ions between shoot and root (Gorham, unpublished observations).

In both species sucrose and glycinebetaine accumulation contributed to osmotic adjustment in the leaves. In contrast, no significant increase in proline was observed in either species, which would support the view that proline accumulation is associated with severe water stress and poor osmotic adjustment[70]. The intracellular model (Fig. 1) would predict that glycinebetaine would accumulate particularly in apical tissue during salt stress, and this is seen to be the case from Table 7. Since the shoot primordia analysed in this experiment contained a considerable amount of vacuolated tissue, the actual cytoplasmic concentrations could be much higher (see also ref. [22]).

Table 7. Glycinebetaine distribution in *Leymus sabulosus* (mol m^{-3} plant water)

	Control	NaCl (250 mol m^{-3})
Shoot primordia	8	40
Youngest leaves	8	14
Inner leaf sheaths	7	11
Outer leaf sheaths	3	6
Emerging leaf	4	14
Mature leaf	6	13
Roots	4	9

An influence of transpiration rate on ion accumulation might be inferred from the data for two lines of *Triticum dicoccoides* shown in Table 8. Since these lines are not necessarily isogenic for characters other than transpiration, proof of this relationship awaits the results of further experiments conducted over a range of atmospheric humidities. As has already been mentioned, transpiration rates during the dark may be a significant proportion of daytime rates (Table 8). A value of about 25% of daytime rates is also shown for data on the halophytic chenopod *Salsola kali*[16] both in control and salt-grown plants. Both stomatal and cuticular transpiration rates decreased in salt-stressed plants of *Leymus sabulosus* and *Elytrigia juncea* (Tables 9 and 10), despite an increase in stomatal frequency (Table 10). One day after removal of the stress the daytime transpiration rates of

Table 8. Transpiration and leaf ion contents of *Triticum aestivum* and *T. dicoccoides*

Parameter	*T. aestivum* (10034)		*T. dicoccoides* (10083)		*T. dicoccoides* (10036)	
	Control	NaCl[a]	Control	NaCl	Control	NaCl
Transpiration ($mg\,g^{-1}\,h^{-1}$)						
Light	96	62	166	47	301	102
Dark	25 (26%)	15 (24%)	29 (18%)	17 (36%)	61 (20%)	29 (28%)
Sap osmotic pressure ($mOsmol\,kg^{-1}$)	443	883	391	799	392	1184
Leaf contents ($mol\,m^{-3}$)						
Potassium	131	91	129	40	152	51
Sodium	5	263	4	291	6	422
Chloride	0	372	0	350	0	488
Glycinebetaine	3	6	2	7	2	6
Na/K ratio	0.04	2.9	0.03	7.3	0.04	8.3

[a] 100 $mol\,m^{-3}$

Table 9. Effect of removal of stress on whole-plant transpiration rates in *Leymus sabulosus*

	Whole-plant transpiration rate ($mg\,g^{-1}$ fr. wt. h^{-1})	
	Control	NaCl ($250\,mol\,m^{-3}$)
1. Stressed		
Light	52.5 + 7.6	14.0 + 2.1
Dark	14.3 + 3.4	5.4 + 0.8
2. One day after removal of salt stress		
Light	78.4 + 10.5	35.8 + 8.0
Dark	16.2 + 3.4	6.8 + 1.9

Table 10. Stomatal numbers and transpiration in *Elytrigia juncea*

	Control	NaCl ($250\,mol\,m^{-3}$)	Sig. level (%)
Stomata/mm cell line	8.5 + 0.30	12.2 + 0.36	0.1
Stomata/mm ridge face	20.0 + 0.70	26.5 + 0.88	0.1
Transpiration ($mg\,g^{-1}$ fr. wt. h^{-1})			
Stomatal	76.7 ± 2.44	30.9 ± 1.80	0.1
Cuticular	22.0 ± 2.59	8.8 ± 1.76	0.2

L. sabulosus had recovered significantly, whereas the rates in the dark had not (Table 9). Reduced cuticular water loss is associated with an increase in the epicuticular wax covering (Table 11), which is the cuticular feature mainly responsible for the impermeability of the leaf surface to water[65]. In spite of the large changes in the rates of transpiration and growth, the water use efficiency of these grasses, expressed in terms of whole plant fresh or dry weight accumulated per gram of water transpired, was only slightly affected by salinity (Table 12). In these terms *E. juncea* transpired less per unit fresh weight gain at $250\,mol\,m^{-3}$ NaCl, and *L. sabulosus* transpired less only in relation to dry weight gain. The differences in water use efficiency between *L. sabulosus* and *E. juncea* were small compared

Table 11. Epicuticular wax of *Leymus sabulosus* and *Elytrigia juncea*

Parameter	NaCl ($mol\,m^{-3}$)	*Leymus sabulosus*	*Elytrigia juncea*
Fresh weight/dry weight	0	5.5	4.0
	250	4.3	3.9
Wax ($mg\,g^{-1}$ fresh wt)	0	1.6	2.0
	250	2.4	2.9
Wax ($mg\,g^{-1}$ dry wt)	0	8.6	8.0
	250	10.5	11.3

Table 12. Water use efficiency in *Leymus sabulosus* and *Elytrigia juncea*

Species	NaCl ($mol\,m^{-3}$)	mg fr. wt. ($g^{-1}\,H_2O$)	mg dry wt. ($g^{-1}\,H_2O$)
Leymus sabulosus	0	7.37	1.07
	250	6.55	1.14
Elytrigia juncea	0	8.27	1.14
	250	12.13	1.73

with the differences in their absolute growth rates. The relative growth rates of both stressed and unstressed *Elytrigia juncea* were, however, slightly higher than those of *L. sabulosus*. In *T. aestivum* water use efficiency, although initially higher (> 2 mg dry weight gain per gram of water transpired) than for the perennial species, decreased dramatically as a result of prolonged exposure to 100 mol m^{-3} NaCl and eventually became negative. An increase in water use efficiency has been reported in *Spartina alterniflora* at 500 mol m^{-3} NaCl[48] but does not occur in *Atriplex halimus*[21].

Conclusions – implications for the development of crops with improved salt tolerance

When considering the usefulness of potentially salt-tolerant agricultural crops it is not appropriate to use survival as the criterion of tolerance. The crop must also produce enough of the desired product (grain, animal feed, biomass or chemicals) to ensure that the exercise is economically viable. Thus we come to the problem of combining high growth rates with salt tolerance. The 'salt includers' are at an obvious advantage here (see O'Leary, this volume), but the uses to which the succulent Chenopodiaceae can be put are limited. On the other hand the 'salt excluding' grasses exhibit drastic reductions in growth rate in response to similar levels of salinity. The choice between these two extremes in any one instance is likely to be determined more by economics than by plant physiology.

The problem that must be taken into account in plant breeding for salinity tolerance (as opposed to screening existing genotypes) is the determination of which of the many mechanisms of tolerance operating in plants will produce a significant increase in tolerance without drastically reducing economic yield. Indeed it is unlikely that improvement of any single mechanism alone would produce the desired result. While we know a considerable amount about various specialized tissues (uptake in low-salt roots for example), we are not yet in a position to integrate the mechanisms operating at the intracellular, tissue and organ levels, especially with respect to changes with time and with fluctuating environmental conditions. It is possible, for example, to envisage feedback regulation of ion uptake at the cellular level, but how the various mechanisms are integrated throughout the plant to strictly control leaf ion levels (in *Elytrigia juncea* for example) requires detailed quantitative information which is not yet available. The interactions between the mass flow of water through the plant (transpiration), ion uptake, turgor regulation, stomatal opening and other determinants of *in vivo* photosynthetic capacity is one area where such information is required[6, 17, 73].

Any potentially salt-tolerant cereal crop must also satisfy agronomic requirements with regard to harvest index, time to flowering, disease resistance, *etc.* [6, 17]. The weighting of these characters for crops grown on marginal salt- or drought-affected soils will not, however, necessarily be the same as would be applicable to Western high-input agriculture. The variety of salt-tolerance mechanisms found in higher plants suggests that there are several possible biological approaches to solving the problems of agriculture on salt-affected soils. These range from the exploitation of existing halophytes, such as *Atriplex* spp.[56] and *Leptochloa* (*Diplachne*) *fusca*[61], to attempts to breed more-salt-tolerant varieties of existing salt-sensitive crops. An understanding of the interactions of salt-exclusion mechanisms with growth rates is essential to evaluating the physiological implications of these two approaches. It is probable that a range of lines will need to be developed to exploit the different conditions existing in various parts of the world.

Acknowledgements The authors would like to acknowledge the financial support of the U.K. Overseas Development Administration.

References

1 Ahmad I and Wainwright S J 1976 Ecotype differences in leaf surface properties of *Agrostis stolonifera* from salt marsh, spray zone and inland habitats. New Phytol. 76, 361–366.

2 Ahmad I, Larher F and Stewart G R 1979 Sorbitol, a compatible solute in *Plantago maritima*. New Phytol. 82, 671–678.

3 Austin R B, Morgan C L, Ford M A and Bhagwat S G 1982 Flag leaf photosynthesis of *Triticum aestivum* and related diploid and tetraploid species. Ann. Bot. 49, 177–189.

4 Ball M C and Farquhar G D 1984 Photosynthetic and stomatal responses of two mangrove species, *Aegiceras corniculatum* and *Avicennia marina*, to long term salinity and humidity conditons. Plant Physiol. 74, 1–6.

5 Bathurst N O 1954 The amino acids of grass pollen. J. Exp. Bot. 5, 253–256.

6 Blum A 1982 Breeding programs for improving crop resistance to water stress. *In* Crop Reactions to Water and Temperature Stresses in Humid, Temperature Climates, Eds. C D Raper Jr. and P J Kramer, pp 263–275. Westview Press, Boulder.

7 Borowitzka L J and Brown A D 1974 The salt relations of marine and halophilic species

of the unicellular green alga Dunaliella. The role of glycerol as a compatible solute. Arch. Microbiol. 96, 37–52.

8 Bradford K J and Hsiao T C 1982 Physiological responses to moderate water stress. *In* Encyclop. Plant Physiol. New Series, Vol 12B. Eds. O L Lange, P S Nobel, C B Osmond and H Ziegler. pp 263–324. Springer-Verlag, Berlin.

9 Champigny M L and Moyse A 1979 Photosynthetic carbon metabolism in wild primitive and cultivated forms of wheat at three levels of ploidy: role of glycolate pathway. Plant Cell Physiol. 20, 1167–1178.

10 Clarkson D T and Robards A W 1975 The endodermis, its structural development and physiological role. *In* Root Structure and Function. Eds. J Torrey and D T Clarkson. pp 415–436. Academic Press, London.

11 Cooil B J, De La Fuente R K and De La Pena R S 1965 Absorption and transport of sodium and potassium in squash. Plant Physiol. 40, 625–633.

12 Cowan I R 1982 Regulation of water use in relation to carbon gain in higher plants. *In* Encyclopaedia of Plant Physiology, New Series, Vol. 12B. Eds. O L Lange, P S Nobel, C B Osmond and H Ziegler. pp 589–613. Springer-Verlag, Berlin.

13 Delane R, Greenway H, Munns R and Gibbs J 1982 Ion concentration and carbohydrate status of the elongating leaf tissue of *Hordeum vulgare* growing at high external NaCl. I. Relationship between solute concentration and growth. J. Exp. Bot. 33, 557–573.

14 Dewey D R 1962 Breeding crested wheatgrass for salt tolerance. Crop Science 2, 403–407. r24.

15 Elzam O E and Epstein E 1969 Salt relations of two grass species differing in salt tolerance. I. Growth and salt content at different salt concentrations. Agrochimica 13, 187–195.

16 Eshel A and Waisel Y 1984 Effect of salt and soil water stress on transpiration of *Salsola kali* L. Plant Cell Environ. 7, 133–137.

17 Fischer R A 1981 Optimizing the use of water and nitrogen through breeding of crops. Plant and Soil 58, 249–278.

18 Flowers T J and Lauchli A 1983 Sodium versus potassium: substitution and compartmentation. *In* Encyclopedia of Plant Physiology, Vol. 15B. Eds. A Lauchli and R L Bielski. pp 651–681. Springer-Verlag, Berlin.

19 Ford C W 1982 Accumulation of O-methyl inositols in water-stressed *Vigna* species. Phytochemistry 21, 1149–1151.

20 Gale J and Zeroni M, 1985 The cost to plant of different strategies of adaptation to stress and the alleviation of stress by increasing assimilation. Plant and Soil 89, 57–67

21 Gale J, Poljakoff-Mayber A 1970 Interrelations between growth and photosynthesis of salt bush (*Atriplex halimus* L) grown in saline media. Aust. J. Biol. Sci. 23, 937–945.

22 Gorham J and Wyn Jones R G 1983 Solute distribution in *Suaeda maritima*. Planta 157, 344–349.

23 Gorham J and Storey R (*In preparation*).

24 Gorham J, Hughes L L and Wyn Jones R G 1981 Low-molecular-weight carbohydrates in some salt-stressed plants. Physiol. Plant. 53, 27–33.

25 Gorham J, McDonnell E and Wyn Jones R G 1984 Salt tolerance in the Triticeae. I *Leymus sabulosus*. J. Exp. Bot. 35, 1200–1209.

26 Gorham J, McDonnell E and Wyn Jones R G (*In preparation*).

27 Gorham J, McDonnell E and Wyn Jones R G 1984 Pinitol and other solutes in salt-stressed *Sesbania aculeata*. Z. Pflanzenphysiol. 114, 173–178.

28 Greenway H and Munns R 1983 Interactions between growth, uptake of Cl^- and Na^+, and water relations of plants in saline environments. II. Highly vacuolated cells. Plant Cell Environ. 6, 575–589.

29 Hajibagheri M A, Hall J L and Flowers T J 1983 The structure of the cutical in relation to cuticular transpiration in leaves of the halophyte *Suaeda maritima* (L.) Dum. New Phytol. 94, 125–131.

30 Hanson A D and Hitz W D 1982 Metabolic responses of mesophytes to plant water deficits. Annu. Rev. Plant Physiol. 33, 163–203.

31 Harvey D M R, Hall J L, Flowers T J and Kent B 1981 Quantitative ion localization within *Suaeda maritima* leaf mesophyll cells. Planta 151, 555–560.

32 Hill A E and Hill B S 1976 Elimination processes by glands: mineral ions. *In* Encyclopedia of Plant Physiology, New Series, Vol. 2B. Eds. U Luttge and M G Pitman. pp 225–243. Springer, Berlin.

33 Holbrook G P, Keys A J and Leech R M 1984 Biochemistry of photosynthesis in species of Triticum of differing ploidy. Plant Physiol. 74, 12–15.

34 Jacoby B 1979 Sodium recirculation and loss from *Phaseolus vulgaris* L. Ann. Bot. 43, 741–744.

35 Jeschke W D 1983 Cation fluxes in excised and intact roots in relation to specific and varietal differences. *In* Genetic Aspects of Plant Nutrition. Eds. M R Saric and Loughman. pp 71–86. Martinus Nijhoff, The Hague.

36 Jeschke W D 1984 K^+-Na^+ exchange at cellular membranes, intracellular compartmentation of cations and salt tolerance. *In* Salinity Tolerance in Plants – Strategies for Crop Improvement. Eds. R C Staples and G H Toenniessen. pp 37–66. Wiley and Sons, New York.

37 Jeschke W D and Stelter W 1976 Measurement of longitudinal profiles in single roots of Hordeum and Atriplex by use of flameless atomic absorption spectroscopy. Planta 128, 107–112.

38 Jeschke W D and Stelter W 1983 Ionic relations of garden orache, *Atriplex hortensis* L.: growth and ion distribution at moderate salinity and the function of bladder hairs. J. Exp. Bot. 34, 796–810.

39 Jeschke W D, Stelter W, Reising B and Behl R 1983 Vacuolar Na/K exchange, its occurrence in root cells of Hordeum, Atriplex and Zea and its significance for K/Na discrimination in roots. J. Exp. Bot. 34, 964–979.

40 Johanson J G and Cheeseman J M 1983 Uptake and distribution of sodium and potassium by corn seedlings. I. Role of the mesocotyl in sodium exclusion. Plant Physiol. 73, 153–158.

41 Johanson J G, Cheeseman J M and Enkoji C 1983 Uptake and distribution of sodium and potassium by corn seedlings. II. Ion transport within the mesocotyl. Plant Physiol. 73, 159–164.

42 Kaiser W M, Weber H and Sauer M 1983 Photosynthetic capacity, osmotic response and solute content of leaves and chloroplasts from *Spinacia oleracea* under salt stress. Z. Pflanzenphysiol. 113, 15–27.

43 Kramer D 1983 The possible role of transfer cells in the adaptation of plants to salinity. Physiol. Plant. 58, 549–555.

44 Larher F, Hamelin J and Stewart G P 1977 L'acide dimethylsulphonium-3-propanoique de *Spartina anglica*. Phytochemistry 16, 2019–2020.

45 Leigh R A and Johnston A E 1983 Potassium concentrations in the dry matter and tissue water of field-grown spring barley and their relationships to grain yield. J. Agric. Sci., Cambridge. 101, 675–685.

46 Lessani H and Marschner H 1978 Relation between salt tolerance and long-distance transport of sodium and chloride in various crop species. Aust. J. Plant Physiol. 5, 27–37.

47 Longstreth D J and Nobel P S 1979 Salinity effects on leaf anatomy. Plant Physiol. 63, 700–703.

48 Longstreth D J and Strain B R 1977 Effects of salinity and illumination on photosynthesis and water balance of *Spartina alterniflora* Loisel. Oecologia 31, 191–9.

49 Lüttge U 1975 Salt glands. *In* Ion Transport in Plant Cells and Tissues. Eds. D A Baker and J L Hall. pp 335–376. North Holland Publishing Co., Amsterdam.

50 Lüttge U 1983 Import and export of mineral nutrients in plant roots. *In* Encyclopedia of Plant Physiology, Vol. 15A. Eds. A Lauchli and R L Beileski. pp 181–211. Springer-Verlag, Berlin.

51 McGuire P E and Dvorak J 1981 High salt tolerance potential in wheatgrasses. Crop Science 21, 702–705.

52 Munns R, Brady C J and Barlow E W R 1979 Solute accumulation in the apex and leaves of wheat during water stress. Aust. J. Plant Physiol. 6, 379–389.
53 Munns R, Greenway H, Delane R and Gibbs J 1982 Ion concentration and carbohydrate status of the elongating leaf tissue of *Hordeum vulgare* growing at high external NaCl. II. Cause of the growth reduction. J. Exp. Bot. 33, 574–583.
54 Munns R, Greenway H and Kirst G O 1983 Halotolerant eukaryotes. *In* Encyclopedia of Plant Physiology, New Series, Vol. 12C. Eds. O L Lange, P S Nobel, C B Osmond and H Ziegler. pp 59–135. Springer, Berlin.
55 Osmond C B 1980 Integration of photosynthetic carbon metabolism during stress. *In* Genetic Engineering of Osmoregulation. Eds. D W Rains and R C Valentine. pp 171–186. Planum Press, New York.
56 Osmond C B, Bjorkman O and Anderson D J 1980 Physiological Processes in Plant Ecology. Towards a Synthesis with Atriplex. Springer, Berlin.
57 Osmond C B, Winter K and Ziegler H 1982 Functional significance of difference pathways of CO_2 fixation in photosynthesis. *In* Encyclopedia of Plant Physiology, New Series, Vol. 12B. Eds. O P Lange, P S Nobel, C B Osmond and H Ziegler. pp 479–547. Springer-Verlag, Berlin.
58 Papp J C, Ball M C and Terry N 1983 A comparative study of the effects of NaCl salinity on respiration, photosynthesis and leaf extension growth in *Beta vulgaris* L. (sugar beet). Plant Cell Environ. 6, 675–677.
59 Passioura J B 1982 Water in the soil-plant-atmosphere continuum. *In* Encyclopedia of Plant Physiology, Vol. 12B. Eds. O L Lange, P S Nobel, C B Osmond and H Ziegler. pp. 5–33. Springer-Verlag, Berlin.
60 Popp M 1984 Chemical composition of Australian mangroves. II. Low molecular weight carbohydrates. Z. Pflanzenphysiol. 113, 411–421.
61 Qureshi R H, Salim M, Abdullah M and Pitman M G 1982 *Diplachne fusca*: an Australian salt-tolerant grass used in Pakistani agriculture. J. Aust. Inst. Agric. Sci. 48, 195–199.
62 Robinson S P, Downton W J S and Millhouse J A 1983 Photosynthesis and ion content of leaves and isolated chloroplasts of salt-stressed spinach. Plant Physiol. 73, 238–242.
63 Rozema J, Van Manen Y, Vugts H F and Leusink A 1983 Airborne and soilborne salinity and the distribution of coastal and inland species of the genus Elytrigia. Acta Bot. Neerl. 32, 447–456.
64 Schleiff U 1984 The effect of soil osmotic and soil matric water potential of the rhizophere on the water uptake by wheat roots. Poster presented at the Third Intern. Workshop on Biosaline Research, March 19–23, 1984, Beer-Sheva, Israel.
65 Schonherr J 1982 Resistance of plant surfaces to water loss: transport properties of cutin, suberin and associated lipids. *In* Encyclopedia of Plant Physiology, New Series, Vol. 12B. Eds. O Lange, P S Nobel, C B Osmond, and H Ziegler. pp 153–179. Springer, Berlin.
66 Schulze E D and Hall A E 1982 Stomatal responses, water loss and CO_2 assimilation rates of plants in contrasting environments. *In* Encyclopedia of Plant Physiology, Vol. 12B. Eds. O L Lange, P S Nobel, C B Osmond and H Ziegler. pp 181–230. Springer-Verlag, Berlin.
67 Shannon M C 1978 Testing salt tolerance variability among tall wheatgrass lines. Agron. J. 70, 719–722.
68 Stelzer R and Lauchli A 1978 Salt- and flooding tolerance of *Puccinellia peisonis*. III. Distribution and localization of ions in the plant. Z. Pflanzenphysiol. 88, 437–448.
69 Stelzer R 1981 Ion localization in the leaves of *Pucinellia peisonis*. Z. Pflanzenphysiol. 103, 27–36.
70 Stewart C R and Hanson A D 1980 Proline accumulation as a metabolic response to water stress. *In* Adaptation of Plants to Water and High Temperature Stress. Eds. N C Turner and P J Kramer. pp 173–189. Wiley and Sons, New York.
71 Stewart G R and Lee J A 1974 The role of proline accumulation in halophytes. Planta Berlin. 120, 279–289.
72 Storey R, Pitman M G, Stelzer R and Carter C 1983 X-ray micro-analysis of cells and cell compartments of *Atriplex spongiosa*. I. Leaves. J. Exp. Bot. 34, 778–794.

73 Turner N C 1979 Drought resistance and adaptation to water deficits in crop plants. *In* Stress Physiology in Crop Plants. Eds. H Mussell and R C Staples. pp 343–372. Wiley-Interscience, New York.

74 Turner N C and Jones M M 1980 Turgor maintainance by osmotic adjustment: a review and evaluation. *In* Adaptation of Plants To Water and High Temperature Stress. Eds. N C Turner and P J Kramer. pp 87–103. John Wiley and Sons, New York

75 Tyree M T and Jarvis P G 1982 Water in tissues and cells. *In* Encyclopedia of Plant Physiology, Vol. 12B. Eds. O L Lange, P S Nobel, C B Osmond and H Ziegler. pp 35–77. Springer-Verlag, Berlin.

76 Wallaart R A M 1980 Distribution of sorbitol in Rosaceae. Phytochemistry 19, 2603–2610.

77 Weatherley P E 1982 Water uptake and flow in roots. *In* Encyclopedia of Plant Physiology, Vol. 12B. Eds. O L Lange, P S Nobel, C B Osmond and H Ziegler. pp 79–109. Springer-Verlag, Berlin.

78 Wyn Jones R G and Gorham J 1983 Aspects of salt and drought tolerance in higher plants. *In* Genetic Engineering of Plants. An Agricultural Perspective. Eds. T Kosuge, C P Meredith and A Hollaender. pp 355–370. Plenum Press, New York.

79 Wyn Jones R G and Storey R 1981 Betaines. *In* Physiology and Biochemistry of Drought Resistance in Plants. Eds. L G Paleg, and D Aspinall. pp 171–204. Academic Press, Sydney.

80 Wyn Jones R G, Storey R, Leigh R A, Ahmad N and Pollard A 1977 A hypothesis on cytoplasmic osmoregulation. *In* Regulation of Cell Membrane Activities in Plants. Eds. E Marre and O Ciferri. pp 121–135. Elsevier/North-Holland Biomedical Press, Amsterdam.

81 Wyn Jones R G and Gorham J 1983 Osmoregulation. *In* Encyclopedia of Plant Physiology, Vol. 12C. Eds. O L Lange, P S Nobel, C B Osmond and H Ziegler. pp 35–58. Springer-Verlag, Berlin.

82 Wyn Jones R G 1981 Salt tolerance. *In* Physiological Processes Limiting Plant Productivity. Ed. C B Johnson. pp 271–292. Butterworth, London.

83 Yeo A R 1981 Salt tolerance in the Halophyte *Suaeda maritima* L. Dum.: intracellular compartmentation of ions. J. Exp. Bot. 32, 487–497.

84 Yeo A R 1983 Salinity resistance: physiology and prices. Physiol. Plant. 58, 214–222.

Note added in proof

The taxonomy of the Triticeae is in a state of flux. The plant referred to as *Elytrigia juncea* above is now also called *Thinopyrum bessarabicum*.

Plant and Soil 89, 41–56 (1985). Ms. BR 03

Physiology of halophytes

T. J. FLOWERS
School of Biological Sciences, University of Sussex, Falmer, Brighton BN1 9QG, Sussex, UK

Key words Halophytes Physiology Salinity Salt tolerance

Summary The cellular basis of salt tolerance in halophytes depends upon the compartmentation of ions necessary for osmoregulation in vacuoles and upon osmotic adjustment of the cytoplasm by compatible solutes. The central role played by Na^+ and Cl^- in osmotic adjustment suggests that the transport of these ions and its regulation must be of primary importance in the physiology of the plant as a whole. There have been few investigations into the regulation of leaf ion concentrations, but such data as are in the literature suggest that limiting xylem Na^+ (and Cl^-) concentrations, together with continued leaf expansion, are particularly important. The role of phloem in retranslocation is uncertain due to lack of data. Decreases in transpiration rate per unit area of leaf help to lower the ion input into leaves. Any linked reductions in photosynthesis appear to be due to decreases in stomatal frequency.

Introduction

As a result of research carried out largely during the last decade [17, 35, 51], it has been possible to deduce the essential features of the cellular basis of salt tolerance in higher plants. For members of the Chenopodiaceae at least, the ability to withstand high external salt concentrations appears to depend on osmotic adjustment, most commonly with Na^+ and Cl^-, but since these are toxic their concentration in the cytoplasm is maintained substantially lower than in the vacuoles (Table 1). Osmotic adjustment in the cytoplasm is achieved with compatible organic solutes [44, 46]. This brief summary of a number of years of experimental work is necessarily neglectful of detail, but highlights the central role played by sodium and chloride in the physiology of the dicotyledonous halophytes. It is their ability to cope with high internal concentrations of Na^+ and Cl^- that separates halophytes from glycophytes. This brief review will then be concentrated on our knowledge of ion transport from root to leaf and its regulation and involvement in growth; often there will be more questions than answers.

Terminology

Plants grown at a constant external NaCl concentration over a period of weeks are considered to be in a 'steady state' in order to contrast them with those recently (up to a few days) subjected to a change in the external salinity.

Table 1. Ion distribution in mesophyll cells of mature leaves of *Suaeda maritima* (L.) Dum. grown at a steady-state salinity of 340 m*M*

Cell compartment	Vol[a] per cell (pl)	Concentration[b] (mol m^{-3})			Relative contents[f]		
		Na^+	K^+	Cl^-	Na^+	K^+	Cl^-
Cytoplasm	9.2	116(136)[c]	19(21)[c]	60(71)[c]	0.03	0.10	0.02
Chloroplasts	1.2	104(208)[d]	22(44)[d]	98(196)[d]	0.003	0.02	0.004
Mitochondria	0.6	–	–	–	–	–	–
Vacuole	67.8	494	20	352	0.94	0.84	0.95
Cell wall	4.4	194(554)[e]	14(40)[c]	138(394)[e]	0.02	0.03	0.02
Total	83.1				35.5[f]	1.6[f]	25.1[f]

a. Data of Hajibagheri[24]
b. Data of Harvey *et al.*[28] : the concentration is in mol m^{-3} of analyzed volume and not mol m^{-3} H_2O.
c. Value in parenthesis assumes 85% (v/v) water content.
d. Value in parenthesis assumes a 55% solute-available space – (ref. [37], p. 86).
e. Based on the water content (35%, w/w) of filter paper at – 1.5 MPa (ref. [50]).
f. Relative to the total content in pmol.
– No data available.

Osmotic adjustment

It has long been known that halophyte cells must have lower water potentials within than outside the plasmalemma to retain cellular water and that the necessary osmotic adjustment in dicotyledons is largely achieved by Na and Cl ions[16]. In the Gramineae, however, K^+ and sugars appear to aid in fulfilling this role[1,20]. The use of organic compounds as vacuolar solutes in succulent halophytes seems to be precluded on energetic grounds[56]. Thus, in succulent halophytes, vacuolar concentrations (based on tissue water) of Na^+ and Cl^- generally exceed the external concentrations, and by a considerable margin where the latter are low (Table 2). This appears to reflect a constitutive ability of halophytes to accumulate high ion concentrations [17]. However, although steady-state Na and Cl ion contents, expressed per unit dry weight, of leaves of some species appear to be relatively constant over higher external NaCl concentrations [16], this is not necessarily so if the leaf concentration is based on the water content. It is, however, very difficult to make any general conclusions concerning the regulation of these ion concentrations (see[11]), because of a lack of information on plant water potential, turgor and ion fluxes.

There is still a dearth of data on ion transport and on net ion fluxes to the shoots of halophytes, largely because too few investigators record data with time[35]. For *Suaeda maritima* plants growing in the

Table 2. The effect of changed external salinity (Na_{out}) on Na and Cl ion concentrations (m*M*) in the leaves (Na_{vac}, Cl_{vac}) of various halophytes

Species and family	Na_{out}	Na_{vac}	Na_{vac}/Na_{out}	Cl_{vac}	Cl_{vac}/Cl_{out}	Reference
Atriplex nummularia	50	348	7.0	314	6.3	4
Chenopodiaceae	200	429	2.1	338	1.7	
	600	870	1.4	834	2.1	
A. nummularia	10	538[a]	[b]	189	19	22
Chenopodiaceae	100	618[a]	6.2	252	2.5	
	300	751[a]	2.5	483	1.6	
A. hastata	100	658	6.6	430	4.3	7
Chenopodiaceae	400	1203	3.0	911	2.3	
A. spongiosa	50	114	2.3	37	0.74	47
Chenopodiaceae	500	598	1.2	341	0.68	
Suaeda maritima	170	380	2.2			54
Chenopodiaceae	340	450	1.3			
S. monoica	50	695	14	392	7.8	47
Chenopodiaceae	500	843	1.7	624	1.2	
Jaumea carnosa	155	166	1.1	162	1.0	45
Asteraceae	600	496	0.83	340	0.57	
Avicennia marina	49	456	9.3	306	5.3[c]	14
Avicenniaceae	123	550	3.7	362	2.5[d]	
	493	956	1.9	631	0.92[e]	
Disphyma australe	25	101	4	80	3.2	36
Aizoaceae	100	222	2.2	157	1.6	
	500	475	0.95	364	0.73	
Elymus oliveri[f]	75	140	1.9			2
Poaceae	100	169	0.56			
Puccinellia maritima[f]	100	112	1.1			2
Poaceae	400	139	0.35			

a, Na^+ plus K^+; b, meaningless; c, Cl_{out} = 58 m*M*; d, Cl_{out} = 145 m*M*; e, Cl_{out} = 580 m*M*; f, unaerated culture but see ref. [10].

steady state in 340 m*M* NaCl, net sodium fluxes to the shoot during the growth of the plant range between 5 and 9 mmol/g dry weight of roots/day and are reasonably correlated with the relative growth rate calculated on the basis of the organic dry weight (Table 3). This net flux is considerably higher than the K^+ fluxes reported for glycophytes (1 to 2 mmol/g d.w. roots/day [39, 40, 56]) and indicates the enhanced transport capabilities of some halophytes as compared with glycophytes. However, for halophytic members of the Graminae, which differ from dicotyledenous plants in terms of net accumulation of Na ions [1, 20] (and Table 2), Na ion fluxes are lower, especially relative to fluxes of K^+ (Table 3). Although net K^+ fluxes are rather low in Suaeda (*cf.* 39),

Table 3. Relative growth rates (RGR) and net ion fluxes (J) to the shoots of three salt-tolerant species. The relative growth rates of Suaeda are based on organic dry matter alone

Species	t (days)	J_{Na} (mmol/g d.w. root/d)	J_K (mmol/g d.w. root/d)	RGR (d^{-1})	J_{Na}/J_K	Reference
Suaeda maritima (340 m*M*)	21–28	6.21	0.65	0.174	9.6	54
	28–35	9.44	0.86	0.249	11.0	
	35–42	8.04	0.55	0.169	14.6	
	42–49	5.09	0.21	0.116	24.2	
	21–49	7.20	0.57	0.177	14.9	
Elymus oliveri (150 m*M*)	3–33	0.48	0.51	0.10	0.94	2
Puccinellia maritima (200 m*M*)	7–63	0.30	0.34	0.10	0.88	2

there may be considerable retranslocation of K^+ from older leaves to the shoots[55], which thus maintain adequate K/Na ratios in the rapidly growing apices [21].

Potassium is evidently required during initiation of protein synthesis, and in most cells cytoplasmic concentrations seem to be about 80–100 m*M*[52]. In mesophyll cells of *S. maritima*, the K^+ concentration is estimated to be only about 20 m*M* (Table 1). This low concentration may be characteristic of mature cells, where Na^+ may substitute for K^+ (see 18), while higher K^+ concentrations exist in meristemic zones [21]. It is, however, apparent from preliminary results in our laboratory that Suaeda microsomes in a heterologous *in vitro* protein synthesis system with wheat postribosomal supernatant are more tolerant of low K^+ and high Na^+ than are wheat ribosomes themselves and can incorporate methionine at 70% of maximal levels (*i.e.* with 120 m*M* potassium acetate) in the presence of 120 m*M* Na^+ and 28 m*M* K^+. The extent to which Na^+ may substitute for K^+ in protein synthesis remains to be fully explored.

For plants growing under natural conditions, rapid osmotic adjustment is presumably essential to cope with fluctuations in the external osmotic potential brought about by rain and drought. It has been calculated from measurements of the initial influx of radioactive ions into *S. maritima* that osmotic adjustment to seawater salinity can be achieved in 24–48 h[17, 56]. In young plants (32d) an increase in the external NaCl concentration from 2 or 200 m*M* to 202 and 400 m*M*, respectively, increased net transport to the shoot to maximum rates of 9 and 11 mmol/g dry weight root/day, respectively, some 6 h after the increase in salinity [9]. Levels of abscisic acid in the shoot also increased over a 24-h period, approximately doubling for plants going from 0 to 200 m*M* NaCl and increasing five-fold when the salinity was increased

Table 4. Salt concentration in a variety of saline environments

Site	Range Na (mM)	Reference
Salt marsh, Hudson Bay, Canada	100–1060	32
Salt marsh, Stiffkey, U.K.	200–910	31
Salt marsh, Georgia, U.S.A.	262–710*	3
Salt marsh, New South Wales, Australia	283–593*	8
Salt desert, Utah, U.S.A.	217–2380*	27
Mangrove swamp, New South Wales, Australia	289–808*	8

*Calculated assuming water with a salinity of 35% has an osmotic pressure of 2.60 MPa, that Na is 30.4% of the S% and that the specific gravity is measured at 25°C to a maximum level of 1.025, (ref. [29]). Chloride concentration would be about 15% higher.

from 200 to 400 mM [9]. It appears unlikely, however, that halophytes have to cope with high water potentials, since dilution of the external solution to produce an osmotic potential greater than − 1.0 MPa seems to be uncommon, from the minimum sodium concentrations recorded in various saline marshes (Table 4). Halophytes are therefore presumably adapted to tolerate fluctuations in the external salinity following rain and drought, but with a minimum Na^+ concentration of about 200 mM (or perhaps 100 mM). A low internal osmotic potential presumably obviates the need for continuous osmotic adjustment under fluctuating salinities and presumably aids the maintenance of positive turgor.

Until recently it has been difficult to estimate turgor pressure in the cells of halophytes. Data on osmotic potentials are relatively easy to obtain and closely reflect changes in the Na^+ and Cl^- concentrations, decreasing with increasing external salinity [16, 17]. This is so not only in whole plants, but in individual leaves [9]. However, good water potential measurements are most difficult to obtain. Those made on excized leaves are probably unreliable because of the high concentrations of salt in cut cells and in cell walls (Table 1). Low water permeabilities make the use of whole leaves in psychometers equally questionable because of excessive equilibration times. Furthermore, the very low water potentials mean very high pressures are required if a pressure bomb is used, and little or no exudate is attainable [42]. However, use of a pressure probe in Wyn Jones' laboratory has shown that turgor is maintained approximately constant in mature leaves of Suaeda at various external salinities [48] (and Fig. 1). Since vacuolar concentrations of Na^+ and Cl^- vary considerably under these conditions, this suggests that a turgor homeostat may operate as outlined by Cram [11]. Even in

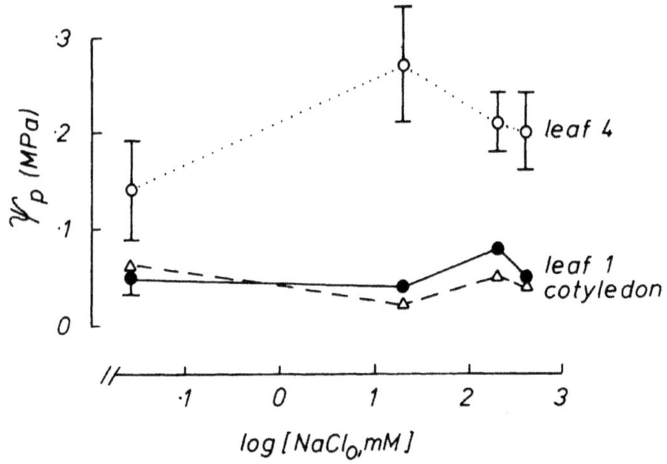

Fig. 1. The turgor pressure in leaf epidermal cells of 27 to 29 day old plants of *Suaeda maritima* grown at various steady-state salinities (mean ± standard deviation). Where no error bar is shown the deviation was too small to illustrate. Leaf age decreases with increasing number. Unpublished data of Clipson and Tomos.

younger leaves (leaf 4), where turgor pressure may reach 0.4 MPa, the variation in turgor is small (less than 0.15 MPa) over a wide range of external salinities (about 1 MPa – Fig. 1).

Factors affecting leaf ion concentrations

In principle, regulation of vacuolar ion concentrations with time, at a steady-state external salinity, could be achieved through balancing import (xylem), export (phloem or salt-glands where present), and volume (i.e. water content). There have, however, been few investigations on the long-term balance of ion concentration within individual leaves. Leaf expansion in Rhizophora, a mangrove species without salt glands, continued throughout the life of the leaf, and the water content per leaf increased with leaf age[5] (and Fig. 2). The amount of sodium per leaf also increased, but its concentration did not change in older leaves. An increase in the fresh weight per leaf was also reported for *Mesembryanthemum crystallinum* grown for over 28 days under saline conditions, although a continuing increase in weight did not occur in older leaves [53] (and Fig. 2). A similar picture appears to hold for leaves of *Suaeda maritima*, although measurements were made over a shorter period (Fig. 2). It is also apparent from work with a pressure probe [48] that the elastic modulus of walls of Suaeda leaves declined with

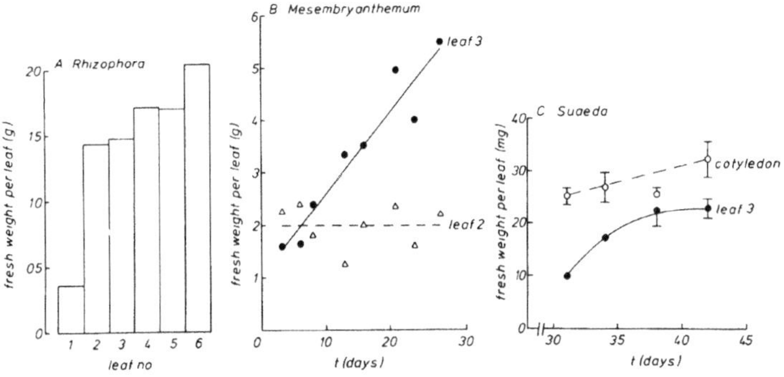

Fig. 2. Changes in leaf size with time for plants growing under saline conditions. **A.** Rhizophora: leaf 1 is the youngest and leaf 6 beginning to senesce. Drawn from the data of Atkinson et al.[5]. **B.** *Mesembryanthemum crystallinum*: leaf 2 is older than leaf 3. Redrawn from [53]. **C.** *Suaeda maritima*; unpublished data of the authors. Mean ± s.e.m. After day 35 the number of live cotyledons decreases.

declining turgor. This suggests that in older leaves with lower turgor pressure, continued leaf expansion by uptake of water would have less effect on that turgor pressure than in younger leaves. If the pattern of leaf expansion described above holds for other species then it may be important to distinguish between phases of rapid and then slow expansion in volume, perhaps with increasing and then constant ion concentrations, respectively. However, definitive statements must await studies of leaf volume in relation to leaf age. The maintenance, with time, of constant leaf Na^+ and Cl^- concentrations is evidently much simplified by the presence of salt glands, and species with such glands may tolerate higher salt concentrations in the xylem than those without glands [35]. Thus, although there is too little data for confident generalization, there is a hint that leaf ion concentrations do increase during the early life of a leaf on a plant growing under steady-state salinity [5, 9, 53]. Volume expansion may occur throughout the life of the leaf, albeit at a changing rate. These factors probably play a significant role in the ion relations during the life of a leaf. What may be particularly important as far as the regulation of turgor pressure is concerned is the balance between ion concentations in vacuole and apoplast. As yet, we can only hypothesize.

The delivery of ions to a leaf is a function of the salt concentration in the xylem and the transpiration rate of that leaf. Ion concentrations in the xylem are perhaps best deduced from changes in the shoot ion contents with time and the transpiration rate of intact plants, since detaching shoots from roots dramatically alters the pressure within

Table 5. Na^{+} concentration in xylem of several halophytes

	Glands	$[Na]_{out}$ (mM)	$[Na]_{xyl}$ (mM)	$[Na]_{xyl}$/$[Na]_{out}$ (%)	Reference
Glaux maritima	+	200	11.6	5.8	41
Armeria maritima	+	200	8.7	4.4	41
Limonium vulgare	+	200	12.2	6.1	41
Spartina anglica	+	200	11.2	5.6	41
Juncus maritima	–	200	7.7	3.9	41
Atriplex litoralis	–	200	5.3	2.7	41
Suaeda maritima	–	200	46.6	23	9
Salicornia virginica	–	210	9.6	4.6	38
S. virginica	–	530	19.0	3.6	38

the xylem (perhaps by some 2.0 MPa for a plant growing in seawater [23]). A large change in pressure is likely to have an effect on ion fluxes [11, 58]. Xylem concentrations of Na^{+} calculated from changes in the shoot content range from about 3 to 10% of the external salinity (Table 5). Although higher concentrations have been reported [23, 42], such high values may be suspect if obtained by the use of pressure, which, when applied to cut shoots [41] and roots [43], may result in values higher than those estimated from the ion and water fluxes. However, measurement of the transpiration rates through weighing or other indirect methods (see 5) may also be subject to considerable technical errors. In a recent study, Na ion flux in *Suaeda maritima*, a halophyte without glands, was deduced from both tracer fluxes and changes in shoot ion content. Transpiration was measured by gas exchange[9]. Xylem concentrations of Na^{+} calculated from net transport agreed well with those based on tracer fluxes and varied with the external concentration (Fig. 3); under optimum conditions for growth, Na^{+} was about 25% of the external salt concentration (200 mM). A particularly notable feature was the apparent increase in xylem Na^{+} concentration at night (Fig. 3). However, water fluxes were very low in the dark, and a 10% change in the plant water content would decrease the apparent xylem concentration by about half. An increase in apoplastic water content would be expected if xylem tensions were reduced during the night [57]. This might lessen the apoplastic salt concentration (Table 1) and hence increase cell turgor.

As already mentioned, the delivery of ions to the shoot is a function of the xylem ion concentrations and the transpiration rate. For many years it has been appreciated that the transpiration decreases as the external steady-state salinity increases. However, as with many other aspects of halophyte physiology, the basis on which the transpiration rate is expressed is all important, since salinity affects fresh and dry

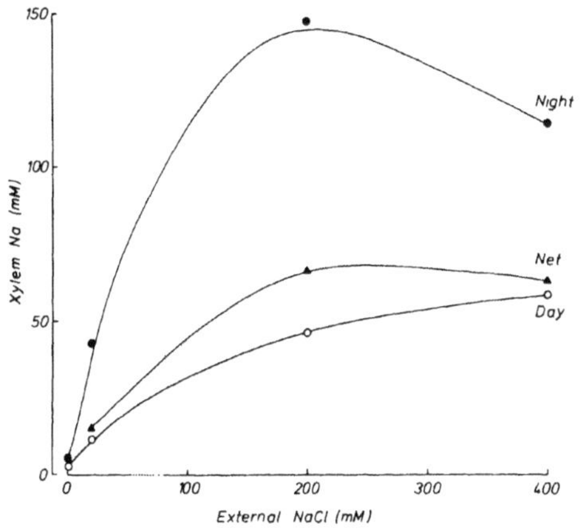

Fig. 3. The effect of steady-state salinity on the xylem concentration of Na^+ calculated from $^{22}Na^+$ transport (•, ○, night and day) and change in shoot ion content (▲). Transpiration rates were measured by gas exchange[9].

weight as well as succulence. It appears that transpiration rates expressed per unit area are depressed at least in the Chenopodiaceae (Fig. 4). For *Suaeda maritima*, the decrease in transpiration is associated with an increase in both stomatal and cuticular resistance (Fig. 5). The increase in cuticular resistance is accompanied by a

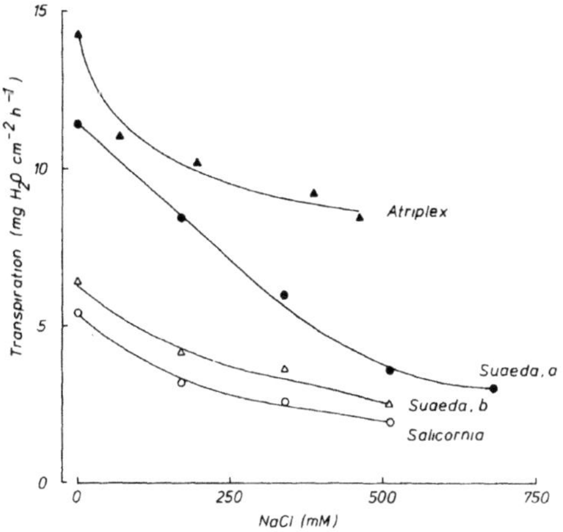

Fig. 4. The effect of increasing steady-state salinity on transpiration in three halophytic members of the Chenopodiaceae. *Atriplex halimus*, saturation deficit (s.d.) 2.5 KPa (drawn from the data in ref. [19]); *Suaeda maritima* (a), s.d. 1.8 KPa (ref. [43]); *Suaeda maritima*, (b), s.d. 0.9 KPa (ref. [13]); *Salicornia europaea* s.d. 0.9 KPa (ref. [13]).

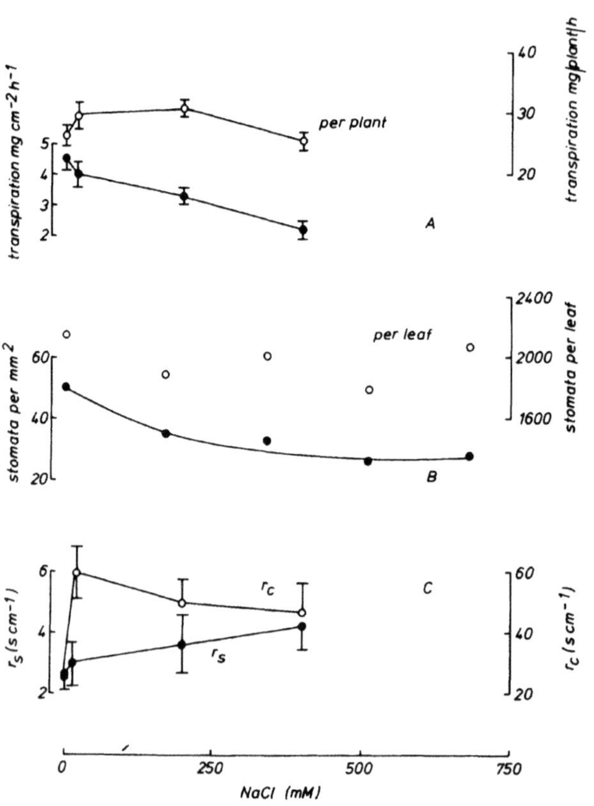

Fig. 5. Transpiration in *Suaeda maritima* at various steady-state salinities. **A.** Transpiration rates [9] expressed per unit area (•) or per plant (○). **B.** Stomatal frequencies[43], per unit area (•) or per unit leaf (○). **C.** Stomatal (•) and cuticular (○) resistances[9]. Vertical bars are standard errors.

1.6-fold increase in the thickness of the cuticle and a 1.8-fold increase in that of the epidermal cell wall under saline conditions, together with a change in the appearance of the epicuticular wax [25]. The change in stomatal resistance parallels exactly a change in stomatal frequency and so does not necessarily imply any increase in resistance to movement through the stomata themselves. Changes in stomatal frequency are, in fact, a consequence of the increase in succulence and have been reported for Salicornia [13] and Distichlis[33]. In *Suaeda maritima* then, transpiration per plant increases under saline conditions (Fig. 5) as a consequence of an increased leaf area. However, ions delivered per unit area or per unit volume of leaf will be reduced, probably as a consequence of the reduced stomatal frequency. Although many general features of the water relations have been deduced, there is still insufficient data to determine the effects of leaf age on water flux and ion balance. Such experiments would now be very timely. Furthermore, the extent of possible differences in transpiration rates between leaves are unknown, as is the extent of bypass flow in halophytes.

There are indications from an anatomical study that the development of the Casparian strip occurs within 3 mm of the root tip in *Suaeda maritima* plants grown under saline conditions and is retarded in the absence of NaCl: 15 mm from the root tip the Casparian strip averages 0.93 μm in length in salt-grown plants but only 0.46 μm in plants raised in nonsaline culture solution [24]. Vacuolation of epidermal and cortical cells is also enhanced by salinity. Although root development is not remarkably more rapid than in salt-resistance glycophytes (see ref. [35], p. 111), early prevention of apoplastic flow would assist in controlling the amount of ions in the shoots. Another feature of *S. maritima* roots is the narrow width of the cortex (3 cells), a feature previously noted in other halophytes [48]. It has been speculated [23] that this may play a role in minimizing ion concentration gradients across the endodermis. However, the model on which these conclusions were based did not specifically include ion uptake by the root, and the true significance of the narrow cortex in these plants must await further work.

There is almost no information concerning phloem Na^+ (and Cl^-) concentrations in halophytes[23] and the possible rates of export of these ions from leaves. Calculations based on the cross-sectional area of phloem in *S. maritima* leaves indicate that Na^+ export from a mature Suaeda leaf might approach 20% of daily import, assuming a phloem Na^+ concentration of 30 m*M*. However, Yeo[66] established that there was little if any apical movement of Na ions from mature leaves of Suaeda into new leaves, a point neglected by Greenway and Munns[23]. The importance of K^+ retranslocation in maintaining adequate K/Na ratios has already been discussed.

The effect of salinity on transpiration per unit area indicates that concomitant effects on photosynthesis may be expected. Such changes may be important in understanding the effects of salinity on growth, in particular, the inhibition of growth at high salinity. Again, determination of the true effect of salinity on the photosynthetic rate is liable to be confounded by the basis of expression, since salinity treatment affects fresh and dry weight and leaf area. Rates per unit area are decreased or are little affected by NaCl. Although there are very few data available, rates per unit of chlorophyll appear either to be unaffected or to increase (Table 6) with increasing external salinity. There is, therefore, little evidence to suggest any major effects of

Table 6. The effect of external salinity (steady state) on net photosynthesis (P_N) expressed per unit area and per unit of chlorophyll in several halophytes. Data are expressed as percent of non-saline controls

Species	$NaCl_{out}$ (mM)	P_N(%)		Reference
		per area	per chlorophyll	
Distichlis spicata	250	71	–	33
	500	60	–	
Aeluropus litorolis	100	–	93	15
Aster tripolium	85	125	–	6
	170	114	–	
	340	98	–	
Atriplex halimus	68	89	–	19
	195	93	–	
	462	64	–	
Suaeda maritima	100	–	125	51
	400	–	125	
	20	90	92	9
	200	80	100	
	400	72	103	

salinity on photosynthesis *per se*, and *in vitro* electron transport from water to ferricyanide is notably tolerant of high Cl^- [12, 26]. There does not appear to be evidence to suggest a decline in photosynthesis *per se* as the cause of growth inhibition at high salinity.

Conclusions

Growth data on halophytes are difficult to evaluate, since such a high proportion of the dry weight consists of ions. Although it is easy to subtract these and compute data on a organic dry matter basis, when there is little stimulation of growth[38,54] the usefulness of this subtraction in evaluating growth and its costs depends upon whether a significant proportion of energy is required for salt transport. The costs of salt transport were considered in some depth in two recent reviews [23, 56]. The consensus is that at high salinities ion transport (through cost or capacity) may limit osmotic adjustment and hence limit growth, presumably through loss of turgor. The rather low turgor pressures recorded for *Suaeda maritima* (Fig. 1) suggest that ion accumulation in cell walls (Table 1) may be an important element of the ion relations in a leaf. Decrease in growth and in leaf longevity at high salinity may be determined by this ion accumulation. It is clear from the number of unknowns entering into the calculations of Yeo[56] and of Greenway and Munns[23] that further attempts at evaluating the energetics of ion transport might be more fruitful if based on an analysis of the carbon and energy budgets, given the problems with the measurement of chemical potential gradients. Such an approach might evaluate the sources of energy for osmotic adjustment – whether photosynthetic or respiratory. It is interesting to note in this respect that large parenchyma cells surrounding the vascular bundles in mature leaves of *Suaeda maritima* contain very few mitochondria and no chloroplasts[24] and may rely on electrical coupling to mesophyll cells for ion transport. It is also noteworthy that chloroplasts of *S. maritima* accumulate starch under optimum growth conditions[24]. Whether this is a consequence of reduced translocation, perhaps induced through phosphate or nitrate deficiency, or simply implies that photosynthesis outstrips demand, is not clear.

An important question relative to this conference is whether the growth rates of halophytes are so low as to preclude their use as crop species. Again, there is rather limited data in the literature on which to base any conclusion, since in most experiments plants are harvested at a fixed time. Certainly the data for Suaeda (Table 3) at its growth optimum do not suggest that the relative growth rates are dramatically limited by salinity, since the maximum value (0.25 day^{-1}; Table 3) is in the range reported for crop species that are not halophytes (0.03 to 0.38 d^{-1})[30]. Long-term values based on organic dry weight (0.18 day^{-1} Table 3, 0.06 g day^{-1} for *Salicornia virginica* [38]) also fall within the range normally reported for glycophytes[30]. It is my view at present, that halophytes do offer a viable alternative[34]. At high salinities their growth is presumably limited not by a single factor but by factors that include a lack of transport capacity, an inbalance in nutrient uptake (*e.g.* K^+, P_i), a limitation on retranslocation of K^+ to growing tips, and excessive ion accumulation in the cell walls of the leaves, which reduces turgor and hence growth. It is clear that the time is right for more detailed experiments aimed at elucidating the regulation of leaf ion concentrations and the overall energy budget of halophytes.

Acknowledgements I would like to acknowledge the help of Maggie Yeo in drawing the figures and Nick Clipson and Nasser Hajibagheri for discussions during the preparation of the manuscript.

References

1 Albert R and Popp M 1977 Chemical composition of halophytes from the Neusiedler Lake region in Australia. Oecologia Berl 27, 157–170.
2 Ansari R 1982 Salt tolerance studies in some grasses. D.Phil. Thesis, University of Sussex.
3 Antlfinger A E and Dunn E L 1983 Water use and salt balance in three salt marsh succulents. Am. J. Bot. 70, 561–567.

4 Ashby W C and Beadle N C W 1957 Studies in halophytes. III. Salinity factors in the growth of Australian salt bushes. Ecology 38, 344–352.

5 Atkinson M R, Findlay G P, Hope A B, Pitman M G, Saddler H D W and West K R 1967 Salt regulation in the mangroves *Rhizophora mucronata* Lam. and *Aegilitis annulata* R.Bl. Aust. J. Biol. Sci. 20, 589–599.

6 Baumeister W and Schmidt L 1962 Uber der Rolle des Natriums im Pflanzlichen Stoffwechsel. Flora 152, 24–56.

7 Black R F 1956 Effect of NaCl in water culture on the ion uptake and growth of *Atriplex hastata* L. Aust. J. Biol. Sci. 9, 67–80.

8 Clarke L D and Hannon N J 1969 The mangrove swamp and salt marsh communities of the Sydney district. II. The holocoenotic complex with particular reference to physiography. J. Ecol. 57, 213–234.

9 Clipson N J C 1984 Salt tolerance in the halophyte *Suaeda maritima* D.Phil.Thesis, University of Sussex.

10 Cooper A 1982 The effects of salinity and water logging on the growth and cation uptake of salt marsh plants. New Phytol. 90, 263–275.

11 Cram W J 1980 Chloride accumulation as a homeostatic system: negative feedback signals for concentration and turgor maintenance differ in a glycophyte and a halophyte. Aust. J. Plant Physiol. 7, 237–249.

12 Critchley C 1982 Stimulation of photosynthetic electron transport in a salt-tolerant plant by high chloride concentrations. Nature London 298, 483–484.

13 Dahesh K 1977 Aspects of the physiology of *Salicornia* spp. D.Phil. Thesis, University of Sussex.

14 Downton W J S 1982 Growth and osmotic relations of the mangrove *Avicennia marina* as influenced by salinity. Aust. J. Biol. Sci. 9, 519–528.

15 Downton W J S and Torokfolvy E 1975 Effect of sodium chloride on the photosynthesis of *Aeluropus litoralis*, a halophytic grass. Z. Pflanzenphysiol. 75, 143–150.

16 Flowers T J 1975 Halophytes. *In* Ion Transport in Plant Cells and Tissues. Eds. D A Baker and J L Hall. pp 309–334. Elsevier/North Holland, Amsterdam.

17 Flowers T J, Troke P F and Yeo A R 1977 The mechanism of salt tolerance in halophytes. Annu. Rev. Plant Physiol. 28, 89–121.

18 Flowers T J and Läuchli A 1983 Sodium versus potassium: substitution and compartmentation. *In* Encyclopedia of Plant Physiology New Series, vol 15 Inorganic Plant Nutrition. Eds. A Läuchli and R L Bieleski. pp 651–681. Springer, Berlin.

19 Gale T and Poljakoff-Mayber A 1970 Interrelations between growth and photosynthesis of salt bush (*Atriplex halimus* L.) grown in saline media. Aust. J. Biol. Sci. 23, 937–945.

20 Gorham J, Hughes L L and Wyn Jones R G 1980 Chemical composition of salt-marsh plants from Ynys Mon (Anglesey): the concept of physiotypes. Plant Cell Environ. 3, 309–318.

21 Gorham J and Wyn Jones R G 1983 Solute distribution in *Suaeda maritima*. Planta Berlin 157, 344–349.

22 Greenway H 1968 Growth stimulation by high chloride concentrations in halophytes. Isr. J. Bot. 17, 169–177.

23 Greenway H and Munns R 1983 Interactions between growth, uptake of Cl^- and Na^+, and water relations of plants in saline environments. II. Highly vacuolated cells. Plant Cell Environ. 6, 575–589.

24 Hajibagheri M A 1983 Physiological and ultrastructural aspects of salt-tolerance in the halophyte *Suaeda maritima* (L.) Dum. D.Phil.Thesis, University of Sussex.

25 Hajibagheri M A, Hall J L and Flowers T J 1983 The structure of the cuticle in relation to cuticular transpiration in leaves of the halophyte *Suaeda maritima* (L.) Dum. New Phytol. 94, 125–131.

26 Hajibagheri M A, Harvey D M R and Flowers T J 1984 Photosynthetic oxygen evolution in relation to ion contents in the chloroplasts of *Suaeda maritima*. Plant Sci. Lett. 34, 353–362.

27 Hansen D J, Dayanandar P, Kaufman P B and Brotherson J D 1976 Ecological adaptations

of salt marsh grass, *Distichlis spicata* (Gramineae), and environmental factors affecting its growth and distribution. Am. J. Bot. 63, 635–650.
28 Harvey D M R, Hall J L, Flowers T J and Kent B 1982 Quantitative ion localization wtihin *Suaeda maritima* leaf mesophyll cells. Planta Berlin 151, 555–560.
29 Harvey H W 1966 The Chemistry and Fertility of Sea Waters. Cambridge University Press, Cambridge.
30 Hunt R 1978 Plant Growth Analysis. Arnold, London.
31 Jefferies R L 1977 Growth responses of coastal halophytes to inorganic nitrogen. J. Ecol. 65, 847–865.
32 Jeffries R L, Jensen A and Bazoly D 1983 The biology of the annual *Salicornia europaea* agg. at the limits of its range in Hudson Bay. Can. J. Bot. 61, 762–773.
33 Kemp P R and Cunningham G L 1981 Light, temperature and salinity effects on growth, leaf anatomy and photosynthesis of *Distichlis spicata* (L.) Greene. Am. J. Bot. 68, 507–516.
34 Malcolm C V 1969 Use of halophytes for forage production on saline wasteland. J. Aust. Inst. Agric. Sci. 35, 38–49.
35 Munns R Greenway, H and Kirst G O 1983 Halotolerant eukaryotes. *In:* Encyclopedia of Plant Physiology New Series, Volume 12C. Physiological Plant Ecology III. Eds. O L Lange, P S Nobel, C B Osmond and H Ziegler. pp 59–135. Springer Verlag, Berlin.
36 Neales T F and Sharkey P J 1981 Effect of salinity on growth and on mineral and organic constituents of the halophyte *Disphyma australe* (Soland) J M Black. Aust. J. Plant Physiol. 8, 165–179.
37 Nobel P S 1983 Biophysical Plant Physiology and Ecology. p. 86, W H Freeman. San Francisco.
38 Ownbey R S and Mahall B E 1983 Salinity and root conductivity: differential response of a coastal succulent halophyte, *Salicornia virginica*, and a weedy glycophyte *Raphanus sativus*. Physiol. Plant. 57, 189–195.
39 Pitman M G 1975 Whole Plants. *In* Ion Transport in Plant Cells and Tissues. Eds. D A Baker and J L Hall. pp 267–308. North Holland, Amsterdam.
40 Pitman M G 1982 Transport across plant roots. Q. Rev. Biophys. 15, 481–554.
41 Rozema H, Gude H, Bijl F and Wesselman H 1981 Sodium concentrations in xylem sap in relation to ion exclusion, accumulation and secretion in halophytes. Acta. Bot. Neerl. 30, 309–311.
42 Scholander P F, Bradstreat E D, Hammel H T and Hemmingsen E A 1966 Sap concentrations in halophytes and some other plants. Plant Physiol. 41, 529–532.
43 Siadat-Pour A 1978 Physiological studies on *Suaeda maritima* (L.) Dum. D. Phil. Thesis, University of Sussex.
44 Stewart G R, Larher F, Ahmad I and Lee J A 1977 Nitrogen metabolism and salt-tolerance in higher plant halophytes. *In* Ecological Processes in Coastal Environments. Eds. R L Jefferies and A J Davy. pp. 211–227. Blackwell, Oxford.
45 St. Omer L and Schlesinger W H 1980 Regulation of NaCl in *Jaumea carnosa* (Asteraceae), a salt marsh species, and its effect on leaf succulence. Am. J. Bot. 67, 1448–1454.
46 Storey R and Wyn Jones R G 1977 Quaternary ammonium compounds in plants in relation to salt resistance. Phytochemistry 16, 447–453.
47 Storey R and Wyn Jones R G 1979 Responses of *Atriplex spongiosa* and *Suaeda monoica* to salinity. Plant Physiol. 63, 156–162.
48 Tomos A D and Wyn Jones R G 1982 Water relations in the epidermal cell of the halophyte *Suaeda maritima. In* Biophysics of Water. Eds. F Franks and S Mathias. pp 327–331. Wiley, New York.
49 Waisel Y 1972 Biology of Halophytes. Academic Press, New York.
50 Wiebe H H 1966 Matric potential of several plant tissues and biocolloids. Plant Physiol. 41, 1439–1442.
51 Wyn Jones R G, Storey R, Leigh A, Ahmad N and Pollard A 1977. A hypothesis on cytoplasmic osmoregulation. *In* Regulation of Cell Membrane Activities in Plants. Eds. E Marré and O Ciferri. pp 121–136, Elsevier/North Holland, Amsterdam.

52 Wyn Jones R G, Brady C J and Speirs J 1979 Ionic and osmotic relations in plant cells. *In* Recent Advances in the Biochemistry of the Cereals. Eds. D L Laidman and R G Wyn Jones. pp 63–103, Academic Press, London.

53 von Willert, D J 1974 Einfluss von NaCl und Blattalter auf die CO_2-Dunkelfixierung bei Halophyten – Vergleichende Untersuchung bei *Mesembryanthemum crystallinum* und *Atriplex spongiosa.* Z. Pflanzenphysiol. 72, 64–74.

54 Yeo A R 1974 Salt tolerance in the halophyte *Suaeda maritima* (L.) Dum. D.Phil. Thesis, University of Sussex.

55 Yeo A R 1981 Salt tolerance in the halophyte *Suaeda maritima* (L.) Dum: intracellular compartmentation of ions. J. Exp. Bot. 32, 487–497.

56 Yeo A R 1983 Salinity resistance: physiologies and prices. Physiol. Plant. 58, 214–222.

57 Zimmermann M H 1983 Xylem Structure and the Ascent of Sap. Springer-Verlag. Berlin.

58 Zimmerman U 1977 Cell turgor pressure regulation and turgor pressure mediated transport processes. *In* Integration of Activity in the Higher Plant. Ed. D H Jennings. Symp. Soc. Exp. Biol. pp 117–154 Cambridge University Press, Cambridge.

Plant and Soil 89, 57–67 (1985). Ms. BR 04

The cost to plants of different strategies of adaptation to stress and the alleviation of stress by increasing assimilation

J. GALE and M. ZERONI
The Jacob Blaustein Institute for Desert Research, Ben-Gurion University of the Negev, Sede Boqer Campus, 84990 Israel

Key words Adaptation Carbon balance Carbon dioxide Environmental stress Heat stress Maintenance respiration Nitrogen fixation Salinity stress

Summary The fitter of two species that use different strategies to overcome the same stress may be the one that expends the least resources to cope with this stress. However, this concept has proven difficult to quantify. It is proposed here that the increase in maintenance respiration in response to stress factors such as high temperature, salinity or a high-oxygen atmosphere (one indirect effect of which is nitrogen deficiency) may provide a measure of the cost of adaptation, in terms of expenditure of assimilated carbon. A corrolary to this is that, where it can be shown that an adaptive strategy results in the expenditure of assimilates, adaptation may be enhanced by increasing carbon assimilation. Results are presented supporting the hypothesis and its corrolary.

Introduction

When Charles Darwin published his 'Origin of the Species' in 1859, questions relating to the mechanism and criteria of natural selection remained open. In 1864 Herbert Spencer, in his 'Principles of Biology' contributed the soon-famous term 'survival of the fittest'. However, the phrase is a tautology and begs the question of the definition of 'fitness' and also the question 'fit for what?' The latter, which is more philosophical than scientific, may be variously answered for plants as 'production of maximum growth', 'longevity', 'maximum reproduction' or, in terms of the presently popular socio-biological hypothesis, 'survival of the genes'. For many agricultural crops and in this paper, 'maximum growth' is the criterion of interest.

A species must adapt to survive and reproduce under conditions of competition and environmental stress. Plants accomplish this by a multiplicity of strategies. These strategies (*e.g.* production of thorns or of alkaloids to deter predators, or secretion of salts to overcome salinity) require expenditure of resources. An implicit paradigm of modern environmental physiology is that when two species adapt equally effectively to the same stress, but by different strategies, the 'fitter' of the two will be the one that expends the least resources (*e.g.* [18]). The resources available to the plant may be water, minerals or photosynthetic assimilates. The latter may be diverted to adaptive structures such as thorns or may be oxidized by respiration to meet energy requirements, for example, for the active secretion of excess salts against an electrochemical gradient.

A major difficulty in assessment of the cost of adaptation is measurement of resource allocation during oxidation of assimilates to provide energy. It will be shown below how, where the major adaptive strategy involves energy expenditure, this expenditure may be estimated from the overall carbon balance of the plant and, in particular, from the derived 'maintenance respiration' (Rm) [9, 10].

If a mechanism of adaptation indeed requires the expenditure of assimilates, then the amount available for normal growth (expressed as dry weight increment) will be reduced. It follows that the retardation of growth by such a stress should be reduced if assimilate production is increased. This has been demonstrated for a number of plants[15], and further evidence is given below for commercial roses.

This paper is an integration and reinterpretation of data published in several previous papers [4, 5, 13, 14, 15, 16] and of new data presented here.

Measurement of maintenance respiration

The definition of Rm can include energy produced to meet a number of different requirements. It is therefore important to define the method used for its measurement. Most methods for estimating Rm are based either on the rate of respiration of 'starved' plants or on the partitioning of the plant's carbon balance[11]. When Rm is obtained from measurements of the 24-hour input and efflux of carbon dioxide, this should include CO_2 exchange by both the tops and roots[7]. This has been the method used in the studies mentioned above.

Measurement of diurnal CO_2 exchange patterns in plants will typically yield curves as shown in Figure 1. Maintenance respiration may be considered to be the steady, low rate of respiration that occurs when assimilates produced during the previous light period have been exhausted by anabolic processes. This occurs after about 8–36 hours in the dark, depending upon previous rates of photosynthesis and temperature.

An alternative method developed by McCree [9, 10] is based on study of CO_2 exchange for several days, on each of which the rate of photosynthesis is changed by, for example, exposing the plant to a different light intensity. The total diurnal CO_2 uptake and efflux can be obtained by integration of CO_2 exchange curves as shown in Figure 1. The data are then plotted as shown in Figure 2. The fraction of respiration that is a linear function of the rate of photosynthesis is considered to be that associated with secondary metabolism of the photosynthetic assimilates (Ra in Fig. 2). The extrapolated rate of

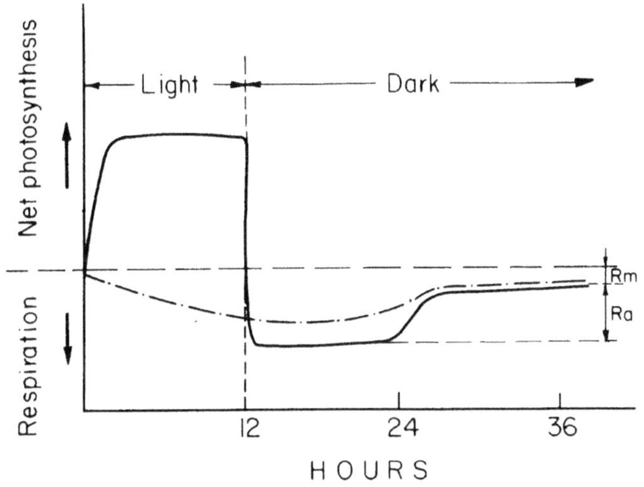

Fig. 1. Typical diurnal plant carbon dioxide exchange, extended to 36 hours in the dark (diagrammatic). Rm – maintenance component of respiration; Ra – anabolic component of respiration; ——— tops; —·—·—·— roots. Redrawn from ref.[13].

respiration at zero photosynthesis is taken to be a measure of Rm (Fig. 2). This method is valid when the experiment is carried out under conditions in which the total dry weight of the plant does not change significantly. When assimilates are used to produce new plant

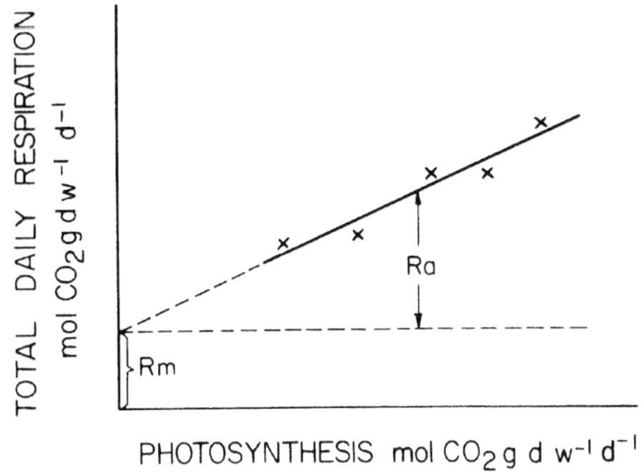

Fig. 2. Derivation of maintenance respiration from diurnal carbon dioxide uptake and efflux. Rm and Ra as in Figure 1.

material or are converted to storage materials, the R/P plots will not be linear and the Rm value cannot then be reliably obtained by this method.

Measurements of CO_2 exchange can be made on plants exposed to various stresses (salt, heat, *etc.*) and Rm plotted as a function of the stress. If the R/P curve is not linear a first estimate of the cost of adaptation to the stress factor can be obtained directly from the daily R/P ratio without partitioning into Ra and Rm.

The cost of adaptation to salt

Schwarz and Gale [13] measured the maintenance respiration of the moderately salt-tolerant *Xanthium strumarium* when it was exposed to low levels of salinity. Measurements were made after osmotic adaptation had taken place, under conditions in which there was little change in overall plant dry weight. A typical response curve is shown in Figure 3. The initial increase in Rm with increasing salt concentration has been interpreted as a result of the demand for assimilate-derived energy for coping with salinity. Similar curves were also obtained for *Phaseolus vulgaris* and *Atriplex halimus*, where the peaks occurred at salt (NaCl) levels typical of the salt tolerance of each species (low for Phaseolus and high for Xanthium and Atriplex: −0.17, −0.7 and −0.9 MPa, respectively). The peaks coincided in each species with the salinity level at which mesophyll resistance to carbon dioxide fixation

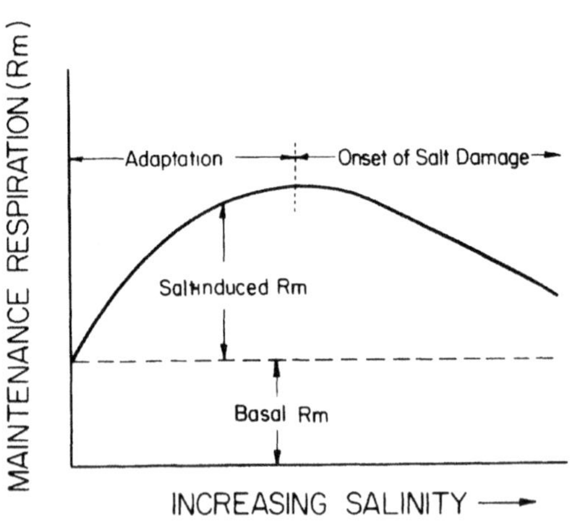

Fig. 3. The cost of adaptation as expressed by an increase in maintenance respiration in *Xanthium strumarium* plants exposed to low levels of salinity. Adapted from ref.[13].

begins to increase [4]. The subsequent decrease in Rm was considered to indicate the onset of salt damage.

Schwarz and Gale[13] calculated that, for *X. strumarium* plants grown in a culture solution having an osmotic potential of −0.5 MPa due to added NaCl, the increased use of assimilates for maintenance was responsible for about 24% of the reduction in dry weight growth (as compared to controls in nonsaline culture solutions).

The cost of adaptation to a combination of heat and salt stress

High temperatures have long been reported to decrease plant tolerance to salt while at the same time increasing the rate of respiration (*e.g.* [3]). Schwarz and Gale [14] found that, although the maintenance respiration rate of *Xanthium strumarium* doubled under saline conditions, the additional increase in response to temperature was relatively less in the saline than in the control plants. This could be interpreted as evidence that respiration is unable to meet the increased energy demands of withstanding a combination of both salt and heat stress. That the increased rate of respiration in response to heat is at least partly a result of increased maintenance demand was demonstrated for *X. strumarium* by Gale[5].

The cost of adaptation to a combination of salt and nitrogen deficiency

Plants cannot absorb gaseous nitrogen directly. Furthermore, soil nitrates, which are easily leached, are often present at suboptimal concentrations, and N_2 fixation by bacteria into a form that can be absorbed by plants is inhibited by oxygen. Consequently, from the point of view of nitrogen nutrition, the present high-oxygen atmosphere is a stress factor. One strategy for obtaining nitrogen that has evolved in plants is symbiosis between nitrogen-fixing *Rhizobium* sp. bacteria and the roots of leguminous plants. The construction of oxygen-excluding nodules, the supply of energy-rich assimilates to the bacteria, and the processes involved in nitrogen reduction require the expenditure of energy which, as for salt adaptation, is derived from the respiration of assimilates [12]. Whether this strategy for obtaining nitrogen or the uptake and subsequent reduction of nitrates requires more energy has been debated for some time [12].

Shone and Gale [16] studied the carbon balance of alfalfa (*Medicago sativum*) grown either in the presence of NO_3^- and absence of *Rhizobium* sp. (NO_3^- plants) or with Rhizobium but without NO_3^- (N_2 plants). To this was added the additional stress factor of salt (NaCl). When alfalfa was grown under conditions causing low levels of carbon assimilation (short days, long nights), overall respiration was increased and photosynthesis was decreased in response to salinity, as reported by Schwarz and Gale [13] for other plants. Furthermore, the respiration/photosynthesis ratio was about 20% higher in the nonsaline N_2 plants than those grown with NO_3^-. When plants were grown with increasing levels of salt, the R/P ratio rose much more rapidly in the N_2 than in the NO_3^- plants (Fig. 4), mainly as a result of increased respiration.

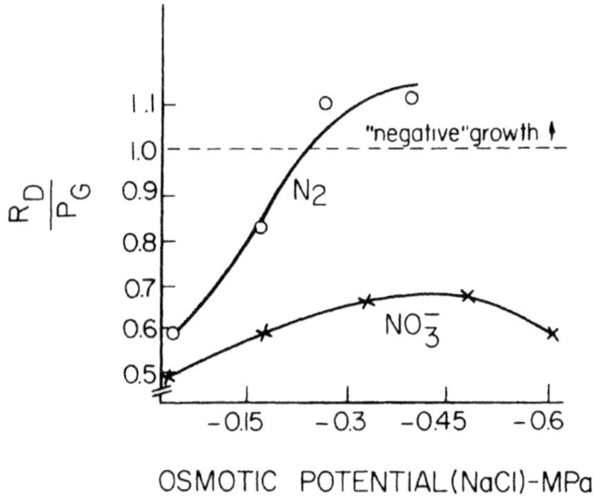

Fig. 4. The effect of salinity on the respiration/photosynthesis ratio in *Medicago sativa* seedlings grown with NO_3^- or with N_2-fixing Rhizobium bacteria. R_D – total dark respiration; P_G – gross photosynthesis. Recalculated from data of Shone and Gale[16].

The combined energy requirements (as evinced by respiration) of symbiotic N_2 fixation and of maintenance under conditions of salinity rapidly raised the R/P ratio above 1.0, *i.e.* to a level at which the plants lost dry weight. Under the short-day/long-night conditions of this experiment, the cost to the plant of coping with both salt and nitrogen fixation *via* the N_2 pathway led to a non-sustainable situation. Shone and Gale[16] were unable to differentiate Rm from R_D in their experiments since the R/P relationship was not linear (See above and Fig. 2).

Raising stress tolerance by increasing the availability of assimilates

As shown above, salt and heat stress and symbiotic N_2 fixation may each cause a measurable increase in plant respiration, presumably in response to an increased demand for energy. This serves as a drain on the available assimilates that would otherwise be used for growth. This is particularly detrimental in salt-stressed plants, where photosynthesis is often limited by the rate of diffusion of CO_2 through stomates, which close partially in response to salt [4]. It follows that any induced increase in the rate of assimilation (photosynthesis) may enhance the ability of the plant to maintain its usual growth rate while at the same time coping with stress. An increase in carbon assimilation may be obtained by growing plants in a high-CO_2 atmosphere.

To test this corollary to the stress/Rm hypothesis Schwarz and Gale [15] grew four plant species (*Phaseolus vulgaris, Xanthium strumarium, Zea mays* and *Atriplex halimus*) with or without salt in the growth media at high or low levels of ambient CO_2. In all four species the relative growth rate increased in response to the addition of CO_2. Moreover, in all four species the relative response was greater in the salinized than in the control plants.

Recently, we carried out a similar study on the response of Sonia roses to salt at different levels of CO_2. In this case the plants were exposed to a mixture of salts simulating the salinity of a brackish desert well. Data on two of the most important commercial criteria for evaluating roses, branch length and flower weight, obtained in a representative experiment are presented in Table 1. In the absence of salt, CO_2 supplementation had a very large effect on both parameters. However, as noted above for other plant species, at low to medium concentrations of salt the relative response of salt-stressed plants was even greater. In an environment of 1200 $\mu l\ l^{-1}$ CO_2 the roses grew

Table 1. The effect of CO_2 supplementation on the response of roses to a mixed salt medium. (*Rosa hybrida* var. Sonia)

Salt concentration (mg l^{-1}, TSS)	[CO_2] ($\mu l\ l^{-1}$)	Branch length (cm)	Flower weight (g)
Normal irrigation water (250)	300 ± 20	26.6 ± 0.6	11.8 ± 0.04
	600 ± 30	42.2 ± 0.2	13.3 ± 0.1
	1200 ± 40	77.2 ± 2.9	23.4 ± 0.9
25% water from brackish well (860)	300 ± 20	21.7 ± 0.9	10.6 ± 0.2
	600 ± 30	45.0 ± 0.8	13.9 ± 0.3
	1200 ± 40	66.6 ± 2.1	28.6 ± 0.7
50% water from brackish well (1700)	300 ± 20	18.4 ± 0.5	8.9 ± 0.1
	600 ± 30	43.9 ± 0.6	12.6 ± 0.3
	1200 ± 40	66.3 ± 1.5	20.7 ± 0.53
75% water from brackish well (2600)	300 ± 20	15.2 ± 0.2	8.6 ± 0.35
	600 ± 30	33.8 ± 0.6	9.6 ± 0.3
	1200 ± 40	54.7 ± 0.8	14.7 ± 0.3

TSS – total soluble salts. N varied from 10 to 28. Figures are means ± S.E. Plants were grown continuously in growth chambers, with conditions simulating the winter greenhouse climate in Israel.

almost as well with 1700 mg l^{-1} TSS as the controls. (This is some three to four times the maximum accepted salt level for growing roses.) At the highest level of salt (2600 ppm) the roses suffered direct toxicity, which could not be alleviated by supplementary CO_2. If we assume that CO_2 supplementation increased photosynthesis and hence the amount of assimilates, this result supports the hypothesis presented above.

A similar effect was recently found by Oren and Gale (unpublished) for alfalfa. The addition of NaCl to the culture solution to give an osmotic potential of -0.2 MPa reduced the relative growth rate (RGR) from 0.072 to 0.056. However, under high-CO_2 conditions the RGR of salt-stressed alfalfa was almost as high as that of the controls, 0.085 and 0.092, respectively (Fig. 5).

Conclusions

The measurement of maintenance respiration is not technically simple, nor is the significance of the measured Rm altogether clear[11]. Energy derived from respiration may be expended for many purposes, such as pumping ions against an electrochemical gradient, turnover of damaged proteins and organelles, or the diversion of nutrients to N_2-fixing symbionts. It is a matter of definition as to whether all these processes should be included in the term 'maintenance'. The present methodology only separates the respiration associated with assimilate processing from that occurring in the absence of unprocessed assimilates. It is the latter which has been called maintenance respiration. Another complication in interpretation is that there is some evidence that the alternative cyanide-resistant pathway [17] may be activated under stress conditions [8]. While this may not affect conclusions regarding the amount of assimilates expended to supply energy for adaptive mechanisms, it does not allow calculation of energy allocation unless the relative amounts of assimilate passing through each pathway can be measured or calculated.

In many instances the allocation of resources to an adaptive strategem does not involve increased energy expenditure as expressed by an increase in Rm as defined above. For example, the allocation of assimilates to the formation of sclerenchymatic tissue would be expressed in the Ra term and not by an increase in Rm (see Fig. 2). Furthermore, in this case there may be no immediate reduction in growth in terms of total dry weight, even though the diversion of assimilates from leaves to thorns would eventually reduce growth.

However, where increased energy expenditure is a major component

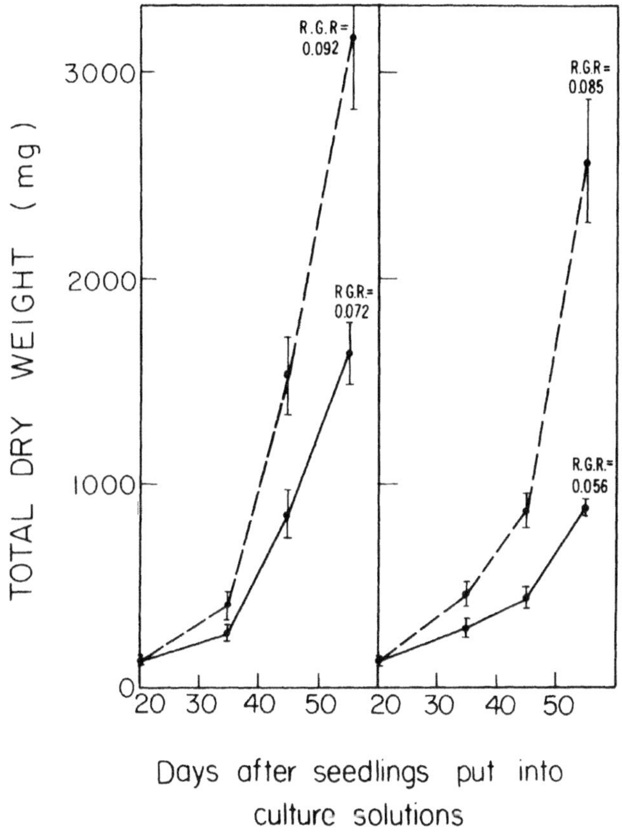

Fig. 5. Response of *Medicago sativa* to salt under two levels of CO_2. Left – grown in normal culture solution; right – grown in culture solution to which NaCl was added to give an osmotic potential of -0.2 MPa; ---- 1200 $\mu l\,l^{-1}$ CO_2; —— ambient (320 $\mu l\,l^{-1}$) CO_2; each point = average ± S.E.; N = 8. (from C. Oren and J. Gale, unpublished).

of an adaptive strategy, measurement of Rm may provide a criterion for the evaluation of the cost of this strategy in terms of resources expended. This was demonstrated above for adaptation to salinity and for the mode of nitrogen acquisition. In these instances we may be able to distinguish the 'fitter' of two equally efficacious evolutionary strategies by finding the one which makes the smaller demand on assimilate resources. This was shown for the case of nitrogen acquisition by either the NO_3^- absorption or Rhizobium symbiosis pathways, under conditions of salt stress (Fig. 4).

For low-level salinity stress in Xanthium, the allocation of assimilates to energy production for adaptive processes accounted for about a quarter of the growth reduction [13]. The other 3/4 of the reduction in growth appeared to be mainly a result of an increase in stomatal resistance to CO_2 uptake. Evidence for this is that the effect of low-level salt stress (resulting in decreased growth rate) can be much reduced by growing plants in a high-CO_2 atmosphere (Ref. [15] and Fig. 5). This finding may be of considerable practical importance in controlled-environment agriculture in deserts where nonsaline water is not available (Table 1 and Ref. [6]).

References

1 Ashi S M and Powers W L 1938 Salt tolerance of plants at various temperatures. Plant Physiol. 12, 767–789.

2 Barnes A and Hole C C 1978 A theoretical basis of growth and maintenance respiration. Ann. Bot. 42, 1217–1221.

3 Fukai S and Salisbury J H 1977 Responses of subterranean clover communities to temperature. II. The effects of temperature on dark respiration rate. Aust. J. Plant Physiol. 4, 159–167.

4 Gale J 1975 Water balance and gas exchange of plants under saline conditions. *In* Plants in Saline Environments. Eds. A Poljakoff-Mayber and J Gale. pp 168–185, Springer-Verlag, Berlin.

5 Gale J 1982 Evidence for essential maintenance respiration of leaves of *Xanthium strumarium* at high temperature. J. Exp. Bot. 33, 471–476.

6 Gale J and Zeroni M 1984 Cultivation of plants in brackish water in controlled-environment agriculture. *In* Salinity Tolerance in Plants. Ed. R C Staples. pp 363–380, Wiley, New York.

7 Hansen G K and Jensen C R 1977 Growth and maintenance respiration in whole plants, tops and roots of *Lolium multiflorum*. Physiol. Plant. 39, 155–164.

8 Lambers H, Blacquiere T and Stuirer C E E 1980 Energy metabolism of *Plantago coronopus* L. as affected by NaCl. *In* Abstracts of II Congress FESPP. Ed. E Vieitez. pp 454–455, Santiago de Compostela, Spain.

9 McCree K J 1970 An equation for rate of respiration of white clover plants grown under controlled conditions. *In* Prediction and Measurement of Photosynthetic Productivity. Ed. I Setlik, pp 221–229, Pudoc, Wageningen.

10 McCree K J 1974 Equations for the rate of dark respiration of white clover and grain sorghum, as function of dry weight, photosynthesis rate and temperature. Crop Sci. 14, 509–514.

11 Ruget F 1981 Respiration de croissance et respiration d'entretien: methodes de mesure, comparaison des resultats. Agronomie 1, 601–610.

12 Ryle G J A, Powell C E and Gordon A J 1981 Assimilate partitioning in red and white clover either dependent on N_2 fixation in root modules or utilizing nitrate nitrogen. Ann. Bot. 47, 515–523.

13 Schwarz M and Gale J 1981 Maintenance respiration and carbon balance of plants at low levels of sodium chloride salinity. J. Exp. Bot. 32, 933–941.

14 Schwarz M and Gale J 1983 The effect of heat and salinity stress on the carbon balance of *Xanthium strumarium*. *In* Effects of Stress on Photosynthesis. Eds. R Marcelle, H Clijsters and A van Poucke, pp 325–331, Martinus Nijhoff/Dr. W. Junk, The Hague, The Netherlands.

15 Schwarz M and Gale J 1984 Growth response to salinity at high levels of carbon dioxide. J. Exp. Bot. 35, 193–196.

16 Shone M G T and Gale J 1983 Effect of sodium chloride stress and nitrogen source on respiration, growth and photosynthesis in lucerne (*Medicago sativa* L.). J. Exp. Bot. 34, 1117–1125.
17 Solomos T 1977 Cyanide-resistant respiration in higher plants. Annu. Rev. Plant Physiol. 28, 279–87.
18 Townsend C R and Calow P 1981 Physiological ecology: an evolutionary approach to resource use. Blackwell Scientific, Oxford.

Plant and Soil 89, 69–81 (1985). Ms. BR 05

Mechanisms of response to salinity in halotolerant microalgae

JOHAN A. HELLEBUST
Department of Botany, University of Toronto, Toronto, Ontario, Canada, M5S 1A1

Key words *Chlamydomonas pulsatilla* *Dunaliella tertiolecta* Glycerol Microalgae Nutrient transport Osmoregulation Salt tolerance *Stichococcus bacillaris*

Summary A limited number of organic solutes are used by microalgae to adjust their internal osmotic pressure in response to changing external salinities. Glycerol and proline are used by the most extremely halotolerant algae. Only glycerol allows growth at salinities approaching saturation. In addition to organic osmoregulatory solutes, inorganic ions also play an important role in osmoregulation. The ability of microalgae to maintain intracellular ions at levels compatible with metabolic functions may set upper limits for their salt tolerance. Requirements for NaCl in the external medium for nutrient transport may define the lower salinity limits for growth observed for some euryhaline algae.

Osmotic upshocks generally cause severe temporary inhibition of photosynthesis in euryhaline microalgae. Extensive osmotic downshocks have little effect on photosynthesis in microalgae with strong cell walls, while wall-less species appear to be more sensitive. Rapid glycerol synthesis takes place in response to increased external salinity in *Chlamydomonas pulsatilla* both in light and dark. Starch supplies carbon for glycerol synthesis in the dark and also during the initial periods of inhibition of photosynthesis in the light. Turnover of osmoregulatory solutes such as glycerol and isofloridoside may be an important aspect of the osmoregulatory mechanism.

At salinities beyond the growth limit for the green flagellate *Chlamydomonas pulsatilla*, resting spores are formed that enable this alga to survive extreme salinities.

Introduction

A considerable number of unicellular algal species are considered euryhaline, since they can adapt to varying external salinity. However, the salinity range over which active growth takes place differs greatly among species (Tables 1 and 2). The osmoregulatory characteristics of microalgae have been studied extensively over the past decade[6, 23, 27, 35]. It is clear that a rather limited number of organic solutes are of key importance in adjusting intracellular osmotic pressure to maintain required turgor or volume control for growth while permitting an ionic cellular environment compatible with metabolic reactions. Information about biochemical mechanisms for the control of cellular levels of osmoregulatory organic solutes is available only for *Poterioochromonas malhamensis*[26, 27, 28], a chrysophyte flagellate that adjusts to a rather low and narrow salinity range, and for species of the genus Dunaliella[2, 3, 6, 16], wall-less green flagellates that grow over a wide salinity range, in some cases reaching salt saturation (Table 2).

Some halotolerant microalgae require a minimum level of external

Table 1. Salinity ranges compatible with growth and organic osmoregulatory solutes of microalgae from freshwater and soil habitats

Species	Osmoregulatory solute	Salinity range (% ASW)	Reference
Freshwater			
Poterioochromonas malhamensis	isofloridoside	0– 40	39
Scenedesmus obliquus	sucrose	0– 70	40
Chlorella emersonii	proline/sucrose	0–100	36
Cyclotella meneghiniana	proline	0–100	39
Soil			
Klebsormidium flaccidum	sucrose/glutamate	0– 60	9, 10
Klebsormidium sterile	glutamate/sucrose	0–140	9, 10
Porphyridium cruentum	glycine/proline/alanine	0–100	13
Stichococcus chloranthus	sorbitol/proline	0–140	9, 10

salt, while many euryhaline species show little or no NaCl requirement for growth (Tables 1 and 2). The role of inorganic ions in osmoregulation of microalgae is uncertain, since technical difficulties in determining ion contents of cells in highly saline media have yielded conflicting values[1, 12, 17, 18, 35]. It is, however, clear that changes in concentrations of organic osmoregulatory solutes are often not sufficient to account for osmotic adaptations to increasing salinity[6, 9, 27, 37], suggesting that inorganic ions may play an important role in osmoregulation in microalgae. In macroalgae, which possess large vacuoles, inorganic ions play a dominant role in osmotic adjustment[23, 25].

While all of the extremely halotolerant microalgae (Table 2) are truly wall-less species, other microalgae, which show low or intermediate halotolerance, possess cell walls of widely differing mechanical

Table 2. Salinity ranges compatible with growth and organic osmoregulatory solutes of microalgae from brackish water, tidepool, marine, salt pond and lake habitats

Species	Osmoregulatory solute	Salinity range (% ASW)	Reference
Brackish, tidepool, marine			
Stichococcus bacillaris	sorbitol/proline	0– 140	8
Cyclotella cryptica	proline/glutamate	10– 150	32
Platymonas suecica	mannitol	10– 300	24
Nannochloris bacillaris	proline	0– 300	7, 9
Chlamydomonas pulsatilla	glycerol	0– 300	31
Chlorella autotrophica	proline	0– 400	1
Dunaliella tertiolecta	glycerol	5– 700	30
Salt pond and lake			
Asteromonas gracilis	glycerol	100– 900	3
Dunaliella bardawil	glycerol	200– 800	3
Dunaliella salina	glycerol	200–1000	3
Dunaliella viridis	glycerol	60–1000	5

properties. Since the absence of a cell wall, or its relative elastic strength when present, is important to the extent and rate of change in cell volume that takes place when microalgae are subjected to rapid increase in external water potential[35, 42], these characteristics may be expected to influence the effect of such treatments on physiological processes of the algae.

The present report deals with studies of the effects of salinity up- or down-shocks on representatives of halotolerant microalgae lacking a cell wall (*Dunaliella tertiolecta*), possessing a weak (*Chlamydomonas pulsatilla*), or a strong cell wall (*Stichococcus bacillaris*). Furthermore, the reasons for the lack of a salt requirement by *C. pulsatilla* and the absolute requirement for NaCl by *D. tertiolecta* for growth[33] were probed in an investigation of the effect of salt concentration on inorganic nutrient transport in the two species. The effect of salinity on changes in inorganic and organic solute contents contributing to osmotic balance in *C. pulsatilla*, and also the relative role of CO_2 photoassimilation and starch degradation in regulation of glycerol levels in this green flagellate, were determined. Beyond the salinity range over which growth takes place, the formation of salt-tolerant resting stages is reported.

Materials and methods

Algae and culture conditions

Stichococcus bacillaris Nägeli (strain UTEX 314; Culture Collection at the University of Texas), *Dunaliella teriolecta* Butcher (Woods Hole Oceanographic Institution Culture Collection, Woods Hole, Mass.), and *Chlamydomonas pulsatilla* Wollenweber (isolated by author from marine supralittoral rockpool near St. Andrew's, New Brunswick) were cultured in an artificial seawater medium (ASW) formulated by McLachlan[34]. The media were enriched with f/2 nutrients and vitamins[21], except for *C. pulsatilla* medium, for which nitrate was replaced with 2 m*M* NH_4Cl. The cells were grown at 18°C on a 12/12 hour light/dark cycle (24 W m^{-2} from 'Daylight' fluorescent tubes). Media of different salinities (10–200% ASW) were made by varying the level of major ions, maintaining their original proportions, and maintaining a constant level of nutrients, bicarbonate and vitamins. Cells were transferred from one medium to another by 10 min centrifugation at 3000*g* and 18°C, removal of the supernatant, and immediate resuspension in the new medium.

Photosynthesis experiments

For all experiments cells from the exponential growth phase were harvested by centrifugation at 1000*g* and 18°C for 10 min and were resuspended at a cell density of about 1×10^5 cells ml^{-1} in nutrient-containing ASW media of different salinities. The experimental system contained 1 μCi $NaH^{14}CO_3$ at a concentration of 2 m*M* in a 5-ml cell suspension. After 30 min incubation at 20°C and 24 W m^{-2} light intensity, cells were harvested by filtration through 0.45 μm pore-size Millipore filters, and the filters were washed twice with 1-ml portions of the same unlabelled medium. The filters were dried and their radioactivity determined by scintillation counting[41]. Similar incubations of dark samples were carried out, and their activities subtracted from the light samples before calculation of photosynthetic rates.

Synthesis of glycerol from photoassimilated $^{14}CO_2$ and starch

Replicate 10-ml samples of *C. pulsatilla* cell suspensions were prepared in media of various salinities and incubated in the light (24 $W m^{-2}$) or dark for different time periods. One set contained unlabelled 2 m*M* $NaHCO_3$, while another contained 10 μCi $NaH^{14}CO_3$ (2 m*M*). Four 10-ml samples were harvested by centrifugation at each time interval. The cells of one duplicate set were extracted for 3 h with 3 ml aqueous K-phosphate buffer (60 m*M*, pH 4.6), and the starch content of the extract was determined enzymatically[36]. Another duplicate set of cell pellets was extracted with 2 ml boiling H_2O for 30 min and the glycerol content of the extract determined enzymatically[4]. Duplicate ^{14}C-labelled samples were similarly harvested by centrifugation, the cell pellets extracted with 2 ml boiling water for 10 min, and the extracts dried down in a stream of N_2 at 60°C. The residue was taken up in 100 μl of water, and 10-μl aliquots were spotted on paper chromatograms. Chromatographic techniques and determinations of radioactivity in glycerol identified by co-chromatography and autoradiography have been described earlier[22].

Uptake of phosphate and methylamine

Replicate 2-ml cell suspensions of *D. tertiolecta* and *C. pulsatilla* from exponential cultures in 10% ASW were prepared by ASW media at different levels of major ion concentrations. The incubation medium designated 0% ASW (Table 3) contained the same major ion concentrations as 0.5% ASW medium, but NaCl was omitted. For methylamine uptake, ammonium was not added, and [^{14}C]methylamine was present at a final concentration of 10 μ*M*. For *ortho*-phosphate uptake, $NaH_2^{32}PO_4$ was added to a final concentration of 2.5 μ*M*. Uptake rates were determined for the initial 5 min period according to a procedure described earlier for methylamine uptake[41]. All incubations were done in the light (24 $W m^{-2}$), and both methylamine and phosphate were taken up linearly for more than 10 min.

Ion analysis

Intracellular ion concentrations were calculated for ion analysis of centrifuged cell pellets. Extracellular volume and cell water content of the centrifuged pellets were determined as described previously[1]. [^{14}C]Mannitol was added to known volumes of resuspended pellets to a final concentration of 10 m*M* to avoid errors due to adsorption in determinations of extracellular space. The cell pellets were suspended in 10% HNO_3, allowed to rest overnight, and then centrifuged at 7000*g* for 10 min. The supernatant was used for ion determinations. Na^+, K^+ and Mg^{2+} were determined by atomic absorption spectroscopy, and Cl^- by titration[11].

Results

Effects of salinity up- and down-shock on photosynthesis

Although *Dunaliella tertiolecta* grows at 10% ASW, this salinity is considerably below that for optimal growth of this marine flagellate[33]. It can be observed (Fig. 1) that after transfer of cells grown in 10% ASW to 50 or 100% ASW medium, there is a substantial increase in the rate of photoassimilation of $^{14}CO_2$. However, a transfer to 200% ASW results in an almost complete inhibition of photosynthesis during the initial 30 min following resuspension of the cells. There was also little or no recovery of photosynthesis over the subsequent 90 min (results not shown). *D. tertiolecta* cells grown at 50% ASW showed maximum photosynthesis when resuspended at the same salinity, and considerable photosynthetic inhibition occurred following transfer to higher or lower salinities. Cells grown at 200% ASW showed a small stimulation

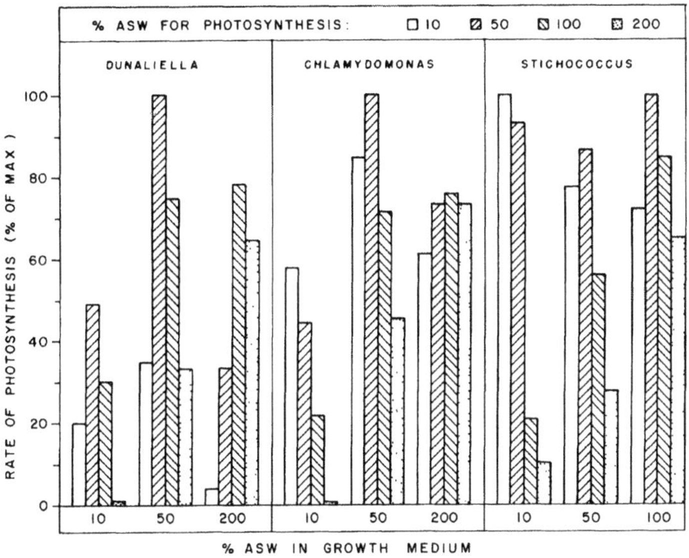

Fig. 1. Effects of osmotic up- and down-shocks on rates of CO_2 photoassimilation by *Dunaliella tertiolecta, Chlamydomonas pulsatilla*, and *Stichococcus bacillaris*.

of photosynthesis upon transfer to 100% ASW, but strong inhibition upon transfer to lower salinities. This was especially evident for cells resuspended in 10% ASW, resulting in a 94% inhibition of photosynthesis, and little recovery over a 2-h period following the downshock (data not shown).

Chlamydomonas pulsatilla cells grown at 10% ASW, a salinity close to the optimum for growth of this flagellate[31], showed marked reductions in photoassimilation of CO_2 following transfer to higher salinities (Fig. 1). As was the case for *D. tertiolecta*, 10% ASW-grown cells almost completely lost their photosynthetic capacity after an osmotic upshock to 200% ASW. Again, 50% ASW-grown cells showed optimal photosynthesis at this salinity, and relatively modest inhibitions of photosynthesis upon resuspension in lower or higher salinities. In striking contrast to the behavior of 200% ASW-grown *D. tertiolecta* cells, *C. pulsatilla* cells adapted to this high salinity showed only a small (about 15%) decrease in photosynthesis during the initial 30-min period following resuspension in 10% media.

Stichococcus bacillaris showed almost the same responses as *C. pulsatilla* to osmotic up- and down-shocks (Fig. 1). However, the inhibition of CO_2 photoassimilation following transfer of 10% ASW-grown cells to 200% ASW media was somewhat less pronounced than for the two flagellates. *Stichococcus bacillaris* was especially resistant to severe osmotic down-shocks. When 100% ASW-grown cells were resuspended in 10% ASW media there was even a small (11%) enhancement of CO_2 photoassimilation.

Effect of salinity on growth and on nutrient uptake

Chlamydomonas pulsatilla grows rapidly in the complete absence of NaCl, the major salt in seawater (Table 3). In agreement with this observation, there was rapid uptake of both methylamine, an ammonium analog, and *ortho*phosphate in the absence of NaCl. The uptake of methylamine, furthermore, showed no significant stimulation by added NaCl. In contrast, phosphate uptake was markedly stimulated by increasing NaCl concentrations. Maximum phosphate uptake occurred at $\geqslant$ 5% ASW (20 m*M* NaCl), at which the rate was approximately five times that in the absence of NaCl.

Dunaliella tertiolecta did not grow at low salinities, and also showed no methylamine or phosphate uptake in the absence of NaCl (Table 3). The minimum salinity at which growth occurred was at 10% ASW, while rapid growth occurred at 25% ASW. Increased NaCl concentrations strongly enhanced both methylamine and phosphate uptake.

Osmoregulation

The contents of both inorganic and organic solutes in *Chlamydomonas pulsatilla* increased at higher external salinity (Table 4). At 50% ASW the contribtuion of inorganic ions to internal osmotic pressure was considerably higher than that of organic solutes, while the total osmolarity of intracellular solutes balanced quite closely that of the external medium. Glycerol was the major internal organic solute at all salinities, and showed a striking nonlinear increase in concentration with increased salinity of the medium. At 200% ASW, glycerol contributed almost 60% of the total internal osmotic pressure. The concentration of amino acids increased only moderately, and the total amino acid pool contributed in only a minor way to regulation of internal osmotic pressure.

Chlamydomonas pulsatilla grown in 33% ASW media and then resuspended in the same medium (Table 5; control), maintained an almost constant glycerol content during 60 min incubation under light (L; 24 W m^{-2}) or dark (D) conditions, while significant amounts of carbon from photoassimilated $^{14}CO_2$ entered the almost-static glycerol pool (Table 5). There was a 14% increase in starch content in the light, and a 3% decrease in the dark. The cells transferred from 33 to 100% ASW media showed an almost equally rapid increase in glycerol content

Table 3. Effects of salinity on growth and nutrient uptake by *Chlamydomonas pulsatilla* and *Dunaliella tertiolecta**

% ASW	NaCl (m*M*)	*Chlamydomonas pulsatilla* Growth constant** (ln units day^{-1})*	*C. pulsatilla* P-uptake (fmol cell^{-1} min^{-1})	*C. pulsatilla* MA-uptake (fmol cell^{-1} min^{-1})	*Dunaliella tertiolecta* Growth constant** (ln units day^{-1})*	*D. tertiolecta* P-uptake (fmol cell^{-1} min^{-1})	*D. tertiolecta* MA-uptake (fmol cell^{-1} min^{-1})
0	0	0.95	0.28	4.3	0	0	0
0.5	2	1.04	0.84	4.5	0	0	0
2	7	1.02	1.25	4.8	0	0.01	0.05
5	20	0.74	1.55	5.1	0	0.07	2.1
10	40	0.70	1.58	5.0	0.45	0.18	4.0
25	100	0.67	1.63	4.7	0.95	0.52	4.7

*P, *ortho*phosphate; MA, methylamine

**Growth constant = ln 2/doubling time in days

Table 4. Effects of external salinity on concentrations of intracellular solutes in *Chlamydomonas pulsatilla*

Solute	Concentration of intracellular solutes* (mOsm)		
	50% ASW	100% ASW	200% ASW
K^+	112	134	128
Na^+	67	158	366
Mg^{2+}	20	22	26
Cl^-	191	286	351
Amino acids	24	39	50
Glycerol	98	290	1,030
Total intracellular solutes	512	989	1,951
External solutes (ASW)	476	890	1,750

*Concentrations of intracellular solutes were calculated from intracellular concentrations (mmol/kg cell H_2O) using an osmotic coefficient of 0.9 for monovalent ions, 0.15 for Mg^{2+}, and published values for osmotic activities of glycerol at various concentrations.

(about 3-fold) in the dark as in the light. The rate of photoassimilation of CO_2 dropped to about one-third of its original level during the initial 15 min following osmotic upshock, then increased to over one-half of the pre-shock rate after 60 min. The starch content of the cells declined to about one-half of its original level after 60 min in the light, while in the dark it dropped by about 75% during the same time period. Degradation of starch in cells kept in the dark could fully account for the rapid synthesis of glycerol during the 60 min dark period. In cells kept in the light following upshock, both photoassimilated CO_2 and starch breakdown contributed to glycerol synthesis. During the initial 15 min most of the carbon for glycerol synthesis must have come

Table 5. Changes in rate of photosynthesis and in glycerol and starch contents of *Chlamydomonas pulsatilla* after transfer from 33 to 100% ASW under light (L) and dark (D) conditions

Time after transfer (min)	Rate of photosynthesis ($fmol\ cell^{-1}\ min^{-1}$)	Glycerol from CO_2 ($fmol\ cell^{-1}$)	Glycerol content ($fmol\ cell^{-1}$)		Starch content ($fmol\ cell^{-1}$)	
	L	L	L	D	L	D
33% ASW (Control)						
0	–	–	160	160	508	508
15	49	21	151	158	–	–
30	51	42	164	143	–	–
60	50	60	171	155	577	490
33 to 100% ASW						
0	–	–	160	160	508	508
15	15	44	376	381	377	306
30	20	100	433	460	292	168
60	26	324	527	506	265	126

from starch breakdown, while during the last 30 min of the 60-min experimental period considerably more carbon entered glycerol from $^{14}CO_2$ photoassimilation than the net observed increase in glycerol during this time interval, suggesting glycerol turnover.

Spore formation

Formation of resting cells (hypnospores and aplanospores) by *C. pulsatilla* was observed when cells were incubated in high salinity media (Table 6). While very few such spore forms were seen in media of 250% ASW or lower salinity, after 11 days of incubation in 300 and 350% ASW media, more than 50% of the cells appeared as resting stages: oval green cells with thick walls (hypnospores), or spherical orange-yellow cells with thick walls (aplanospores). The latter spore type survived complete evaporation of seawater media, and then germinated after incubation for several days in 100% ASW media (results not shown).

Table 6. Formation of resting cells (hypnospores and aplanospores) by *Chlamydomonas pulsatilla* at different salinities*

	% Resting cells			
Time (days)	200% ASW	250% ASW	300% ASW	350% ASW
2	0	0	< 1	< 1
3	< 1	< 1	< 1	< 1
4	< 1	2	10	27
8	3	3	43	51
11	2	2	56	64

* Inoculum from 250% ASW culture.

Discussion

A comparison of the salinity ranges allowing cell growth with the key organic osmoregulatory solutes used for salinity adaptation in a variety of microalgae (Tables 1 and 2) shows that only proline and glycerol permit growth at very high salinity. However, it is also clear that species that utilize proline or glycerol as osmoregulatory solutes show widely different salinity tolerances. Clearly, there must be distinct differences in the osmoregulatory mechanisms of the species utilizing the same organic solutes that result in such striking differences in salt tolerance. It is possible that the biochemical regulatory mechanisms are sufficiently different to set different upper limits for proline and glycerol accumulation. However, it is equally possible that differences in the regulation of ion transport in response to varying salinities result in different salinity tolerance ranges for growth. Significant ion accumulation has been reported for several microalgae as a function of increasing salinity[1, 12, 18, 19] (Table 4), which may lead to inhibition of growth and metabolism[6, 23, 35, 42]. The effects of high external salinities on membrane functions such as nutrient transport are also not known, and may impose upper salinity limits for certain species.

It is clear that the lower limit of salinity for growth of *Dunaliella tertiolecta* is at least in part caused by a salt (presumably Na^+) requirement for nutrient transport (Table 3). *Chlamydomonas pulsatilla*, which grows in NaCl-free media, has no absolute NaCl requirement for phosphate or methylamine transport. Since the centric diatom *Cyclotella cryptica*, which does not grow at salinities below 10% ASW (Table 2), has a Na^+ requirement for transport of methylamine[41] and organic substrates[25], it seems likely that the lower salinity limits compatible with growth of euryhaline species may be due to ion requirements for nutrient transport.

The ability of euryhaline microalgae to tolerate rapid up- and downshocks in salinity varies considerably. A comparison of three species (*Dunaliella tertiolecta*, a small naked flagellate; *Chlamydomonas pulsatilla*, a weak-walled flagellate 10–20 times the size of *D. tertiolecta*; and *Stichococcus bacillaris*, a small non-motile alga with a strong cell wall) showed several interesting similarities and differences in behavior. All three species show a strong photosynthetic inhibition in response to transfer from low to high salinity (Fig. 1), in agreement with findings for several other microalgae of various types[6, 7, 14, 29].

The photosynthetic responses of the three microalgae to osmotic downshocks were quite different. *Dunaliella tertiolecta* grown at high salinities showed a striking loss of photosynthetic capacity upon down shock from 200 to 10% ASW (Fig. 1). This is in agreement with earlier reports for this flagellate[15] and for the green flagellate *Platymonas suecica*[24]. *Stichococcus bacillaris* and *Chlamydomonas pulsatilla*, on the other hand, showed little or no decrease in photosynthesis upon a similar downshock (Fig. 1). The same observation was made for the strong-walled green alga *Nannochloris bacillaris*[7]. The reason why the weak-walled flagellate *Chlamydomonas pulsatilla* is so much more resistant to osmotic downshocks than *D. tertiolecta* is not clear, but it is possible that the small surface/volume ratio of the large *Chlamydomonas* species coupled with a low hydraulic conductivity (unpublished observations) make it less vulnerable to drastic hypoosmotic shocks.

Although several reports suggest almost linear increases in the concentrations of organic osmoregulatory solutes with increasing external salinity, such increases in organic solutes are usually insufficient to account for the required increase in internal osmotic pressure[6, 9, 27, 35, 36, 37]. *Chlamydomonas pulsatilla* adjusts its internal osmotic pressure in response to increasing salinity chiefly by increases in glycerol concentration (Table 4). However, at 200% ASW, only about 60% of the required internal osmotic pressure is accounted for by glycerol, with other organic solutes and mainly inorganic ions accounting for the difference. The method used in the present work for correcting for ions in the extracellular space of cell pellets with [^{14}C]-mannitol can lead to overestimates of cellular Na^+ and Cl^- at very high external salinities. Since the cell wall of Chlamydomonas does not appear to be able to withstand significant turgor pressures[20], one would expect that the internal osmotic pressure should nearly equal that of the external medium. This holds true for cells grown at 50 and 100% ASW, but the cells grown at 200% ASW show a difference between internal and external osmotic pressure of about 200 mOsm (Table 4). This is probably due to an overestimation of NaCl of about 100 m*M*. Nevertheless, it is clear that inorganic ions play a significant role in osmoregulation of *C. pulsatilla*, in agreement with earlier findings for Dunaliella species[12, 17, 18, 19] and for *Chlorella autotrophica*[1].

Chlamydomonas pulsatilla adjusts its glycerol content in response to an osmotic upshock almost equally rapidly in darkness as in light (Table 5). While the source of carbon for glycerol synthesis in darkness, as expected, appears to be starch degradation, starch appears to supply most of the carbon for glycerol also during the initial phase of rapid glycerol synthesis in the light. It is also clear that a considerable amount of glycerol turnover takes place during steady-state conditions at 33% ASW and when the steady state is approached after an upshock to 100% ASW (Table 5). Such rapid turnover of an osmoregulatory solute has been observed for *Poterioochromonas malhamensis*[26] and indicates that both synthetic and degradative pathways for osmoregulatory solutes are involved in osmoregulatory mechanisms[26].

Even though *Chlamydomonas pulsatilla*, in contrast to other glycerol-accumulating green flagellates studied so far, grows actively only to an upper salinity limit of 300% ASW (Table 2), it reacts to salinities at or beyond its growth range by forming resting cells or spores (Table 6). These resting cells can withstand salinities up to salt saturation and even complete evaporation of the medium. Presumably, excessive intracellular ion concentrations prevent growth and trigger the transformation of motile flagellates into non-motile, thick-walled spores. Similar survival of cells at salinities beyond their active growth range has been observed also for *Chlorella autotrophica*[1].

Acknowledgements This work was supported by Natural Sciences and Engineering Research Council of Canada Grant A 6032. I wish to thank Mr. Y H Lin for expert technical assistance.

References

1 Ahmad I and Hellebust J A 1984 Osmoregulation in the extremely euryhaline marine microalga *Chlorella autotrophica.* Plant Physiol. 74, 1010–1015.
2 Ben-Amotz A and Avron M 1972 Photosynthetic activities of the halophilic alga *Dunaliella parva.* Plant Physiol. 49, 240–243.
3 Ben-Amotz A and Avron M 1980 Osmoregulation in the halophilic algae Dunaliella and Asteromonas. *In* Genetic Engineering of Osmoregulation. Eds. D W Rains, R C Valentine and A Hollaender. pp 91–99. Plenum Press, New York.
4 Bergmeyer U 1970 Methoden der Enzymatischen Analyse, Vol. II. pp 1267–1372. Verlag Chemie, Weinheim.
5 Borowitzka L J, Kessly D S and Brown A D 1977 The salt relations of Dunaliella. Arch. Microbiol. 113, 131–138.
6 Brown A D and Borowitzka L J 1979 Halotolerance of Dunaliella. *In* Biochemistry and Physiology of Protozoa. Eds. M Levandowsky and S H Hunter. pp 139–190. Academic Press, New York.
7 Brown L M 1982 Photosynthetic and growth responses to salinity in a marine isolate of *Nannochloris bacillaris* (Chlorophyceae). J. Phycol. 18, 483–488.
8 Brown L M and Hellebust J A 1980 The contribution of organic solutes to osmotic balance in some green and eustigmatophyte algae. J. Phycol. 16, 265–270.
9 Brown L M and Hellebust J A 1980 Some new taxonomic characteristics applied to *Stichococcus bacillaris* (Chlorophyceae). Can. J. Bot. 58, 1405–1411.
11 Cattle M 1935 Estimation of small amounts of chlorine in plant tissue. New Phytol. 34, 151–154.
12 Ehrenfeld J and Coussin J-L 1984 Ionic regulation of the unicellular green alga *Dunaliella tertiolecta*: response to hypertonic shock. J. Membr. Biol. 77, 45–55.
13 Gilles R and Pequeux A 1977 Effect of salinity on the free amino acid pool of the red alga *Porphyridium purpureum* (= *P. cruentum*). Comp. Biochem. Physiol. 57A, 183–185.
14 Gilmour J, Hipkins M F and Boney A D 1984 The effect of osmotic and ionic stress on the primary processes of photosynthesis in *Dunaliella tertiolecta.* J. Exp. Bot. 35, 18–27.
15 Gilmour J, Hipkins M F and Boney A D 1984 The effect of decreasing the external salinity on the primary processes of photosynthesis in *Dunaliella tertiolecta.* J. Exp. Bot. 35, 28–35.
16 Gimmler H and Moller E-M 1981 Salinity-dependent regulation of starch and glycerol metabolism in *Dunaliella parva.* Plant Cell Environ. 4, 367–375.
17 Gimmler H and Schirling R 1978 Cation permeability of the plasmalemma of the halotolerant alga *Dunaliella parva.* II. Cation content and glycerol concentration of the cells as dependent upon extracellular NaCl concentrations. Z. Pflanzenphysiol. 87, 435–444.
18 Ginzburg M 1981 Measurements of ion concentrations and fluxes in *Dunaliella parva.* J. Exp. Bot. 32, 321–332.
19 Ginzburg M 1981 Measurements of ion concentrations in *Dunaliella parva* subjected to hypertonic shock. J. Exp. Bot. 32, 333–340.
20 Guillard R R L 1960 A mutant of *Chlamydomonas moewussii* lacking contractile vacuoles. J. Protozool. 7, 262–268.
21 Guillard R R L and Ryther J H 1962 Studies of marine planktonic diatoms. I. *Cyclotella nana* Hustedt, and *Detonula confervaceae* (Cleve) Gran. Can. J. Microbiol. 8, 229–239.
22 Hellebust J A 1965 Excretion of some organic compounds by marine phytoplankton. Limnol. Oceanogr. 10, 192–206.
23 Hellebust J A 1976 Osmoregulation. Annu. Rev. Plant Physiol. 27, 485–505.
24 Hellebust J A 1976 Effect of salinity on photosynthesis and mannitol synthesis in the green flagellate *Platymonas suecica.* Can. J. Bot. 54, 1735–1741.
25 Hellebust J A 1978 Uptake of organic substrates by *Cyclotella cryptica* (Bacillariophyceae): effects of ions, inophores and metabolic and transport inhibitors. J. Phycol. 14, 79–83.

26 Kauss H 1974 Osmoregulation in Ochromonas. *In* Membrane Transport in Plants. Eds. U Zimmermann and J Dainty. pp 90–94. Springer-Verlag, Berlin.
27 Kauss H 1978 Osmotic regulation in algae. Progr. Phytochem. 5, 1–27.
28 Kauss H 1981 Sensing of volume changes by Poterioochromonas involves a Ca^{2+}-regulated system which controls activation of isofloridoside-phosphate synthase. Plant Physiol. 68, 420–424.
29 Kessly D S and Brown A D 1981 Salt relations of Dunaliella. Transitional changes in glycerol content and oxygen exchange reactions in water stress. Arch. Microbiol. 129, 154–159.
30 Latorella A H and Vadas R L 1973 Salinity adaptation by *Dunaliella tertiolecta*: I. Increase in carbonic anhydrase activity and evidence for a light-dependent Na^+/H^+ exchange. J. Phycol. 9, 273–277.
31 Le Gresley S M L 1979 Osmoregulation in a euryhaline green flagellate. M. Sc. Thesis, Department of Botany, University of Toronto.
32 Liu M-S and Hellebust J A 1976 Effect of salinity changes on growth and metabolism of the marine centric diatom *Cyclotella cryptica*. Can. J. Bot. 54, 930–937.
33 McLachlan J 1960 The culture of *Dunaliella tertiolecta* Butcher – a euryhaline organism. Can. J. Microbiol. 6, 367–379.
34 McLachlan J 1964 Some considerations of the growth of marine algae in artificial media. Can. J. Microbiol. 10, 769–782.
35 Munns R, Greenway H and Kirst G O 1983 Halotolerant eukaryotes. *In* Encyclopedia of Plant Physiology, N. S. Vol. 12C. Physiological Plant Ecology III. Eds. O L Lange, P S Nobel, C B Osmond and H Ziegler. pp 59–135. Springer-Verlag, Berlin.
36 Setter T L and Greenway H 1979 Growth and osmoregulation of *Chlorella emersonii* in NaCl and neutral osmotica. Aust. J. Plant Physiol. 6, 47–60.
37 Setter T L and Greenway H 1983 Changes in the proportions of endogenous osmotic solutes accumulated by *Chlorella emersonii* in the light and dark. Plant Cell Environ. 6, 227–234.
38 Schobert B 1974 The influence of water stress on the metabolism of diatoms. I. Osmotic resistance and proline accumulation in *Cyclotella meneghiniana*. Z. Pflanzenphysiol. 74, 106–120.
39 Schobert B, Untner E and Kauss H 1972 Isofloridoside und die Osmoregulation bei *Ochromonas malhamensis*. Z. Pflanzenphysiol. 67, 385–398.
40 Wetherell D F 1963 Osmotic equilibrium and growth of *Scenedesmus obliquus* in saline media. Physiol. Plant. 16, 82–91.
41 Wheeler P A 1980 Use of methylammonium as an ammonium analogue in nitrogen transport and assimilation studies with *Cyclotella cryptica* (Bacillariophyceae). J. Phycol. 16, 328–334.
42 Wyn Jones R G and Gorham J 1983 Osmoregulation. *In* Encyclopedia of Plant Physiology, N.S. Vol. 12C. Physiological Plant Ecology III. Eds. O L Lange, P S Nobel, C B Osmond and H Ziegler. pp 35–38. Springer-Verlag, New York.

Plant and Soil 89, 83–94 (1985). Ms. BR 06

The effect of compatible solutes on proteins

L. G. PALEG, G. R. STEWART* and R. STARR
*Department of Plant Physiology, Waite Agricultural Research Institute, University of Adelaide, Adelaide, South Australia 5064, Australia; and *Department of Botany, Birkbeck College, University of London, Malet Street, London WCIE 7HX, UK*

Key words Compatible solutes Environmental stresses Polyethylene glycol Proline Proteins and enzymes

Summary The ability of the compatible solute, proline, to affect the behavior of proteins has been examined in many different systems by many researchers. In the present study of protein solvation, proline has been shown to prevent or diminish, in a concentration-dependent manner, the glutamine synthetase-precipitating ability of polyethylene glycol (PEG). The effects of PEG concentration and molecular weight are reduced by proline, and the interaction is strongly affected by pH.

PEG causes precipitation of many proteins, and the ability of proline to reduce the precipitation of two non-enzymatic conjugated proteins, alfalfa mosaic virus and an ^{3}H-testosterone/antiserum complex, was also examined. Proline was effective in reducing the PEG-induced precipitation of both proteins. Virus precipitation by PEG and its alleviation by proline are influenced by pH. The increased virus-precipitating effect of PEG in the presence of salt (NaCl) is also alleviated by proline. The precipitation of the radioimmune complex by PEG is diminished by proline and by a mixture of free amino acids.

These results indicate the generality of the three-way interaction between proline, protein and PEG. They may be of importance for extraction of proteins from biological systems and in studies of enzyme inactivation or protein denaturation in a cytoplasmic milieu. The results suggest that the protective effects of some amino acids are at least additive and are consistent with the conclusion that the compatible solutes protect protein-containing systems against the unfavorable consequences of dehydration and other stresses, by increasing the tendency of the system to maintain the *status quo*.

Introduction

One of the best-known responses of plants to drought and other stresses is the accumulation of very soluble low-molecular-weight solutes such as proline [3], glycine betaine [30], sucrose [24], sorbitol [2], and mannitol [9]. The role of these compounds during stress is the subject of considerable controversy [11, 25] and interest since they accumulate to very high levels (0.1 to 0.3 *M* or more) [17, 27]. In the case of proline, three separate stress-induced control processes seem to be associated with its accumulation: 1) loss of feedback control of proline synthesis [4], 2) stimulation of proline synthesis from glutamate [5], and 3) inhibition of proline oxidation[26]. The testing of high concentrations of these compounds on isolated enzyme systems has shown that they have essentially no deleterious effects [6, 22] and, in fact, that they protect various enzyme systems against a range of perturbing effects. This has been shown for salt[22], heat[20], pH[1] and chemical[7] perturbations, with isolated organelles[19] as well as enzymes. No demonstration has been published, however, of the ability of compatible solutes to alleviate the perturbing effects, if any, of dehydration on enzyme activity. Recently, the effects of proline and glycine betaine on enzyme solvation were described. These solutes and others prevented or diminished the ability of polyethylene glycol (PEG) to precipitate a key enzyme of nitrogen metabolism, glutamine synthetase [21].

The present paper discusses some of the characteristics of the proline/PEG/glutamine synthetase interaction and the generality of the three-way interaction between proline, PEG and different types of proteins.

Materials and methods

Preparation and assay of enzyme

Glutamine synthetase from barley (*Hordeum distichum* L.) was extracted and partially purified as described earlier[15]. The transferase assay procedure was as indicated previously [21, 23].

PEG-induced precipitation of glutamine synthetase

Tris-HCl (0.05 *M*) at pH 6.5 (except as indicated) was used as the solvent for all reagents. Precipitation with PEG was carried out in a total volume of 1.0 ml in Eppendorf tubes which were chilled and whose contents were thoroughly agitated. After enzyme addition, the tubes remained chilled for 30 min before being centrifuged for 10 min at room temperature. The supernatants were decanted and, after draining, the tubes were wiped free of any remaining supernatant. The precipitated enzyme remaining in the tubes was redissolved in 0.2 ml 0.05 *M* Tris-HCl at pH 6.5 and assayed.

Extraction and purification of alfalfa mosaic virus (AMV)

The procedures followed were those of van Vloten-Doting and Jaspers [29], except that 6000 instead of 20,000 MW PEG was used.

PEG-induced precipitation of AMV and its assay

The virus was precipitated by techniques developed earlier [31], except that precipitation and resuspension were in a 0.1 *M* KH_2PO_4/KOH buffer, and a final volume of 1 ml was held on ice for 2 h prior to centrifugation in Eppendorf tubes. The resuspended viral pellet was assayed spectrophotometrically by determining the absorbance at 260 nm.

Preparation of steroid complex and antiserum

The conjugation of ^{3}H-testosterone and bovine serum albumin was carried out by M. D'Occhio according to Thorell and Larson[28], and antiserum was prepared and supplied by R. I. Cox of the CSIRO Division of Animal Production, Hormone Assay Development Group, Blacktown, NSW, Australia.

PEG-induced precipitation of the steroid/antiserum complex

PEG-induced precipitation of the radioimmune complex was carried out as for a normal radioimmunoassay[28].

Results

Effects of PEG and proline on the solubility of glutamine synthetase

As the concentration of PEG (6000 MW) is increased above 10%, an increasing amount of glutamine synthetase is precipitated (Fig. 1). The effect reaches a maximum at about 25% PEG, at which concentration about 75% of the enzyme is precipitated. In the presence of 1 *M* proline at pH 6.5 the effect of intermediate PEG concentrations is mitigated, *i.e.*, it is still possible to precipitate the enzyme, but higher concentrations of PEG are required.

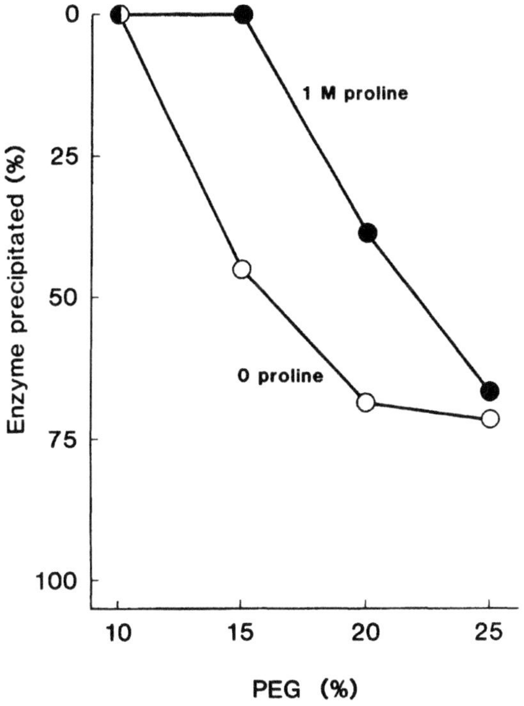

Fig. 1. Precipitation of glutamine synthetase by different concentrations of PEG (6000 MW, pH 6.5) in the absence and presence of 1 *M* proline.

This is clearer, perhaps, when the effect of proline concentration is compared at 12.5 and 17.5% PEG (Fig. 2). At 17.5% PEG, about 75% of the enzyme is precipitated, and increasing proline concentration does not have a large influence, decreasing precipitation to about 55% at 1 *M* proline. When the PEG concentration is 12.5%, however, enzyme precipitation is reduced to nil by 1 *M* proline. Thus, it is apparent that

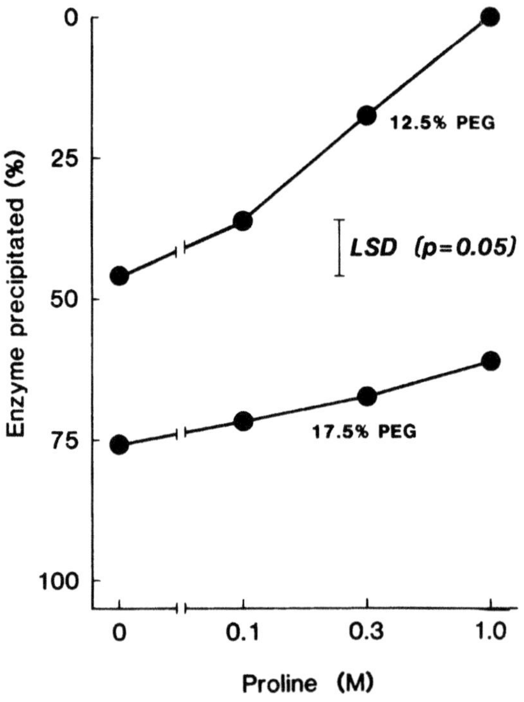

Fig. 2. Effect of proline concentration on the precipitation of glutamine synthetase by 12.5 and 17.5% PEG (6000 MW, pH 6.5).

the effects of the compatible solute are concentration-dependent, as are the effects of PEG.

The effects of lower-molecular-weight PEG polymers are illustrated in Fig. 3. As the molecular weight is decreased from 6000 (Fig. 1) to 4000 and to 1000, the enzyme-precipitating ability of a given concentration (%, wt/vol) of the polymer decreases. The figure also indicates that the ameliorating effect of 1 *M* proline is greatest when the concentration or molecular weight of PEG does not exceed a medium range.

The enzyme/PEG interaction is markedly sensitive to pH, with low (< 20%) concentrations of PEG causing considerable precipitation at pH 5.5 (Fig. 4). At pH 7.5, however, no enzyme is precipitated until 15% PEG is exceeded, although the effect of 25% PEG is significantly greater at pH 7.5 than at pH 5.5. The effect of 1 *M* proline is also affected by pH, in that it more effectively reduces PEG-induced enzyme precipitation at pH 7.5 than at pH 5.5.

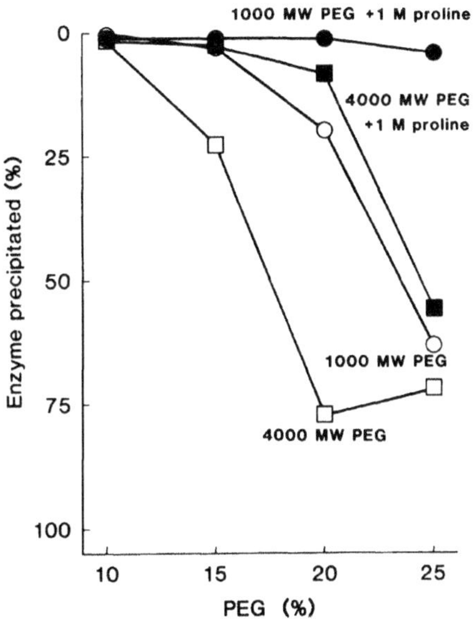

Fig. 3. Effect of 1 *M* proline on glutamine synthetase precipitation by different concentrations of 1000 and 4000 MW PEG at pH 6.5.

Effects of PEG and proline on the solubility of AMV

AMV is an RNA-containing virus that has between 80 and 82% protein. The use of PEG in its purification has been the subject of several studies, and standardized procedures have been developed, as indicated in the Materials and methods section.

The pH sensitivity of the virus/PEG/proline interaction (Fig. 5) is similar to that shown above for the interaction involving glutamine synthetase (Fig. 4), although the concentrations of PEG required to cause precipitation of AMV are lower than those required for the enzyme (Fig. 1). AMV is more effectively precipitated at the lower pH, and 1 *M* proline is active at both pH values in decreasing the PEG-induced precipitation of the virus.

The addition of 0.1 *M* salt (NaCl) enhanced the ability of PEG to precipitate the virus (Fig. 6), so that 7.5% PEG was able to cause nearly complete precipitation. Proline (1 *M*) not only reduced the effect of PEG (Fig. 6A), but also alleviated the NaCl-enhanced effect of PEG (Fig. 6B), thus demonstrating a tendency to maintain the pre-stress *status quo* in the face of two simultaneous 'stresses'.

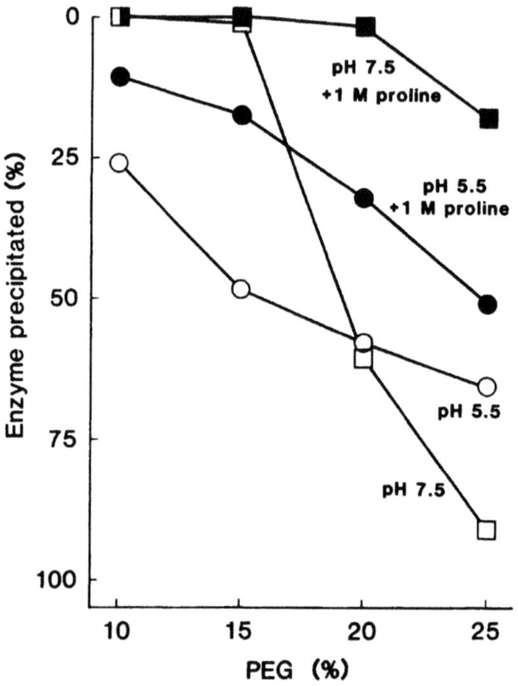

Fig. 4. Effect of 1 *M* proline on glutamine synthetase precipitation by different concentrations of PEG (6000 MW) at pH 5.5 and 7.5.

Effects of PEG and proline on the solubility of ^{3}*H-testosterone/antiserum complex*

Sensitive and specific assays for hormones and other ligands, by means of antibodies generated in rabbits or sheep, *etc.*, have been in use for some time. PEG is routinely employed to precipitate the hormone/antibody complex during the assay procedure [28]. Because the binding of the radioactive ligand is an equilibrium phenomenon, the highest sensitivity of the assay is achieved when about 50% of the radioactivity is bound, which is an important aspect of the methods. This amount of radioactivity is precipitated by about 15 to 20% PEG (6000 MW) (Fig. 7), and 1 *M* proline decreases the PEG-induced effect, particularly at low and intermediate polymer concentrations.

Thus, whether the proteins are uncomplexed soluble enzymes, viral proteins bound to RNA, or antibodies complexed with radioactive steroids, they can be precipitated by PEG (at the appropriate molecular weight, concentration and pH). Further, regardless of the system,

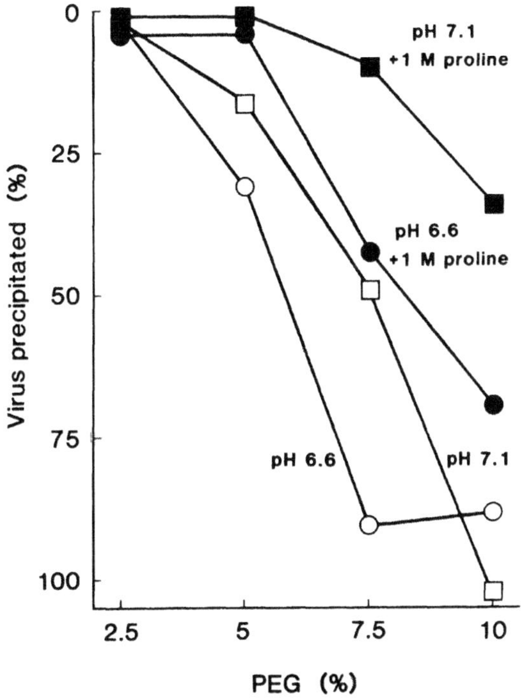

Fig. 5. Effect of 1 *M* proline on alfalfa mosaic virus precipitation by different concentrations of PEG (6000 MW) at pH 6.6 and 7.1.

proline (at the appropriate concentration and pH) can alleviate the precipitation of the protein.

The ability of a mixture of five amino acids, each at 0.2 *M*, to protect the steroid/antiserum complex from precipitation by PEG, in comparison with the influence of 1 *M* proline, is shown in Fig. 8. The amino acid mixture was as protective as proline alone up to a PEG concentration of about 12.5%. Above that percentage of PEG the mixture appeared even more effective than proline, suggesting either that one or more of the amino acids in the mixture may be even more active than proline in 'protecting' the solubility of the hormone/protein complex, or that, in combination, the amino acids may be more active than they are singly.

Discussion

The protein-precipitating action of PEG has been ascribed to its

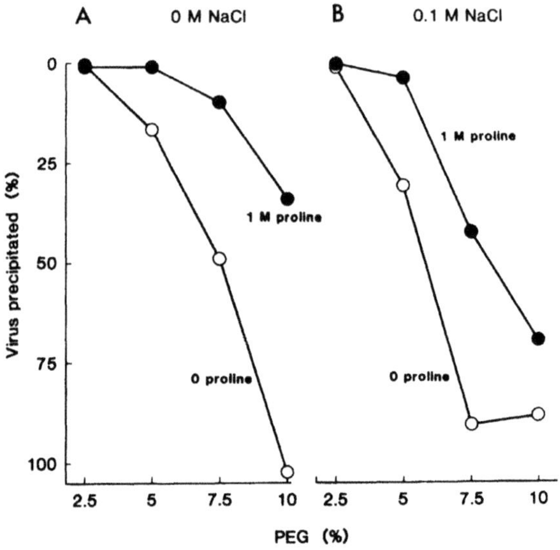

Fig. 6. Effect of 0.1 *M* NaCl and 1 *M* proline on the PEG (6000 MW)-induced precipitation of alfalfa mosaic virus at pH 7.1.

ability to sterically exclude the protein from regions of the solvent occupied by the polymer [13, 16]. The interaction between the protein and the polymer depends on a number of parameters, such as the size of both components, the concentration of the PEG polymer (but not of the protein [18, 21]), the pH, and the presence or absence of other solutes like proline or salt, *etc.* The end result, however, is a PEG-induced thermodynamic destabilization of the protein, which, in effect, decreases the amount of protein remaining in solution. The ability of PEG to reduce the solubility of proteins is widespread, and extends from soluble, uncomplexed proteins to those complexed with RNA, with radioactive steroids, or with other materials [8, 12]. The effect of proline on the protein/PEG interaction is also dependent on a number of similar factors, but under appropriate conditions proline reduces the precipitation of the protein. In essence, therefore, proline reduces the perturbing or destabilizing influence of the polymer and at least partially returns the solution to the thermodynamic condition it was in before PEG was added. It is not clear how proline brings this about, since it has not yet been determined whether the ability of PEG to precipitate the protein is reduced, or whether the protein has an increased tendency to remain in solution under unfavorable conditions.

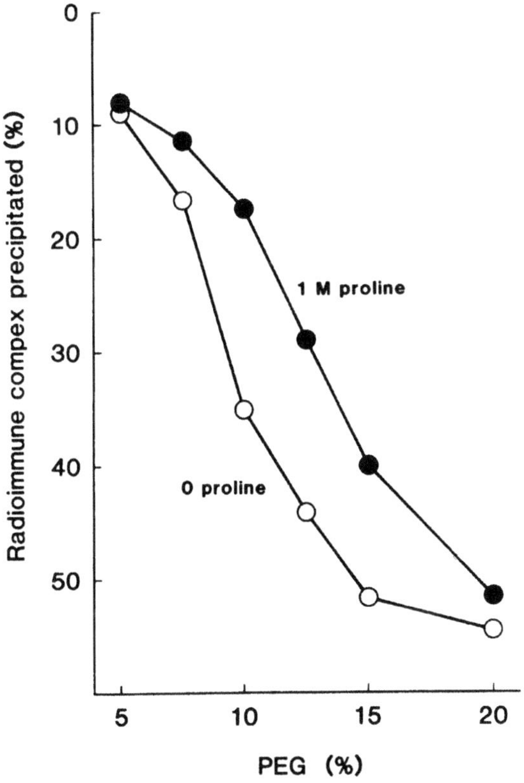

Fig. 7. Precipitation of ^{3}H-testosterone/antiserum complex at pH 7.5 by different concentrations of PEG (6000 MW) in the absence and presence of 1 *M* proline.

Proline is not the only amino acid or compatible solute to decrease the effect of PEG on proteins[21]. Because of the effects of solvation on enzyme activity [10, 14], and the likelihood that natural polymeric macromolecules exist in cells and cause effects under conditions of reduced water availability (*e.g.*, drought) that are similar to those of PEG, it seems likely that our system has important parallels with and implications for the consequences of biological dehydration and the protective role played by proline. With the demonstration that proline has the ability to protect (*i.e.*, maintain the pre-stress *status quo*), to a greater or lesser extent, protein-containing systems against 'dehydration', the effects of all the major environmental stresses have now been shown to be alleviated by proline. Thus, a conceptual basis for the adaptive significance of *in vivo* accumulation of proline and

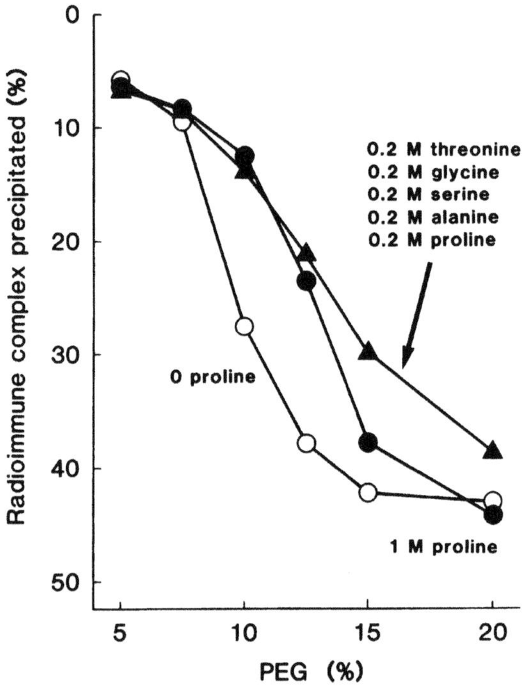

Fig. 8. Precipitation of ^{3}H-testosterone/antiserum complex at pH 7.5 by different concentrations of PEG (6000 MW) in the absence and presence of 1 *M* proline or a mixture of proline, serine, glycine, alanine and threonine, each at 0.2 *M*.

other compatible solutes has been demonstrated, although similar effects with intact plants have yet to be reported.

Acknowledgements The authors thank the Royal Society, the Nuffield Foundation and the Barley Industry Research Council for grants to L G Paleg for support of the research. We also thank M O'Dea for valuable technical assistance, Dr. O W Barnett of Clemson University, Clemson, S. C. who, while on leave in the Department of Plant Pathology, Waite Agricultural Research Institute, graciously provided the alfalfa mosaic virus and the techniques for working with it, and Dr. M D'Occhio of the Department of Animal Sciences, Waite Agricultural Research Institute who generously provided the materials and methods for examining the effects of proline in the radioimmunoassay system.

References

1 Ahmad I, Larher F, Mann A F, McNally S F and Stewart G R 1982 Nitrogen metabolism of halophytes. IV. Characteristics of glutamine synthetase from *Triglochin maritima* L. New Phytol. 91, 585–595.

2 Ahmad I, Larher F and Stewart G R 1979 Sorbitol, a compatible osmotic solute in *Plantago maritima*. New Phytol. 82, 671–678.

3 Aspinall D and Paleg L G 1981 Proline accumulation: physiological aspects. *In* Physiology and Biochemistry of Drought Resistance in Plants. Eds. L G Paleg and D Aspinall. pp 205–241. Academic Press, Sydney.

4 Boggess S F, Aspinall D and Paleg L G 1976 Stress metabolism. IX. The significance of end-product inhibition of proline biosynthesis and of compartmentation in relation to stress-induced proline accumulation. Aust. J. Plant Physiol. 3, 513–525.

5 Boggess S F, Stewart C R, Aspinall D and Paleg L G 1976 Effect of water stress on proline synthesis from radioactive precursors. Plant Physiol. 58, 398–401.

6 Borowitzka L J 1981 Solute accumulation and regulation of cell water activity. *In* Physiology and Biochemistry of Drought Resistance in Plants. Eds. L G Paleg and D Aspinall. pp 97–130. Academic Press, Sydney.

7 Bowlus R D and Somero G N 1979 Solute compatibility with enzyme function and structure: rationales for the selection of osmotic agents and end-products of anaerobic metabolism in marine invertebrates. J. Exp. Zool. 208, 137–152.

8 Fried M and Chun P W 1971 Water-soluble non-ionic polymers in protein purification. Methods Enzymol. 22, 238–248.

9 Gorham J, Hughes L I and Wyn Jones R G 1981 Low-molecular-weight carbohydrates in some salt-stressed plants. Physiol. Plant. 53, 27–33.

10 Greaney G S and Somero G N 1979 Effects of anions on the activation thermodynamics and fluorescence emission spectrum of alkaline phosphatase: evidence for enzyme hydration changes during catalysis. Biochemistry 18, 5322–5332.

11 Hanson A D, Nelsen C E and Everson E H 1977 Evaluation of free proline accumulation as an index of drought resistance using two contrasting barley cultivars. Crop Sci. 17, 720–726.

12 Ingham K C 1977 Polyethylene glycol in aqueous solution: solvent perturbation and gel filtration studies. Arch. Biochem. Biophys. 184, 59–68.

13 Lee J C and Lee L L Y 1981 Preferential solvent interactions between proteins and polyethylene glycols. J. Biol. Chem. 256, 625–631.

14 Low P S and Somero G N 1975 Protein hydration changes during catalysis: a new mechanism of enzymic rate enhancement and ion activation/inhibition of catalysis. Proc. Nat. Acad. Sci. USA 72, 3305–3309.

15 McNally S F, Orebamjo T O, Hirel B and Stewart G R 1983 Glutamine synthetase isoenzymes of *Striga hermonthica* and other angiosperm parasites. J. Exp. Bot. 34, 610–619.

16 Miekka S I and Ingham K C 1980 Influence of hetero-association on the precipitation of proteins by poly (ethylene glycol). Arch. Biochem. Biophys. 203, 630–641.

17 Munns R, Brady C J and Barlow E W R 1979 Solute accumulation in the apex and leaves of wheat during water stress. Aust. J. Plant Physiol. 6, 379–389.

18 Na G C and Timasheff S N 1981 Interaction of calf brain tubulin with glycerol. J. Mol. Biol. 151, 165–178.

19 Nash D, Paleg L G and Wiskich J T 1982 The effect of proline, betaine and some other solutes on the heat stability of mitochondrial enzymes. Aust. J. Plant Physiol. 9, 47–57.

20 Paleg L G, Douglas T J, van Daal A and Keech D B 1981 Proline and betaine protect enzymes against heat inactivation. Aust. J. Plant Physiol. 8, 107–114.

21 Paleg L G, Stewart G R and Bradbeer J W 1984 Proline and glycine betaine influence protein solvation. Plant Physiol. 75, 974–978.

22 Pollard A and Wyn Jones R G 1979 Enzyme activities in concentrated solutions of glycine betaine and other solutes. Planta 144, 291–298.

23 Rhodes D, Rendon G A and Stewart G R 1975 The control of glutamine synthetase activity in *Lemna minor* L. Planta 125, 203–210.

24 Setter T L and Greenway H 1979 Growth and osmoregulation of *Chlorella emersonii* in NaCl and neutral osmotica. Aust. J. Plant Physiol. 6, 47–60 and corrigendum in Aust. J. Plant Physiol. 6, 569–572.

25 Singh T N, Aspinall D and Paleg L G 1972 Proline accumulation and varietal adaptability to drought in barley: a potential metabolic measure of drought resistance. Nature New Biol. 236, 188–190.

26 Stewart C R, Boggess S F, Aspinall D and Paleg L G 1977 Inhibition of proline oxidation by water stress. Plant Physiol. 59, 930–932.

27 Stewart G R and Lee J A 1974 The role of proline accumulation in halophytes. Planta Berlin 120, 279–289.

28 Thorell J I and Larson S M 1978 Radioimmunoassay and Related Techniques, C V Mosley Co., St. Louis.

29 van Vloten-Doting L and Jaspers E M J 1972 The uncoating of alfalfa mosaic virus by its own RNA. Virology 48, 699–708.

30 Wyn Jones R G and Storey R 1981 Betaines. *In* Physiology and Biochemistry of Drought Resistance in Plants. Eds. L G Paleg and D Aspinall. pp 172–204. Academic Press, Sydney.

31 Yamamoto K R, Alberts B M, Benzinger R, Lawhorne L and Treiber G 1970 Rapid bacteriophage sedimentation in the presence of polyethylene glycol and its application to large-scale virus purification. Virology 40, 734–744.

Section 2

Production of micro and macroalgae with saline water

Plant and Soil 89, 97–106 (1985). Ms. BR 07

Blue-green algae (cyanobacteria): prospects and perspectives

R. H. REED*, S. R. C. WARR*, D. L. RICHARDSON*, D. J. MOORE*
and W. D. P. STEWART**
**Department of Biological Sciences, and **A.F.R.C. Research Group on Cyanobacteria,
The University of Dundee, Dundee, DD1 4HN Scotland, UK*

Key words Blue-green algae Cyanobacteria Osmotic responses Salinity tolerance Spirulina

Summary Photosynthetic, prokaryotic blue-green algae (cyanobacteria) occur in a wide range of natural habitats of diverse ionic composition and, as such, represent an important source of biological material for biosolar energy conversion programs using saline water. The gas-vacuolate, filamentous Spirulina is grown in 'seminatural' culture in Lake Texcoco, Mexico, as a major source of single-cell protein for animal nutrition. Pilot-scale trials in other areas of the world have also demonstrated the suitability of blue-green algae, including Spirulina, for growth under brackish conditions. The carbohydrate accumulation profiles of blue-green algae differ in isolates from freshwater, marine and hypersaline habitats, with a trend towards sucrose or trehalose accumulation in stenohaline freshwater strains grown in media containing NaCl, while euryhaline and marine forms frequently accumulate glucosylglycerol. Many halotolerant isolates from hypersaline habitats accumulate glycinebetaine in response to osmotic stress. This knowledge may provide scope for future improvement in the N_2 fixation rates of blue-green algae in saline media, using betaine-accumulating N_2-fixing strains in preference to other, salt-sensitive isolates.

Introduction

The blue-green algae (Cyanophyta, Cyanobacteria) occupy a unique taxonomic position, since they combine an autotrophic mode of growth that is common to eukaryotic plant cells with a metabolic system that is generally regarded as bacterial, rather than plant-like. These prokaryotic organisms thus offer the growth potential of microbial cells together with the light-harvesting capabilities of plant cells, making them ideal candidates for biosolar energy conversion programs. A further advantage of blue-green algae over eukaryotic plant cell types is the capacity of some strains to fix atmospheric N_2 [34].

There is historical evidence that blue-green algae were harvested, dried and eaten by the Aztecs of Tenochtitlan (Mexico City) at the time of the Spanish Conquest[15]. African tribes in the vicinity of Lake Chad have also been reported to consume blue-green algae[20]. Except for the Far East, these reports form the only record of the traditional use of microbial biomass as a food for human consumption[14]. The two species that have been utilized as a food source are both filamentous, nonheterocystous forms, namely *Spirulina maxima* (*geitleri*) and *Spirulina platensis*, and recent research studies have concentrated on the potential of these two blue-green algae in large-scale culture systems.

Mass production of Spirulina

The major plant for large-scale commercial production of Spirulina, on Lake Texcoco in Mexico, is operated by the Sosa Texcoco SA Company. *S. maxima* grows naturally in the waters of L. Texcoco, and the 'seminatural' culture methods used at the plant involve raceway ponds filled with lake water and supplemented with fertilizer to increase biomass production[31]. The plant has now been operating for a decade, with a reported daily production of more than 2 tons of dried product. An expansion program to produce 'several thousand tons yearly' is forecast within the next few years[31].

Extensive investigations have also been conducted in other countries on the possibility of large-scale cultivation of Spirulina. Most of these studies have used *S. platensis* rather than *S. maxima*.

Under the All-India Coordinated Project on Algae, several research groups in India have evaluated the production technology of *S. platensis*. The results of the research group at the Central Food Technological Research Institute (CFTRI), Mysore have recently been published[35]. Mass outdoor cultivation at CFTRI employed PVC tanks holding up to 1.4 m^3 of growth medium, with a surface area of up to 11 m^2 (depth 20–25 cm). With a modified culture medium based on that described by Zarrouk[35], but containing less $NaHCO_3$ (0.45%) and with rural wastes (cattle dung and urine) and sea salt to provide additional N, P, and trace elements, the algae could be harvested after 8–10 days with a 4-fold increase in algal biomass over this period. The temperature regime at Mysore (subtropical and not subject to marked seasonal fluctuations) appears to favour the mass cultivation of *S. platensis*, which grows best within the range 20–37°C. Yields of *S. platensis* from CFTRI, Mysore are comparable to those quoted for *S. maxima* (L. Texcoco), at 8–12 $g\,m^{-2}\,d^{-1}$ (ref. [31, 35]). Contamination of culture ponds by other algae (especially Chlorella) and bacteria has also been reported[35]. One possible reason for this problem lies in the reduced $NaHCO_3$ level of the CFTRI growth medium – it is recognized generally that Spirulina grows as a quasi-monoculture under conditions where both the pH and total salt concentration are high (*i.e.* > pH 10.0 and > 3.0% salt)[14].

The growth of *S. platensis* in brackish water unsuitable for irrigation has been investigated in recent years in Israel[28, 36]. In the Negev Desert, the algae have been grown in shallow PVC-lined channels up to 300 m long (20–25 cm deep), with agitation to increase the final population density[29]. The production of *S. platensis* varied seasonally, with almost no growth during the period from December to February, due to the low temperatures during the day[28]. Contamination by *Chlorella*

Table 1. Spirulina from various locations: gross chemical compositions (% dry weight)

Component	Texcoco[31] (*S. maxima*)	CFTRI, India[35] (*S. platensis*)	NBRI, India[32] (*S. platensis*)
Protein	⩽ 70.0	55–65	50–55
Carbohydrate	16.5	10–15	18–20
Lipids	7.0	2–6	6–9
Total	⩽ 93.5	≃ 80	≃ 80

vulgaris (and its grazers) occurred during the winter months when the growth of *S. platensis* was reduced and also in media where the $NaHCO_3$ content was artificially lowered below 1.6%.

Chemical composition of Spirulina

Table 1 shows the major constituents of *S. platensis* and *S. maxima*. Proteins form the bulk of the dry weight, with values that are higher than those for other blue-green algae and eukaryotic plant cells[9]. The amino acid composition of Spirulina protein is similar in many respects to that of other microorganisms (Table 2), being marginally deficient in

Table 2. Amino acid composition of *Spirulina* protein (% total)

Amino acids	Texcoco[31] (*S. maxima*)	CFTRI, India[35] (*S. platensis*)	NBRI, India[32] (*S. platensis*)
Essential			
Isoleucine	6.2	4.0	4.2
Leucine	8.7	8.1	8.8
Lysine	6.0	6.2	4.2*
Methionine	3.3	1.7	1.6*
Phenylalanine	6.0	5.3	3.0
Threonine	6.3	5.7	5.8
Tryptophan	1.7	0.3	–
Tyrosine	–	3.5	2.3
Valine	9.0	6.7	7.3
Non-essential			
Alanine	8.8	10.0	10.4
Arginine	9.0	7.9	4.3
Aspartic acid	9.7	9.3	13.3
Cysteine	1.0	0.7	–*
Glutamic acid	13.4	11.9	11.1
Glycine	–	7.9	8.3
Histidine	1.6	2.3	2.0
Proline	4.5	4.3	5.7
Serine	4.8	4.2	7.7

*marginally deficient by comparison with FAO standards[35]

lysine, methionine and cysteine when compared with FAO standards[35]. However, Spirulina protein is generally considered to be superior to all other plant proteins, with a high annual protein productivity rate per hectare when compared with other plant and animal sources[1]. The remaining non-protein nitrogen content is mainly due to nucleic acids and, in this respect, Spirulina contains $< 5\%$ (dry weight) nucleic acids[31]. Such low values are characteristic of algal, rather than microbial (yeast, bacteria) cells and increase the suitability of Spirulina for human consumption.

An examination of the literature reveals a wide range of values for the lipid content of dried Spirulina (1.5–12% of the dry weight[14]). A feature of the lipid composition of Spirulina that is stressed frequently is the high proportion of unsaturated fatty acids and the predominance of linolenic acid[35].

Carbohydrates account for up to 20% of the total dry weight of Spirulina, with the polyglucoside glycogen forming the major component[13]. Of the low molecular weight carbohydrates, our studies have shown that the heteroside glucosylglycerol is accumulated in cells of *S. platensis* in response to osmotic stress (see below).

The carbon reserve poly-β-hydroxybutyrate has recently been isolated from *S. platensis*[11]. This is only the second record of this compound from a blue-green algal cell (the first report of poly-β-hydroxybutyrate was made by Carr[12], in *Chlorogloeopsis fritschii* grown on sodium acetate). Poly-β-hydroxybutyrate accumulates to maximum levels in stationary-phase cells of *S. platensis*, reaching levels in excess of 5% of the dry weight[11].

The major pigments of Spirulina include chlorophyll *a* (up to 1.5% of the dry weight of *S. platensis* and *S. maxima*), β-carotene, myxoxanthophyll and C-phycocyanin. In animal feeding trials, the presence of these pigments (and especially β-carotene) led to colour enhancement of egg yolks, chicken flesh[35] and ciprinid fish scales[31].

Salinity tolerance of Spirulina

Growth of *S. platensis* in a 'minimal' medium containing $NaHCO_3$ (1.68%) together with varying amounts of NaCl was studied by Riccardi *et al.*[26, 27]. They showed that addition of NaCl at 1.87% or more results in a 50% decrease in growth, with little evidence of growth at 5.25%. Our observations, using a strain of *S. platensis* from L. Chad, showed that this alga can grow in BG11 medium supplemented with $NaNO_3$ (0.15%), as described by Rippka *et al.*[30], over a range of external sea salt concentrations from 0.0 to 7.0%. Growth under these conditions is slower than in the presence of excess $NaHCO_3$. These findings demonstrate that the salinity tolerance of *S. platensis* may be increased under conditions where the external $NaHCO_3$ level is reduced. However, the reduction in growth observed as a result of lowering the external $NaHCO_3$ content might be expected to lead to problems of contamination in field experiments with high NaCl-low $NaHCO_3$ growth media.

Osmotic responses of Spirulina

The mechanism of low molecular weight carbohydrate production in blue-green algal cells subjected to salt stress is now well established. The first demonstration of an osmotic role for an organic molecule in a blue-green algal cell was made by Borowitzka *et al.*[7], who showed that an isolate of the marine unicellular *Synechococcus* sp. could synthesize the heteroside glucosylglycerol[18] in response to increasing salt in the medium. Subsequent studies showed that other low molecular weight organic solutes may be involved in osmotic adjustment, including sucrose[4], trehalose[23], and glycinebetaine[22]. We have recently investigated the low molecular weight carbohydrate profiles of *S. platensis* grown in low salt and high salt media (up to 7.0% sea salt). Natural-abundance ^{13}C nuclear magnetic resonance spectroscopy of cells and cell extracts showed glucosylglycerol to be the only major low molecular weight organic component–spectra show the nine characteristic resonances due to glucosylglycerol (at $\delta = 61.6$, 61.7, 62.6, 70.8, 72.7, 73.1, 74.2, 79.9 and 98.9, measured with respect to 3-(trimethylsilyl)propane sulfonic acid (Na^+ salt)[24]), with no other resonances being observed. When assayed by gas-liquid chromatography, the intracellular concentration of glucosylglycerol was found to vary linearly with increasing external salinity up to 5.25% sea salt with a maximum glucosylglycerol concentration of 243.3 $mmol\,kg^{-1}$ (dry weight). Trehalose was also present in cells grown at low salinity, accounting for more than 40% of the total low molecular weight carbohydrate fraction in cells grown at 28°C in a medium containing sea salt at 1.75% (Table 3). Growth at higher temperatures increased the concentration of trehalose in cells of *S. platensis*, which reached levels in excess of one third of the total low molecular weight carbohydrates in cells grown in full-strength seawater medium (3.5%) at 37°C (*cf.* Table 3).

Osmotic responses of other blue-green algae

Several recent studies have presented data on the carbohydrate accumulation profiles of blue-green algae under conditions of osmotic

Table 3. Low molecular weight carbohydrates of *Spirulina platensis* grown at 28°C at several salt concentrations

Sea salt ($g\,dm^{-3}$)	Glucosylglycerol ($mmol\,kg^{-1}$)*	Trehalose ($mmol\,kg^{-1}$)*
17.5	33.9	27.2
35.0	207.5	26.1
52.5	243.3	7.9

* values (3 replicates) are expressed in terms of dry wt.

stress. The studies of Blumwald and coworkers[4, 5, 6] showed that sucrose may be a major organic solute in freshwater blue-green algae, while Mackay *et al.*[21] demonstrated that glucosylglycerol is more frequently recorded from marine strains. Our research studies lend some support to this general pattern, with the majority of stenohaline freshwater strains showing an accumulation of either sucrose or trehalose under osmotic stress, while the trend is towards glucosylglycerol accumulation in euryhaline and marine isolates[25]. We have recently studied four unicellular blue-green algae isolated from hypersaline habitats (*Synechocystis* spp. and *Dactylococcopsis salina*), and found that the quaternary ammonium compound glycinebetaine was accumulated in response to increasing external salt concentration[24]. Quaternary ammonium compounds (of as yet unspecified structure) have also been recorded for other halotolerant isolates of filamentous and unicellular forms[21].

Glycinebetaine has been reported as an intracellular solute in halophytes[38] and in the photosynthetic bacterium *Ectothiorhodospira halochloris*[16], which suggests that quarternary ammonium compounds may play a fundamental role in the increased halotolerance of these organisms, allowing them to survive and grow at higher salt concentrations than is possible for organisms that do not accumulate quarternary ammonium compounds. If this is the case, then future screening programs should investigate the organic solutes of blue-green algae under conditions of osmotic stress, to establish whether strains that accumulate glycinebetaine are generally more tolerant of extreme hypersaline conditions. These strains could then be grown in high salt media (and, in this respect, it is worth remembering that the betaine-accumulating unicell *Synechocystis* sp DUN52 is able to grow in media containing up to 20% sea salt[22]), to minimize the problems of contamination by other organisms, as is the case for Dunaliella[3]. It is tempting to suggest that a salt-tolerant, betaine-accumulating, N_2-fixing strain would have considerable potential in this respect, bearing in mind the protective effect of glycinebetaine and other quaternary ammonium compounds on nitrogenase activity in bacterial cells, which has recently been demonstrated by Bouillard and Le Rudulier[8]. Since the nitrogenase of both freshwater and marine blue-green algae is particularly sensitive to the external salt concentration[17, 23, 37], this avenue of research may offer scope for the improvement of N_2 fixation under highly saline conditions (*i.e.* > 3.5%). However, it is also worth noting that the sucrose-accumulating marine strain *Anabaena* CA is one of the fastest-growing N_2-fixing blue-green algae, growing optimally in full-strength seawater[33] (3.5% sea salt), suggesting that high sucrose levels may not inhibit nitrogenase in seawater-grown cells of this isolate.

Future prospects

Pilot-scale trials for the production of Spirulina to be used as a protein supplement for poultry have been carried out with encourgaging preliminary results. Chickens showed similar weight gain when Spirulina protein replaced groundnut protein or fish protein concentrate over a 12-week growth period[35]. It seems likely that production of Spirulina protein will be geared to animal feeding programs in the immediate future.

More than 50 'health food' products containing Spirulina are manufactured in capsule, tablet and powder form[31]. Spirulina has been claimed to help in the treatment of obesity[31]. Venkataraman[35] has also reported that Spirulina contains a factor that lowers blood cholesterol levels, although this component has not yet been identified. The more extreme claims of the health food industry have yet to be evaluated, and it seems likely that Spirulina consumption in the West will remain limited, though lucrative, until these claims are evaluated.

The recent isolation of mutants of *S. platensis* that overproduce the amino acid proline are of interest[26]. Proline-overproducing mutants excrete this amino acid into the external medium and contain higher intracellular levels than wild-type *S. platensis*[27]. The possibility of amino acid production using mutant strains of blue-green algae has yet to be studied, although this has been considered by other workers who have proposed schemes to obtain proline from salt-stressed Chlorella[19]. It is clear that proline-overproducing mutants of *S. platensis* are more tolerant of hypersaline stress than wild-type cells, and Riccardi *et al.*[27] have suggested that strains of Spirulina may be selected for enhanced NaCl tolerance by screening for resistance to proline analogs. Similar procedures might be expected to yield NaCl-tolerant mutants of other blue-green algae (including N_2-fixing forms). Since one of the unique features of cyanobacteria is the ability of filamentous, heterocystous forms to fix N_2 while growing photoautotrophically under aerobic conditions, we believe that future research could benefit from studies directed at maximizing N_2 fixation rates in these strains under conditions of salinity stress.

Our studies have indicated that many blue-green algae efficiently interconvert low molecular weight organic osmotica and glycogen, with little loss of organic osmotica to the bathing medium upon 'downshock'[23, 37]. Preliminary studies of cells entrapped in alginate have shown that minimal amounts of carbohydrates (typically 1–12% of the intracellular concentration) are released to the medium upon 'downshock', suggesting that this technique is unlikely to yield carbohydrates in sufficient quantities to be commercially exploited.

The presence of the organic storage compound poly-β-hydroxybutyrate in blue-green algae grown in saline media[12] may find future commercial applications, since it is a biodegradable substance characterized by a number of physical properties related to conventional plastics[10]. Poly-β-hydroxybutyrate is also accumulated in marine heterotrophic bacteria when the carbon supply is non-limiting for growth[2]. However, the photosynthetic conversion of inorganic carbon into poly-β-hydroxybutyrate by blue-green algae has an added advantage, since no additional source of fixed carbon is required.

Acknowledgements RHR currently holds a Royal Society Research Fellowship in Subjects Supporting Biotechnology.

References

1 Aaronson S and Dubinsky Z 1982 Mass production of algae. Experientia 38, 36–40.
2 Baumann P and Baumann L 1981 The marine gram-negative eubacteria: genera Photobacterium, Beneckea, Alteromonas, Pseudomonas and Alcaligenes *In* The Prokaryotes, Vol II. Eds. M P Starr, H Stolp, H G Trüper, A Balows and H G Schlegel. pp 1302–1331. Springer, Berlin.
3 Ben-Amotz A and Avron M 1983 Accumulation of metabolites by halotolerant algae and its industrial potential. Annu. Rev. Microbiol. 37, 95–119.
4 Blumwald E and Tel-Or E 1982 Osmoregulation and cell composition in salt adaptation of *Nostoc muscorum*. Arch. Microbiol. 132, 168–172.
5 Blumwald E, Mehlhorn R J and Packer L 1983 Studies of osmoregulation in salt adaptation of cyanobacteria with ESR spin-probe techniques. Proc. Nat. Acad. Sci. USA 80, 2599–2602.
6 Blumwald E, Mehlhorn R J and Packer L 1983 Ionic osmoregulation during salt adaptation of the cyanobacterium *Synechococcus* 6311. Plant Physiol. 73, 377–380.
7 Borowitzka L J, Demmerle S, Mackay M A and Norton R S 1980 Carbon-13 nuclear magnetic resonance study of osmoregulation in a blue-green alga. Science 210, 650–651.
8 Bouillard L and Le Rudulier D 1983 Nitrogen fixation under osmotic stress: enhancement of nitrogenase biosynthesis in *Klebsiella pneumoniae* by glycinebetaine. Physiol. Veg. 21, 447–457.
9 Boyd C E 1973 Amino acid composition of freshwater algae. Arch Hydrobiol. 72, 1–9.

10 Braunegg G, Sonnleitner B and Lafferty R M 1978 A rapid gas chromatographic method for the determination of poly-β-hydroxybutyric acid in microbial biomass. Eur. J. Appl. Microbiol. Biotechnol. 6, 29–37.

11 Campbell J, Stevens S E and Balkwill D L 1982 Accumulation of poly-β-hydroxybutyrate in *Spirulina platensis*. J. Bacteriol. 149, 361–363.

12 Carr N J 1966 The occurrence of poly-β-hydroxybutyrate in the blue-green alga *Chlorogloea fritschii*. Biochim. Biophys. Acta 120, 308–310.

13 Casu B, Naggi A and Vercelloti J R 1980 Polisaccaridi di riserva della *Spirulina platensis*, estrazione e caratterizzazione *In* Prospettive della coltura di Spirulina in Italia. Cons. Naz. delle Rich. Rome. pp 145–153.

14 Ciferri O 1983 Spirulina, the edible microorganism. Microbiol. Rev. 47, 551–578.

15 Farrar W V 1966 Tecuitlatl: a glimpse of Aztec food technology. Nature London 211, 341–342.

16 Galinski E A and Trüper H G 1982 Betaine, a compatible solute in the extremely halophilic phototrophic bacterium *Ectothiorhodospira halochloris*. FEMS Microbiol. Lett. 13, 357–360.

17 Jones K and Stewart W D P 1969 Nitrogen turnover in marine and brackish habitats. III the production of extracellular nitrogen by *Calothrix scopulorum*. J. Mar. Biol. Ass. UK 49, 701–716.

18 Kollman V H, Hanners J L, London R E, Adame E G and Walker T E 1979 Photosynthetic preparation and characterization of ^{13}C-labelled carbohydrates in *Agmenellum quadruplicatum*. Carbohydr. Res. 73, 193–202.

19 Leavitt R I 1983 Process for the preparation of L-proline by cultivating algae. US Patent No 4383038, US Patent Office.

20 Leonard J 1966 The 1964–65 Belgian trans-Saharan expedition. Nature 209, 126–128.

21 Mackay M A, Norton R S and Borowitzka L J 1983 Marine blue-green algae have a unique osmoregulatory system. Mar. Biol. 73, 301–307.

22 Mohammad F A A, Reed R H and Stewart W D P 1983 The halophilic cyanobacterium *Synechocystis* DUN52 and its osmotic responses. FEMS Microbiol. Lett. 16, 287–290.

23 Reed R H and Stewart W D P 1983 Physiological responses of *Rivularia atra* to salinity: osmotic adjustment in hyposaline media. New Phytol. 95, 595–603.

24 Reed R H, Chudek J A, Foster R and Stewart W D P 1984 Osmotic adjustment in cyanobacteria from hypersaline environments. Arch. Microbiol. 138, 333–337.

25 Reed R H, Richardson D L, Warr S R C and Stewart W D P 1984 Carbohydrate accumulation and osmotic stress in cyanobacteria. J. Gen. Microbiol. 130, 1–4.

26 Riccardi G, Sora S and Ciferri O 1981 Production of amino acids by analog-resistant mutants of the cyanobacterium *Spirulina platensis*. J. Bacteriol. 147, 1002–1007.

27 Riccardi G, Cella R, Camerino G and Ciferri O 1983 Resistance to azetidine-2-carboxylic acid and sodium chloride tolerance in carrot cell cultures and *Spirulina platensis*. Plant Cell Physiol. 24, 1073–1078.

28 Richmond A and Vonshak A 1978 Spirulina culture in Israel. Arch. Hydrobiol. Beich. Ergebn. Limnol. 11, 274–280.

29 Richmond A, Karg S and Boussiba S 1982 Effects of bicarbonate and carbon dioxide on the competition between *Chlorella vulgaris* and *Spirulina platensis*. Plant Cell Physiol. 23, 1411–1417.

30 Rippka R, Deruelles J, Waterbury J B, Herdman M and Stanier R Y 1979 Generic assignments, strain histories and properties of pure cultures of cyanobacteria. J. Gen. Microbiol. 111, 1–61.

31 Santillan C 1982 Mass production of Spirulina. Experientia 38, 40–43.

32 Saxena P N, Ahmad M R, Shyam R and Misra P S 1982 Chemical composition of sewage-grown *Spirulina platensis*. Experientia 38, 1438.

33 Stacey G, Van Baalen C and Tabita F R 1977 Isolation and characterization of a marine *Anabaena* sp. capable of rapid growth on molecular nitrogen. Arch. Microbiol. 114, 197–201.

34 Stewart W D P 1980 Some aspects of structure and function in N_2-fixing cyanobacteria. Annu. Rev. Microbiol. 34, 497–536.
35 Venkataraman L V 1983 Blue-green alga Spirulina. CFTRI Press, Mysore, India.
36 Vonshak A, Boussiba S, Abelovich A and Richmond A 1983 Production of Spirulina biomass: maintenance of monoalgal cultures outdoors. Biotechnol. Bioeng. 25, 341–349.
37 Warr S R C, Reed R H and Stewart W D P 1984 Physiological responses of *Nodularia harveyana* to osmotic stress. Mar. Biol. 79, 21–26.
38 Wyn Jones R G and Gorham J 1983 Osmoregulation *In* Encyclopedia of Plant Physiology Vol 12C. Eds. O L Lange, P S Nobel, C B Osmond and H Ziegler. pp 35–58. Springer, Berlin.

Plant and Soil 89, 107–116 (1985). Ms. BR 08

Cyanobacterial biomass production in saline media

RACHEL GABBAY and ELISHA TEL-OR
Department of Agricultural Botany, The Hebrew University of Jerusalem, Rehovot 76100, Israel

Key words Cyanobacteria Ferredoxin-NADP reductase Malate dehydrogenase Respiration Salt tolerance *Spirulina subsalsa*

Summary The cultivation, growth patterns, and physiological activities of the marine cyanobacterium (blue-green alga) *Spirulina subsalsa* were studied. A comparison of its growth in three different media (diluted seawater, seawater, and seawater + 0.5 *M* NaCl) revealed a faster growth in the hypersaline medium. In the hypersaline medium, the culture was homogeneous, in contrast to the aggregates formed in the lower-salt media. Enzymic analysis of the cells demonstrated selective sensitivity of soluble malate dehydrogenase to sodium ions, while chloride ions or nonionic solutes caused no inhibition. The membrane-associated enzyme ferredoxin-NADP reductase was only partially sensitive to sodium ions. The respiratory enzymes exhibited well-coupled activity, and faster respiration was observed with the preparation from the hypersaline culture.

Introduction

Few marine cyanobacterial genera have been defined and characterized with respect to their salt tolerance[4], and no general concepts have been presented to explain the mechanisms of salt tolerance in marine cyanobacteria. The cyanobacterial cell must have structural, physiological, and biochemical means to survive and proliferate under the low water potential and ionic stress of the marine environment. Osmoregulation by polyols and amino acids in Aphanotheca and by glucosylglycerol in Synechococcus has been described[2]. The physiology[8] and motility[1] of *Spirulina subsalsa*, isolated from a marine lagoon in the Sinai peninsula, has been investigated.

Mass cultivation of marine cyanobacteria has not been reported yet, and it is challenging to assess its feasibility. This report presents our recent studies of *Spirulina subsalsa*.

Materials and methods

Organism and culture

The marine cyanobacterium *Spirulina subsalsa* was isolated by M. Shilo of the Hebrew University from the Bardawill Lagoon in the Sinai peninsula, and was kindly made available to us by the Department of Aquatic Microbiology of the Hebrew University. The organism was cultured in seawater medium (SW) enriched with minerals according to Abeliovich and Gan[1]. Diluted seawater medium (SW/2) was 1:1 dilution of seawater medium with fresh water. Hypersaline medium was prepared by adding sodium chloride to the SW medium to a final concentration of 0.5 *M* (SW + 0.5 *M* NaCl).

The cultures were grown in 250-ml Erlenmeyer flasks containing 100 ml of medium on a gyratory shaker (200 rpm), illuminated with continuous cool-white fluorescent light (5 W m^{-2}) at 26°C, and harvested in the late logarithmic phase of growth.

Cell-free preparations

Crude soluble extracts for enzymic analyses were prepared from cells after harvesting and washing with 0.1 *M* HEPES buffer, pH 7.6. The cells were sonicated twice for 3 min in a Branson Cell Sonicator W140, followed by centrifugation (30,000 *g* for 30 min). The supernatant was used as the crude extract for the analysis of malate dehydrogenase and ferredoxin-NADP reductase.

The pellet was resuspended in 0.1 *M* HEPES buffer, pH 7.6, recentrifuged (1000 *g* for 5 min), and the supernatant, containing the washed membrane fraction, was used for assay of respiration.

The protein content of the cell-free preparations was determined by the Bradford procedure. The chlorophyll *a* content was determined in an 80% acetone extract of sonicated cells, according to MacKinney [5].

Results

Growth pattern of Spirulina subsalsa

Cultivation of marine Spirulina in mineral-enriched seawater always resulted in the formation of large aggregates, leading to a heterogeneous culture in which the cells tended to settle or adhere to the walls of the flasks (Fig. 1). Addition of 0.5 *M* NaCl to the seawater medium resulted in dissociation of the large aggregates into fewer and smaller aggregates.

Fig. 1. *Spirulina subsalsa* cultures in SW/2 (left), SW (center), and SW + 0.5 *M* NaCl (right).

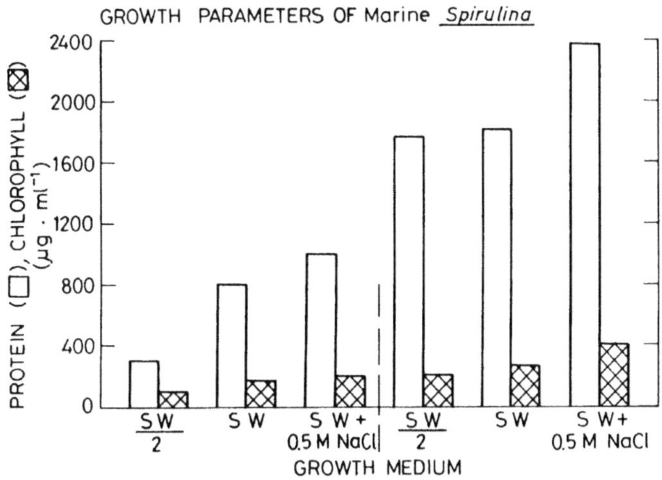

Fig. 2. Growth of marine Spirulina in diluted seawater (SW/2), seawater, and seawater + 0.5 *M* NaCl media. Two different experiments are shown. The cells were harvested after 6 days of growth.

The culture was more homogeneous and diffuse, the intensity of pigmentation higher, and the growth rate faster. Figure 2 shows the relative growth of cells in the three different media, as expressed by the chlorophyll *a* and protein contents of the cells in two different experiments. These cultures were grown in Erlenmeyer flasks as described under Materials and methods. The advantage of hypersaline medium was demonstrated also in 20-liter cultures stirred with a magnetic stirrer or flushed with an air stream, where the addition of 0.5 *M* NaCl again yielded a more homogeneous culture and faster growth.

Scanning electron microscopy (Fig. 3) showed that the filaments in the diluted seawater medium aggregate and form mineral-containing precipitates, while the filaments in the seawater are more dissociated. The extent of dissociation is more complete in the seawater + 0.5 *M* NaCl medium.

The response of enzyme activities to salt

Malate dehydrogenase (MDH) is soluble in the cyanobacterial cytosol and may therefore be affected by solutes in the cell. Figure 4 shows the effect of solutes added to the enzyme assay mixture on the activity of MDH extracted from cells grown in seawater medium. The activity is stable in the presence of the nonionic solute mannitol and of lysyl chloride, while sodium ions with either chloride or glutamate severely inhibited the activity.

A similar pattern of salt sensitivity was observed when the activity of ferredoxin-NADP reductase (FNR) from cells grown in seawater was assayed as diaphorase (NADPH $\xrightarrow{\text{FNR}}$ DCPIP) or as cytochrome *c* reductase (NADPH $\xrightarrow{\text{FNR}}$ ferredoxin$\rightarrow$cytochrome *c*). The results (Fig. 5) demonstrate the sensitivity of the enzyme to sodium ions accompanied by either chloride or glutamate anions, and its insensitivity to mannitol or lysyl chloride in the diaphorase assay.

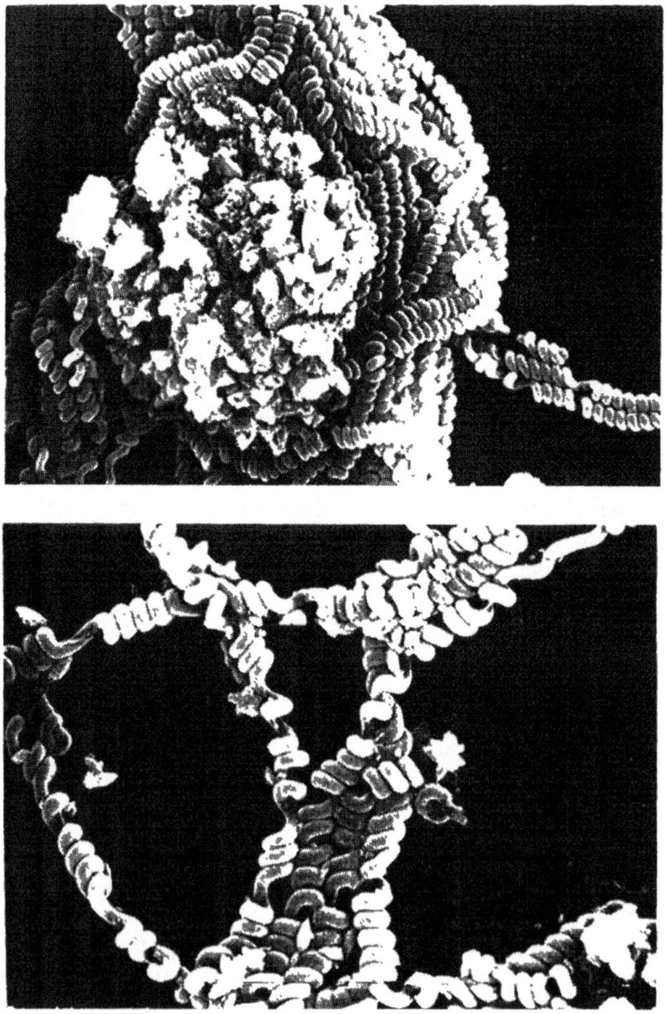

a

b

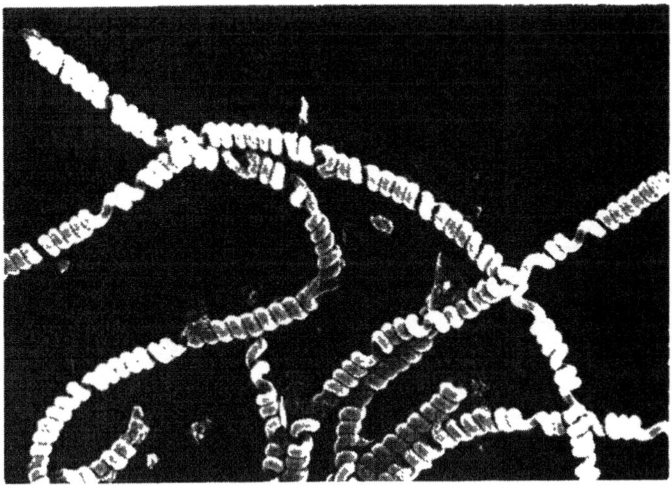

c

Fig. 3. Morphology of marine Spirulina filaments in the three growth media: a, SW/2 (magnification 2000 ×); b, SW (3000 ×); c, SW + 0.5 *M* NaCl (2000 ×).

A comparison of the enzymic activities assayed in extracts of cells grown in the three different media is presented in Figure 6. The activities of both FNR and MDH were highest in the cells cultured in the hypersaline medium. The effects of solutes added to the reaction mixture were similar in all three preparations. The results suggest that FNR and MDH activities in Spirulina cultured in the hypersaline medium may be increased by induction of the synthesis of these enzymes.

The respiratory activities in membrane preparations from the three cultures are shown in Figure 7. The endogenous oxygen reduction activity in the absence of an exogenous electron donor is low and is markedly enhanced by addition of the uncoupler CCCP or in the presence of the phosphorylation substrates ADP, phosphate, and magnesium.

Discussion

A marine Spirulina was employed in these studies to define several physiological and biochemical indicators for its salt tolerance. The growth rate of the cells was enhanced in hypersaline medium, suggesting that the organism is halophilic rather than halotolerant. The microscopic observations showed a dissociation of the filaments from aggregates when the salinity is elevated from the hyposaline to the

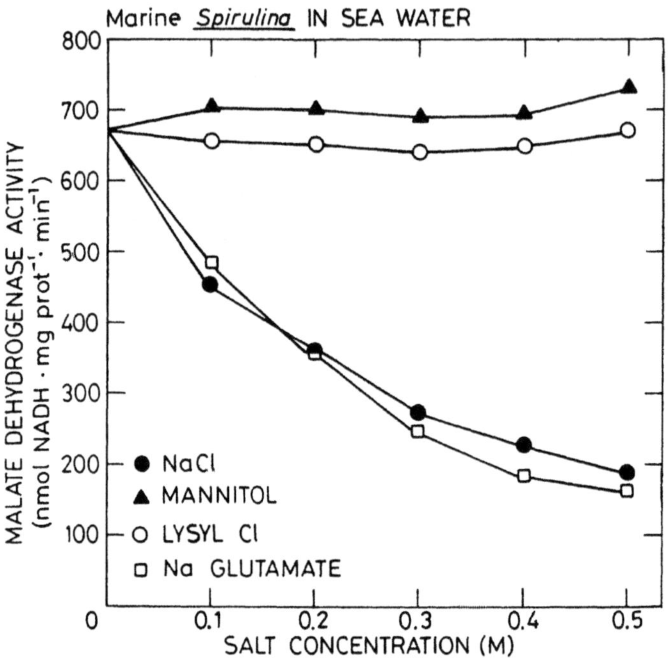

Fig. 4. Response of malate dehydrogenase activity to solutes added to the enzyme assay mixture. MDH activity was assayed in a 1-ml cuvette containing 100 m*M* glycine buffer, pH 10, 6 m*M* NAD^+, 0.6 m*M* malate, and crude cell extract containing 1–2 mg protein. Reduction of NAD^+ was followed by the change in absorbance at 350 nm.

hypersaline medium. The cells in the hypersaline medium are also more exposed to light and may exhibit increased photosynthetic activity, while those in the diluted seawater medium are partially shaded within the aggregates. Our earlier studies[8] had shown the stability of photosynthetic O_2 evaluation to increased salt concentrations. The comparative studies on FNR activities of cells grown at different salt concentrations suggest that salt may induce the synthesis of the enzyme in the cells grown at the higher salt concentration. The enzyme is still sensitive to salt to the same extent as the FNR in the cells grown at lower salt concentration.

Both FNR and MDH are sensitive to sodium ions and the marine Spirulina cells must possess an adequate mechanism to remove sodium ions and counterbalance the osmotic pressure of the medium. The respiratory activity of the cells is well coupled and may provide the ATP required for ion translocation by specific pumps such as a sodium/proton antiporter.

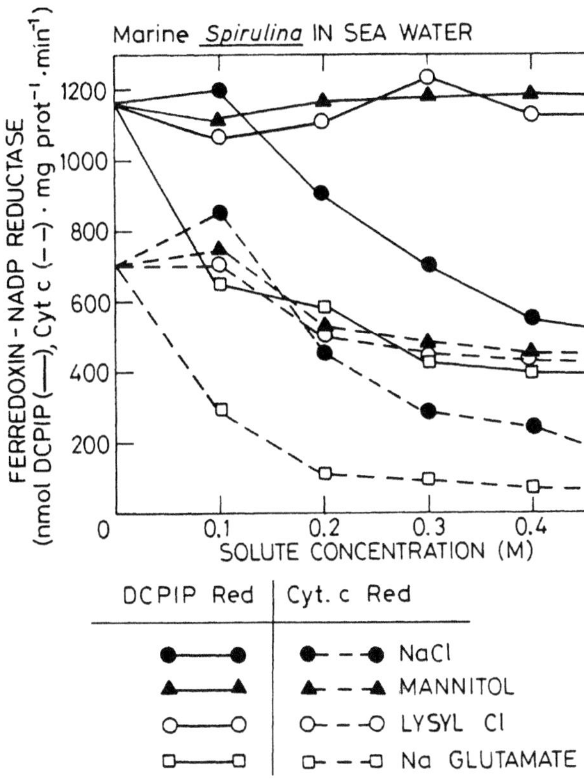

Fig. 5. Ferredoxin-NADP reductase activities and their response to added solutes. Diaphorase activity was assayed in a 1-ml cuvette containing 30 m*M* glycine, pH 9.4, 20 μ*M* DCPIP, 20 μ*M* NADPH, and crude cell extract containing 1–2 mg protein. DCPIP reduction was followed by the change in absorption at 600 nm. Cytochrome *c* reductase was assayed in a 1-ml cuvette containing 30 m*M* HEPES, pH 6.5, 20 μ*M* cytochrome *c* from horse heart, 20 μ*M* NADPH, and crude cell extract containig 1–2 mg protein. Cytochrome *c* reduction was followed at 550 nm.

Further analyses done more recently revealed a ten-fold increase in respiration supported by an exogenous electron donor to the membrane-bound respiratory enzymes. Respiratory and oxidative phosphorylation are very active and generate the energy required when photosynthesis does not function in the dark.

The marine Spirulina cells are assumed to possess a suitable mechanism for osmoregulation. We have not been able to detect the presence of high amounts of glycerol or potassium ions in the cells. Most recently, we have obtained preliminary results on the accumulation

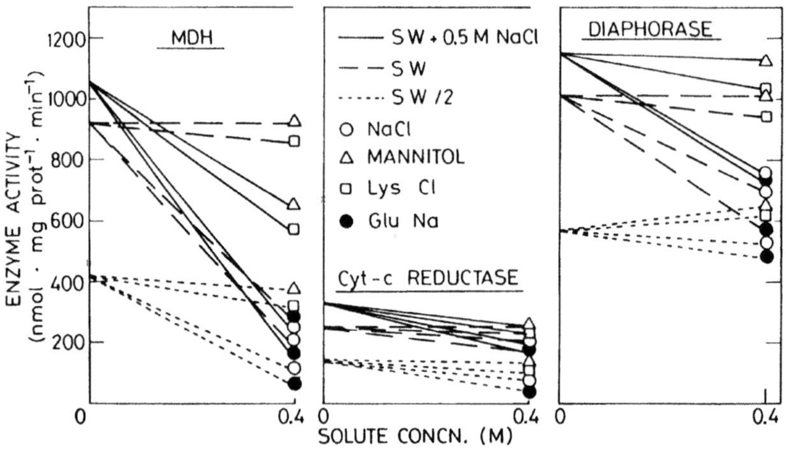

Fig. 6. A comparison of MDH and FNR activities in extracts of cells grown in the three different media.

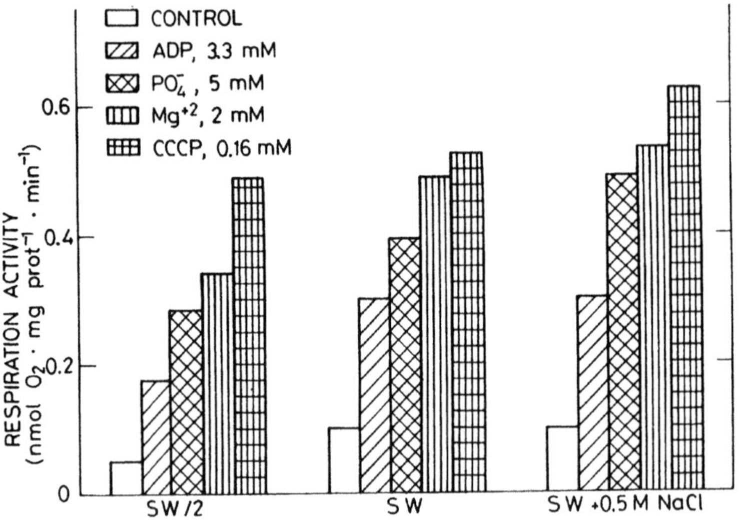

Fig. 7. Respiratory activities of membrane preparations of marine Spirulina. Oxygen reduction was assayed at 25°C with a Clark O_2 electrode (YSI 4004). The reaction mixture, in a final volume of 2.5 ml, contained 0.1 *M* HEPES, pH 7.6, and membrane fraction containing 4–5 mg protein.

of glycinebetaine in the soluble fraction of the cells (analysed by Wyn Jones). The glycinebetaine content of cells grown in the hypersaline medium was three times higher than that of cells grown in the hyposaline medium.

Table 1. Criteria for selection of marine cyanobacteria for mass cultivation

Property	Advantageous	Example
1. Size: length	Filamentous	Trichodesmium, Anabaena CA
diameter	Spiral shape	Spirulina, Oscillatoria
2. Harvestibility	Cells settle due to intracellular compatible solutes, extracellular carbonates, phosphates	Spirulina
3. Salt-dependence	Broad range of salinity:	
	SW → SW + 0.5 *M* NaCl	Spirulina
	SW → ?	Oscillatoria
4. Biomass-loading of the pond	Withstand mixture by	
	gyrotory,	Spirulina
	rotatory,	Oscillatoria
	paddle-wheel devices	
5. Primary productivity (during the day)	High photosynthetic rate, low respiration, high NO_3^- assimilation	
6. Maintenance respiration (during the night)	Ion translocation	Spirulina

Conclusions

Marine plankton organisms are essential components of food chains and may be grown for biomass production in artificial ponds. The above studies on a marine Spirulina and earlier studies of Oscillatoria in Israel[3], Trichodesmium in Japan[6], and marine Anabaena in the USA[7] reveal a promising potential for mass cultivation of marine cyanobacteria. These filamentous strains exhibit a fast growth rate, a broad range of salt tolerance, and suitable dimensions for easy, economical harvesting procedures. These species are physiologically geared for high productivity and for survival under extreme growth conditions. Table 1 provides a summary of the key criteria for mass cultivation of marine cyanobacteria.

Acknowledgements We are grateful to the U.S.-Israel Binational Science Foundation for financial support and to Dr H Melamed-Harel for her collaboration.

References

1 Abeliovich A and Gan J 1982 Factors affecting motility of *Spirulina subsalsa. In* Photosynthetic Prokaryotes: Cell Differentiation and Function. Eds. G C Papageorgiou and L Packer. pp 353–362 Elsevier, New York.

2 Ben-Amotz A and Avron M 1983 Accumulation of metabolites by halotolerant algae and its industrial potential. Annu. Rev. Microbiol. 37, 95–119.

3 Cohen Y, Padan E and Shilo M 1975 Facultative anoxygenic photosynthesis in the cyanobacterium *Oscillatoria limnetica*. J. Bacteriol. 123, 855–861.

4 Fogg G E 1982 Marine Plankton. *In* The Biology of Cyanobacteria. Eds. N G Carr and B W Whitton. pp 491–514 Blackwell, Oxford.

5 MacKinney G 1941 Absorption of light by chlorophyll solutions. J. Biol. Chem. 140, 315–322.

6 Ohki K and Fukita Y 1982 Laboratory culture of the palegic blue-green alga *Trichodesmium thiebautii*: conditions for unialgal culture. Mar. Ecol. Prog. Ser. 1, 185–190.

7 Stacey G, Van Baalen C and Tabita F 1977 Isolation and characterization of a marine Anabaena sp capable of rapid growth on molecular nitrogen. Arch. Microbiol. 114, 197–201.

8 Tel-Or E 1980 Adaptation to salt of the photosynthetic apparatus in cyanobacteria. FEBS Lett. 110, 253–256.

Plant and Soil 89, 117–127 (1985). Ms. BR 09

The potential of production of sulfated polysaccharides from Porphyridium

SHOSHANA (MALIS) ARAD, MICHEL ADDA and EPHRAIM COHEN
The Boyko Institute for Agriculture and Applied Biology. The Institutes for Applied Research, Ben-Gurion University of the Negev, P.O. Box 1025, Beer-Sheva 84110, Israel

Key words Carrageenan Cell wall Marine algae Porphyridium Sulfated polysaccharides

Summary The environmental conditions prevailing in Israel make marine algae an attractive crop for the production of valuable chemicals. A marine species of Porphyridium seems to fit this purpose.

The unicellular red alga Porphyridium is encapsulated by a polysaccharide envelope that is present in the gel state. This polysaccharide is an acidic heteropolymer composed of sulfated sugars. It forms ionic bridges through divalent cations, thus reaching a very high molecular weight. The thickness of the polysaccharide capsule varies according to the phase of growth and the growth conditions. Its outer part dissolves in the growth medium, which becomes progressively more viscous. Sulfated polysaccharides form thermally reversible gels similar to agar and carrageenan, which are usually extracted from marine macroalgae. These gels have been finding increasing use in commercial applications as gelling agents, thickeners, stabilizers, and emulsifiers.

We have done experiments on the cultivation of a marine species of Porphyridium for the production of polysaccharides. This unicellular alga has an advantage over the macroalgae due to its relatively faster growth rate and the possibility to regulate its growth. The potential for production of the polysaccharide, both that dissolved in the external medium and that attached to the cell (including an intracellular fraction), and the effects of growth conditions on productivity were studied in the laboratory. Porphyridium was also cultivated outdoors in seawater in 1-m^2 ponds and its growth potential investigated.

Introduction

General background

The conditions prevailing in Israel, *i.e.* high intensity of solar radiation, relatively high temperatures, large expanses of arid land unsuitable for conventional agriculture, a constant search for new agricultural crops, lack of fresh water, and availability of saline water, make marine algae an attractive potential crop for the production of valuable chemicals. During the past ten years scientists in Israel have gained much experience in microalgal biomass production[10]. As a result of these efforts, a new type of agriculture has been started in the southeastern part of Israel, the Arava. Spirulina is cultivated today at Ein Yahav and near Eilat (by Koor Foods) and sold as a health food, and Dunaliella is cultivated for β-carotene production. Several technological breakthroughs enabled the cultivation of algae as new crops in the Arava and succeeded in putting it 'on the map'.

However, it is obvious that cultivation of only these two algae is not enough to introduce algae into agriculture. It thus seems to us that the search for new suitable algae must continue. Marine algae containing valuable chemicals seem to fit this program.

Phycocolloids of the red algae

Cells of the red algae are generally surrounded by gelled polysaccharides known as D and L sulfated galactans, which belong to the agar family. This unique group of polysaccharides includes a number of structurally different polysaccharides, most of which are heteropolysaccharides[5]. These polysaccharides typically exist in a gel form, which is a stable net structure formed by a solution of polymers. The network, which is composed of solute and solvent molecules, is responsible for the characteristic structure of the gel and determines its physical properties. The biological function of these polysaccharides is still unknown. The gel state is responsible for the hygroscopic character of the polysaccharides, which probably facilitates maintenance of the hydration state of algal cells during periods of drought.

The main uses of the phycocolloids agar and carrageenan are for thickening and gelation. They are used as stabilizers, solubilizers, and thickeners in the food, textile, paint, cosmetics, paper, and pharmaceutical industries (McLachlan, this volume). They are also used in hydrophylic processes in biochemical research and in industry. Some of the polysaccharides are acidic as a result of the presence of carboxyl and sulfate groups. They can therefore be used as cation exchangers. The source of the raw material for phycocolloids, the marine macroalgae, has been severely depleted in the past few years as a result of intensive harvesting and ecological damage to their natural habitats. The combination of depletion of natural resources and the increase in world demand for phycocolloids has resulted in a large increase in their price. These factors have contributed to the need for a search for additional sources. One solution has been the use of sulfated polysaccharides from microbial sources that, similar to agar, form thermally reversible gels. Such polysaccharides are finding increasing applications in industry to provide viscosity, stabilize emulsions, and form useful films[1].

Porphyridium as a source of phycocolloids

We have been testing the cultivation of Porphyridium, which contains a sulfated polysaccharide. The advantages of cultivating Porphyridium rather than red macroalgae (from which agar and carrageenan are usually produced) lie in the fact that it is a unicellular organism having a relatively fast growth rate and in the possibility to control the environmental conditions regulating its growth.

Porphyridium is morphologically the simplest of all the red algae. The structure of the cells has been described[2,4]. Most species are brown since they contain chlorophyll and phycoerythrin, but many strains are blue due to the presence of phycocyanin instead of phycoerythrin[2]. Many strains grow in fresh water, while others grow in seawater[3].

Significant work has been done on *P. aerugineum*, a freshwater species isolated from soil by Ramus and colleagues[4–9]. The cells of Porphyridium are encapsulated in a layer of gelatinous polysaccharide, the size of the capsule depending on the growth conditions. During the logarithmic phase of growth the cell walls are the thinnest, and during the stationary phase they are the thickest. During growth in liquid medium, the viscosity of the medium rapidly increases due to the release of polysaccharide from the cell wall. The polysaccharide dissolved in the growth medium was found to be identical to that extracted from the cell wall. Measurement of the solubility of the polysaccharide from the cell wall in the external medium as a function of its production[5] showed that in the stationary phase of the culture the rate of production is higher than the rate of dissolution. This enables the formation of a thick capsule. Thus, the thickness of the capsule is affected by three factors: the rate of production, degree of solubility, and cell surface area. The involvement of the Golgi apparatus in the production and transfer of polysaccharides to the cell surface was supported by electron microscopy.

The polysaccharide of *P. aerugineum* is composed of xylose, glucose, galactose, and sulfate esters, with several minor additional constitutents, including hexuronic acids[5]. The presence of sulfate esters and uronic acids in the polymer is responsible for its acidic properties. The polysaccharide of *P. aerugineum* is a heteropolymer similar to the sulfated galactans of other red algae.

Effect of light intensity on production

We have started cultivating a marine species of Porphyridium (UTEX 637, referred to here as *Porphyridium* sp) under controlled conditions in 700-ml conical tubes continuously illuminated with a bank of fluorescent lamps supplying about 15 $\mu E\, m^{-2}\, s^{-1}$ at the surface of the cultures. The media were mixed by a stream of sterile air containing 3% CO_2. The temperature was maintained at 24 + 1°C. For experiments the cultures were usually cultivated on artificial seawater (ASW)[3]. Growth on seawater from the Red Sea (Eilat) and the Mediterranean

(Ashkelon) was optimized. The addition of nitrogen, phosphate, and iron together to seawater resulted in a growth rate as high as that obtained in ASW, whereas the addition of any of these elements alone did not support the same rate of growth.

Porphyridium was cultivated under two light intensities, $15\,\mu E\,m^{-2}\,s^{-1}$ (high) and $3.8\,\mu E\,m^{-2}\,s^{-1}$ (low), and growth was assessed. The maximum number of cells under high light intensity was 50×10^6 cells ml^{-1}, and under low intensity 25×10^6 cells ml^{-1}. In both cases the culture reached the stationary phase of growth after 5–6 days. The biomass produced by these cells (Fig. 1) continued to increase after the culture reached the stationary phase of growth. Hence, this biomass increase results from the synthesis of cellular constituents.

Phycoerythrin

The culture under low light intensity looked red, whereas that growing under high light intensity was yellowish. Indeed, the amount of

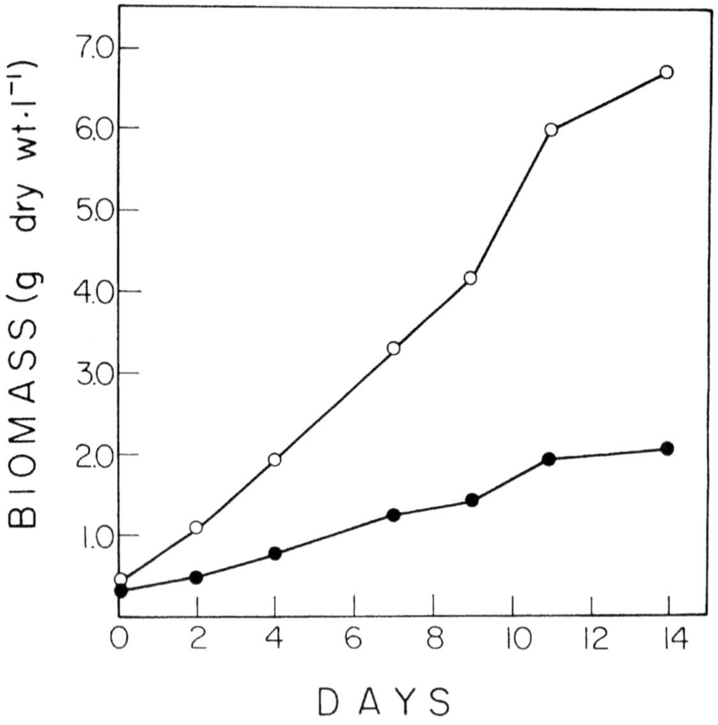

Fig. 1. Biomass production by *Porphyridium* sp grown under high (○) and low (●) light intensity.

phycoerythrin (absorption of the extract at 565 nm) increased with time under low light intensity and decreased with time under high light intensity. These changes in the proteinaceous pigment content were accompanied by parallel changes in the level of protein. Thus, the amount of protein per cell increased in the culture maintained under low light intensity, whereas at high light intensity it decreased, although the cell biomass increased under these conditions.

Carbohydrates

Starch The storage material of the red algae is Floridean starch[2,4]. For analysis the algae were extracted with a mixture of methanol, chloroform, and water (12:5:3), boiled, and treated with amyloglucosidase. Reducing sugars were determined with the Sumner reagent. The amount of Floridean starch increased linearily with culture age at both the high and low light intensities (Fig. 2). As expected, under the high light intensity the amount of starch per cell was more than 3 times higher than under low light intensity, reaching a concentration of $3.5\ g\ l^{-1}$ under the high light intensity.

Sulfated polysaccharides The sulfated polysaccharides were measured with Alcian Blue reagent[6], with carrageenan for the standard. This method is specific for sulfated polysaccharides and does not detect non-sulfated polysaccharides. The total production of sulfated polysaccharides, *i.e.* the attached fraction (intracellular and on the cell wall) and the dissolved fraction, increased continuously with time under both light intensities (Fig. 3). The high-light-grown cells produced almost 3 times more polysaccharides of both fractions (attached and dissolved) than the low-light-grown cells. The amount of attached fraction (intracellular and on the cell wall) was almost two fold that of the dissolved fraction (not shown). Both fractions had similar patterns of change with time, although the amount of attached fraction started increasing at a constant rate before the start of the stationary growth phase (day 4), whereas the dissolved fraction also increased at a constant rate, but starting on the ninth day (not shown).

Each cell produces twice as much attached polysaccharide under high light as under low light intensity. The amount of polysaccharide can be appreciated from measurement of the relative viscosity (Fig. 4) of the medium enriched with biopolymer and of the biopolymer attached to the cell after its extraction and separation from the cells. It is possible that the viscosity of the attached fraction is also affected by other cellular macromolecules that were not separated from the biopolymer.

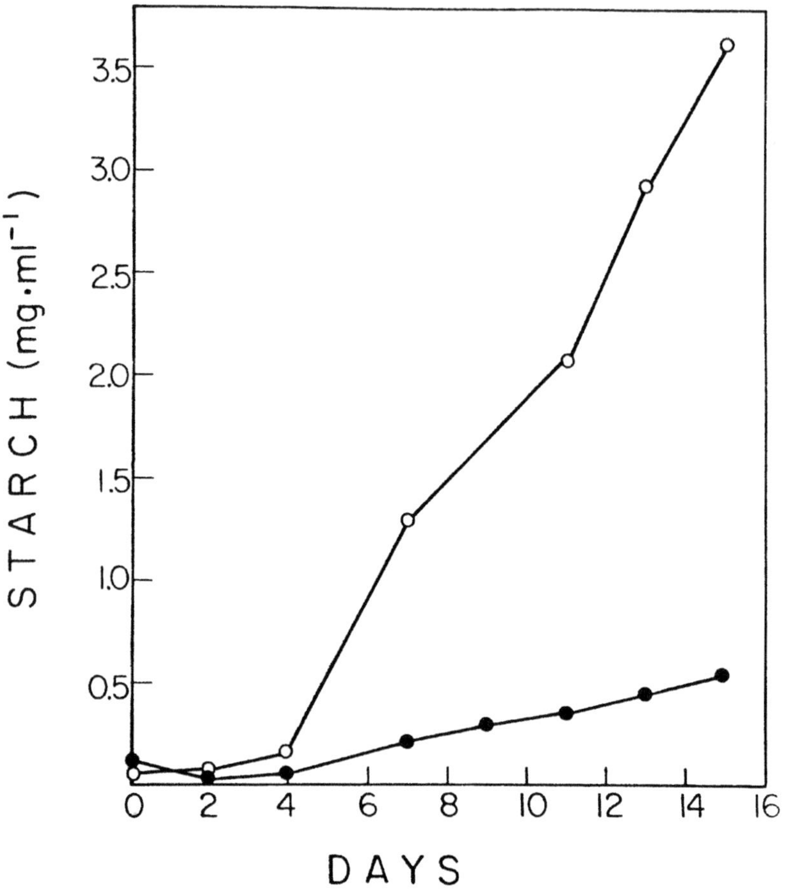

Fig. 2. Accumulation of Floridean starch in cultures of *Porphyridium* sp grown under high and low light intensity. Symbols as in Figure 1.

The medium becomes viscous during growth as the cells become enriched with the viscous polysaccharide.

Summary of light effects

The amounts of the various cellular constituents were calculated as the percent of dry weight (Table 1). The percent of protein decreased during the stationary phase of growth in the high-light-grown cells from 40 to 10%. In the low-light-grown cells there was only a small decrease. The amount of starch increased in high-light-grown cells from 10 to 46% and only to 20% in the low-light cells. The percent

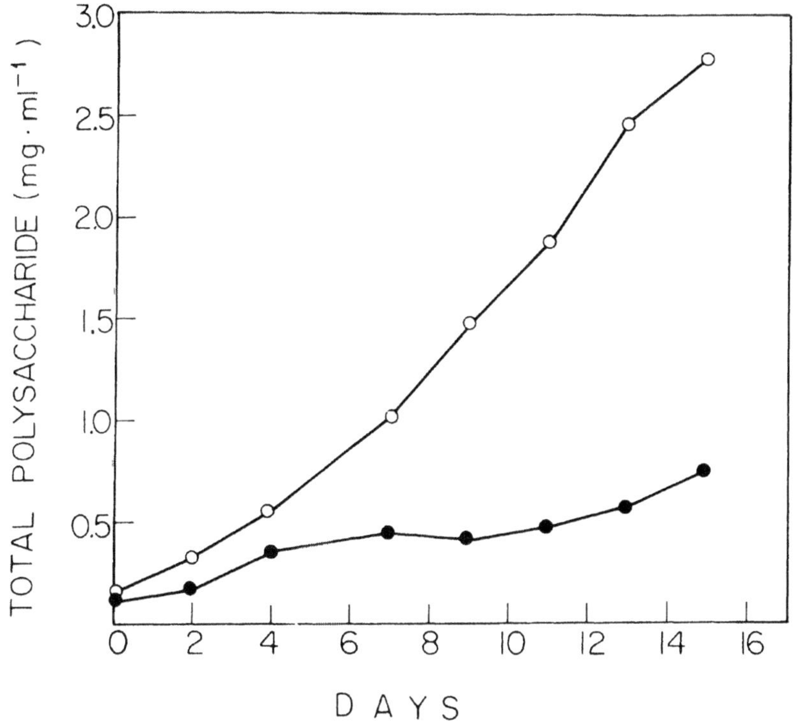

Fig. 3. Production of sulfated polysaccharides by *Porphyridium* sp. grown under high and low light intensity. Symbols as in Figure 1. Total polysaccharides were determined as the sum of the attached (intracellular and cell wall) and the dissolved fractions.

of total polysaccharides, *i.e.* attached and dissolved fractions, increased in high-light-grown cells from 29 to 36% and decreased in the low-light-grown cells. It thus seems that under low light intensity cellular metabolism is directed towards protein accumulation, probably of the pigment phycoerythrin. Under high light intensity, on the other hand, the cell directs its energy towards accumulation of carbohydrates, both starch and the cell wall polysaccharide. Under both light intensities the cell directs similar amounts of carbon towards starch and towards cell wall carbohydrates.

Chemical properties of the sulfated polysaccharide

We have started characterizing the sulfated polysaccharide of *Porphyridium* sp. The polysaccharide was isolated by cetyl pyridinium chloride precipitation, calcium chloride solubilization, and ethanol

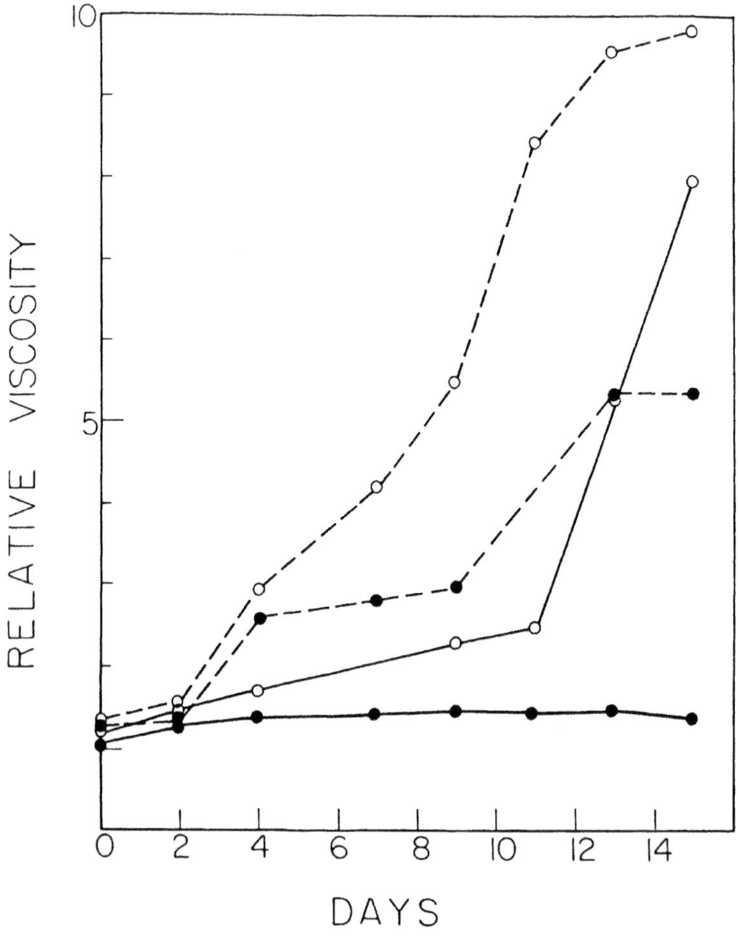

Fig. 4. Viscosity of the cell-free medium of *Porphyridium* sp culture grown under high (○) and low (●) light intensity. Viscosity was measured with an Oswald viscometer: ———, dissolved fraction; – – – –, attached fraction after extraction from the cells.

precipitation. General analysis showed that it contains 8.5% sulfate, 1–2% protein, and the rest as various sugars, which are now being identified after acid hydrolysis. We could not detect 3,6-anhydro-galactose, an important factor in gelling ability. In preliminary studies the molecular weight was found to be over 2×10^6, which is in keeping with the requirements for carrageenan used commercially in the food industry. The viscosity of Porphyridium polysaccharide was higher than that of various commercial carrageenans at the same concentration

Table 1. Protein, starch and polysaccharide contents (% of dry weight) in Porphyridium grown under high and low light intensity

Day	Protein		Starch		Total polysaccharide	
	High	Low	High	Low	High	Low
2	40	41	8	12	29	35
4	39	46	10	8	28	44
7	23	37	39	18	33	36
11	16	34	35	20	31	25
15	10	32	46	20	36	27

(Fig. 5). In this characteristic the Porphyridium polysaccharide is similar to that of Gloipeltis because it contains sulfate esters, thus forming viscous solutions but not strong gels. We are now studying its characteristics in more detail.

Outdoor cultivation

We see that there is a potential for producing sulfated polysaccharides from Porphyridium. But those who have experience with the transfer of algae from controlled conditions to outdoors will appreciate the difficulties that may be encountered in growing an organism releasing so much polysaccharide into solution in open-air cultures while maintaining a monoalgal culture. This problem is being studied by growing *Porphyridium* sp outdoors in 1-m^2 ponds filled with artificial seawater, stirred by a paddle-wheel, and enriched with CO_2 *via* porous tubes.

Porphyridium has been grown outdoors from Sept. 1983 to March 1984, and its growth was measured during various periods of the year (Fig. 6). As expected, growth was faster in the summer than the winter. The maximum density reached in the summer was *ca.* 26×10^6 cells ml^{-1}, with a doubling time of 1–1.3 days. When winter started, growth slowed down, and the maximum cell density was $11–12 \times 10^6$ cells ml^{-1}. In December the doubling time was 4–5 days. It is interesting that the cultures survived under these conditions. In all cases in which growth was intense, the cultures were monoalgal, probably due to the salt concentration in the ASW. However, we should remember that in Beer-Sheva, contamination by other salt-resistant microalgae is not very likely because of the distance (*ca.* 50 km) from the sea. It seems to us that there is potential in the production of sulfated polysaccharide from Porphyridium cultivated outdoors and that, by

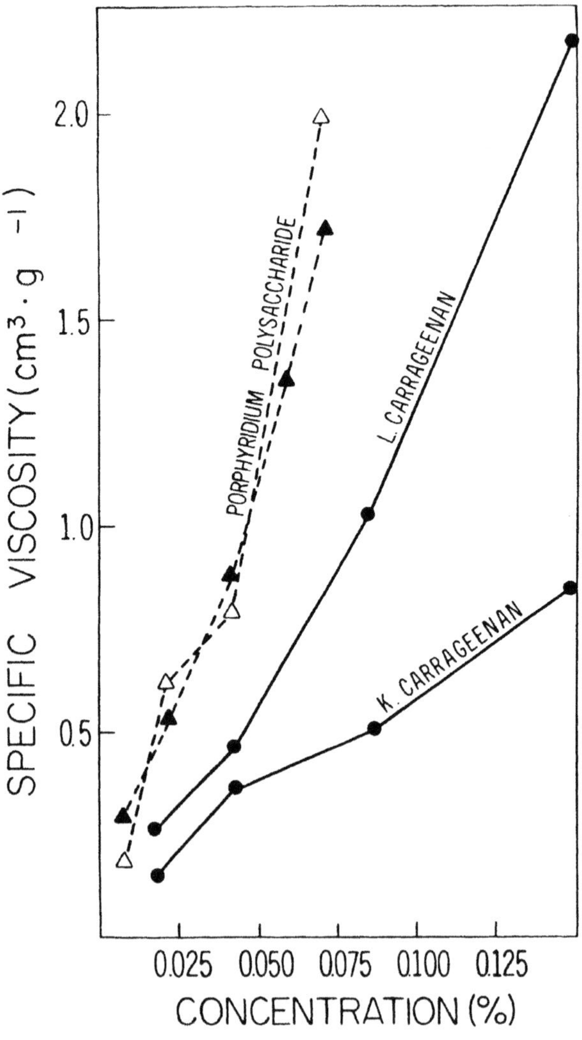

Fig. 5. Viscosity of Porphyridium polysaccharide in comparison with that of lambda and kappa carrageenans.

optimizing the system, productivity could be increased to make the yield economically feasible.

Acknowledgements The authors wish to thank Ms. Dana Friezem, Ms. Victoria Samoiloff and Mr. Avraham Rotberg for their skillful assistance and Dr. Marjorie Tiefert and Ms. Dorot Imber for help with the manuscript.

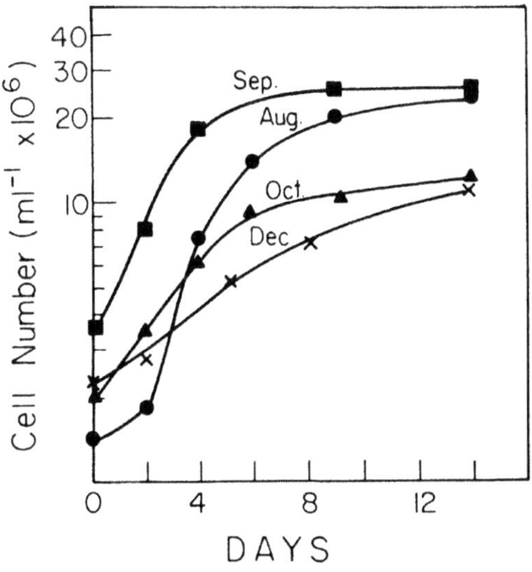

Fig. 6. Growth curves at various seasons of the year of *Porphyridium* sp grown outdoors in Beer-Sheva, Israel.

References

1 Colegrove G T 1983 Agricultural applications of microbial polysaccharides. Ind. Eng. Chem. Prod. Res. Dev. 22, 456–460.
2 Gant E and Conti S F 1965 The ultrastructure of *Porphyridium cruentum*. J. Cell Biol. 26, 365–381.
3 Jones R F, Speer H L and Kury W 1963 Studies on the growth of the red algae *Porphyridium cruentum*. Physiol. Plant. 16, 636–643.
4 Ramus J 1972 The production of extracellular polysaccharide by the unicellular red alga *Porphyridium aerugineum*. J. Phycol. 8, 97–111.
5 Ramus J 1976 Cell surface polysaccharides of the red alga Porphyridium. Plant Cell Wall Polysaccharides. p 333.
6 Ramus J 1977 Alcian Blue: a quantitative aqueous assay for algal acid and sulfated polysaccharides. J. Phycol. 13, 345–348.
7 Ramus J 1980 Algae Biopolymer Production. US Patent No. 4, 236, 349.
8 Ramus J and Groves S T 1974 Precursor-product relationships during sulfate incorporation into Porphyridium capsular polysaccharide. Plant Physiol. 53, 434–439.
9 Ramus J and Robins D 1975 The correlation of Golgi activity and polysaccharide secretion in Porphyridium. J. Phycol. 11, 70–74.
10 Shelef G and Soeder C J Eds. 1981 Algal Biomass: Production and Use. Elsevier/North-Holland, Amsterdam.

Plant and Soil 89, 129–135 (1985). Ms. BR 10

Problems in developing the biotechnology of algal biomass production

AVIGAD VONSHAK and AMOS RICHMOND
The Institute for Desert Research, Ben-Gurion University of the Negev, Sede Boqer 84990, Israel

Key words Arid zones Outdoor algal ponds Population density Saline water *Spirulina platensis*

Summary The effects of environmental conditions (solar irradiance and temperature) and population density on the production of Spirulina biomass with brackish water are reported for cultures grown in outdoor ponds. Higher specific growth rates were observed at lower population densities. Lower growth rates were associated with limitation by light in dense cultures under optimum conditions in the summer. Seasonal variation in productivity was observed. In summer, light was the limiting factor, whereas in winter the low daytime temperature appeared to constitute the major limitation. The oxygen concentration in the culture can serve as a useful indicator of limiting factors and can also be used to estimate the extent of such limitations.

Introduction

The unique environment of many arid and desert areas imposes severe limitations on conventional agriculture, yet offers important advantages for the cultivation of algae, *i.e.* saline water resources, high temperatures, abundant solar irradiation, and large areas of currently unused land.

Saline water

Arid zones are deficient in fresh water. However, saline aquifers are found under many desert areas all over the world. The saline aquifers of the Negev in Israel, which vary in salinity from 2500 to 6000 ppm total dissolved solids, could supply $30 \cdot 10^6$ m^3 of water annually with no depletion[1], and up to three times that amount with only minor depletion. In addition to brackish water, seawater can also be considered for use in algal culture.

Temperature

Algal growth increases exponentially with temperature until an optimum temperature is reached. Daytime temperatures in many desert areas approach an average of 40–50°C in the summer, although the water temperature in the ponds would be a few degrees cooler due to evaporation. The optimum temperature for growth of many warm-water algae is about 35°C. Hence, normal daytime temperatures for hot deserts, which subject many conventional plant species to severe stress, fall in a range that permits maximum yields of algae.

Solar radiation

The average annual levels of solar radiation in hot desert regions are considerably higher than in other areas of the world because of their latitudes and the characteristic absence of cloud cover. High radiation levels offer two advantages for algal cultivation. First, the intensity of solar radiation in warm regions is a major factor that determines the rate of growth of algae, unlike with conventional agricultural crops, where water, soil nutrients, and carbon dioxide are usually the major limiting factors. Second, the high radiation levels should allow the use of solar energy for processing the harvested algae.

Large tracts of land

For maximal absorption of solar radiation, algae must be cultivated in shallow, 10–20 cm deep ponds, which requires large level stretches of vacant land. These are abundant in most desert areas.

Technological problems

Three major technical aspects must be considered in developing commercial systems for the mass cultivation of algae. The first relates to pond construction: its shape and depth, the lining material, and the system for mixing the algae-laden water. The second concerns separation of the algal mass from the medium, and the third relates to dehydration and storage of the harvested algae.

Biotechnological problems

Algae are among the most efficient plants in utilizing solar energy, having the highest output rate in terms of dry weight and protein per unit area[2] when the sole limiting factor for growth is light. Thus, the most important single issue in the practical understanding of biomass production in outdoor cultures is the study of light as the major factor limiting the output rate. Because of self-shading, the significant parameter in studying the effect of illumination is the integrated radiant flux incident on each algal cell. Outdoors, this parameter depends on three factors: (1) the light intensity, (2) the algal population density, and (3) the dark/light cycle to which the average cell in the culture is

Table 1. Percent distribution of incident light throughout the depth of a pond of *Spirulina platensis*. Incident light (100%) = 2300 $\mu E\ m^{-2}\ s^{-1}$

Depth (cm)	Cell concentration (O.D. 560 nm)					
	0.10	0.16	0.22	0.28	0.34	0.40
1	57	53	47	41	37	35
2	48	41	33	29	20	16
3	43	33	28	16	10	3
4	38	38	18	8	3	0
5	29	18	2	14	0	0
6	18	12	0	0	0	0
7	2.4	2	0	0	0	0
15	0	0	0	0	0	0

exposed. The latter is affected by the turbulence, the depth of the medium, and the population density, as will be discussed below.

The interrelationship between light irradiance and population density is presented in Table 1, which shows a typical pattern of distribution of incident light throughout the depth of a pond of *Spirulina platensis*. When the cell density was relatively high (optical density of 0.40 at 560 nm, or 0.5 g dry weight per liter), only the upper 3 cm of the pond, or about 20% of the cells, received light. Accordingly, about 80% of the cells were in almost complete darkness at any given moment. Even when the cell concentration was halved, solar radiation did not penetrate beyond 5 cm, and over 60% of the cells were in complete darkness or at a light intensity below the compensation point at any given instant. Even in cultures of very low cell densities (which exhibit the maximum specific growth rate but cannot be economically maintained in the pond), light penetrated to only about half the pond's depth, leaving half the cell population unilluminated at any given time. Thus in outdoor ponds, the extent of mutual shading, which is a function of the population density and the pond depth, is the major factor determining the amount of solar radiation available to the cells in the culture.

Turbulent flow should be maintained in the pond, so that each cell in the culture is exposed to a light/dark cycle. The cycle may take a few seconds to many minutes to complete as each cell travels back and forth from the upper, illuminated layer of the pond to the lower and much thicker unilluminated layer. The light/dark regime to which each cell in the culture is thus exposed has hardly been investigated. However, when light is the factor limiting growth, the effective light/dark cycle ought to have a crucial effect on the growth rate and the photosynthetic efficiency. The nature of the light/dark regime depends on the intensity and duration of solar irradiance, the depth

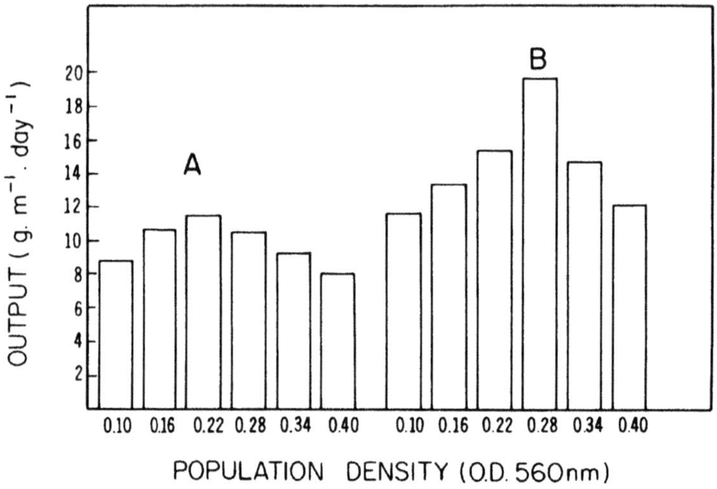

Fig. 1. Effects of cell density and turbulence on output rate in cultures of *Spirulina platensis*. **A**, paddle speed, 15 rpm; **B**, paddle speed, 30 rpm.

of the pond, the population density, and the extent and type of turbulence.

Since the net output of biomass is a product of both the cell density and the specific growth rate, and since these parameters are negatively related (see Fig. 3), then when the system is only light-limited, maximum output may be achieved at some optimum cell density (Fig. 1). The output rate is also clearly affected by the extent of turbulence in the pond (Fig. 1). Increased mixing may impose a more favorable light/dark cycle on the average cell in the culture, resulting in improved photosynthetic efficiency per unit area. The greater the turbulence, the shorter should be the duration of one complete light/dark cycle. In addition, when irradiation is very high, e.g. 250 micro-einsteins m^{-2} s^{-1}, cells located in the uppermost layer of the pond may suffer from overexposure to light. Intense stirring would decrease the duration of this overexposure. Increasing the turbulence clearly increases the growth rate (Fig. 2). The finding that, when stirring is enhanced, the maximum output rate is shifted to a higher cell density (Fig. 1) is consistent with our thesis that greater turbulence improves the light regime for the cells.

The complexity of optimizing outdoor biomass production is shown by the effect of population density on μ (the average specific daily growth rate) throughout the year (Fig. 3). Since μ is substantially affected by temperature, this relationship varies greatly with the

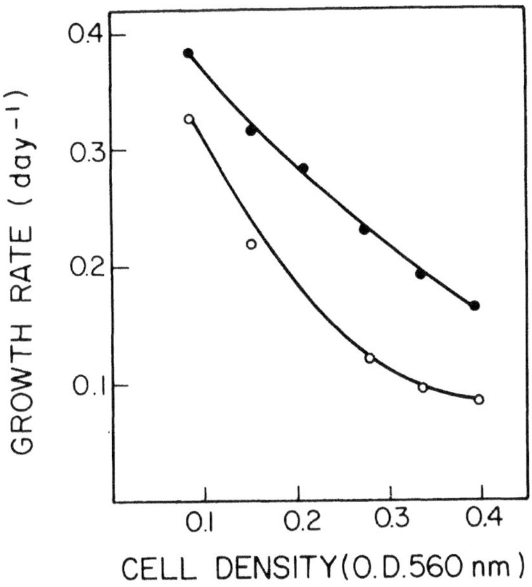

Fig. 2. Influence of cell density and turbulence on the specific growth rate. Upper curve, paddle speed, 30 rpm; lower curve, paddle speed, 15 rpm.

seasons. The more severe the temperature limitation on μ, the smaller is the dependence of μ on the population density. This dependency becomes very low in mid-winter, when the specific growth rate is low (open triangles, Fig. 3). In winter, temperature becomes the main factor limiting the output rate.

A major biological question relevant to the mass production of algae is to what extent is it possible to maintain a monoalgal continuous culture outdoors. In nature, there are some examples of algal species such as Spirulina, which predominate in a body of water as the major photosynthetic species. In small-scale experiments, we found that a continuous culture of Spirulina was readily maintained throughout the summer when the cell density was kept constant by continuously filtering the excess biomass, returning the effluent to the pond, and keeping the volume of the medium constant by adding tap water daily. Carbon dioxide was added to maintain the pH between 9.5 and 9.8, and the nutrient level was maintained by analyzing for PO_4^{3-} and NO_3^- three times weekly and adding appropriate amounts of the entire mineral nutrient mixture, as needed. As long as the temperature in the pond did not fall below 20°C, Spirulina cultures could be kept

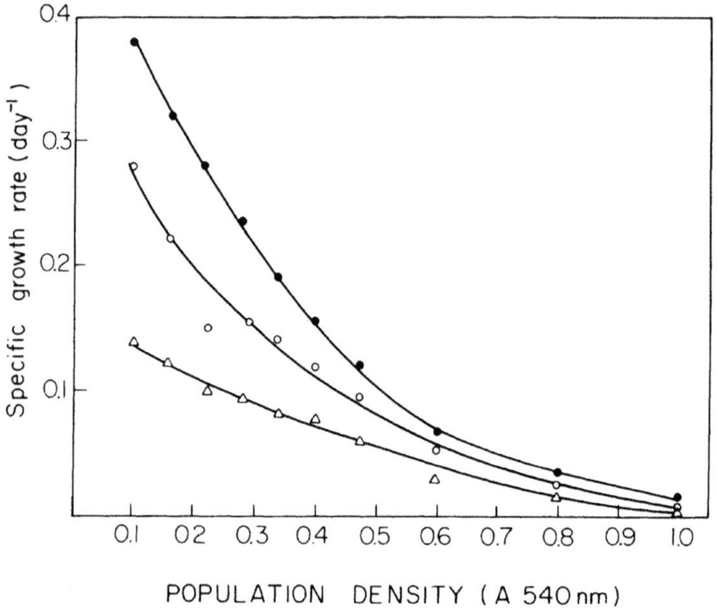

Fig. 3. Effect of cell density on the specific growth rate in August–September (top), May–June (middle), and December–January (bottom).

essentially clean of other organisms. The number of bacterial cells did not increase above $1 \cdot 10^4$ ml^{-1}. In addition, analysis of the daily specific growth rate did not reveal any signs of self-limitation in the pond throughout its continuous operation from April to October.

A crucial requirement for pond maintenance is a way to constantly and readily evaluate the relative performance or 'well-being' of the culture in the pond. We, as well as others, have found that the partial pressure of oxygen in the pond during the daytime serves as a useful tool. Table 2 shows the effects of the intensity of solar irradiance and of temperature on pond oxygen. Data were recorded hourly throughout the year. Most figures represent the average of many scores of observations, 10% of which were deleted from both extremes. All readings were made at 1 PM, when the concentration of oxygen in the pond was usually increasing. The highest oxygen content was recorded at the highest values of temperature and irradiation, thus further illustrating that production of this warm-water algal species depends on both of these environmental parameters.

Table 2. Effects of temperature and the intensity of solar irradiance on oxygen concentration in *Spirulina platensis* ponds

Temperature (°C)	Pond oxygen content (% saturation)					
	Incident light (klux)					
	0–5	5–20	20–40	40–60	60–80	80–100
6–12	71	–	–	–	–	–
12–18	87	95	109	115	145	–
18–24	88	101	113	122	140	157
24–30	94	108	120	131	142	181
30–36	107	–	135	125	155	208

Conclusion

Consistent efforts towards optimization of algal biomass production, harvesting, and product processing, as well as genetic improvement and basic research will lead in time to improved biotechnologies of significant economic importance. This seems especially useful in arid lands, where agricultural productivity is particularly low and cultivation of conventional crops is severely handicapped. In these areas, algaculture in brackish or sea water has distinct advantages.

References

1 Issar A 1981 The role of non-replenishable aquifers in development projects in arid regions. *In* Settling the Desert. Eds. L Berkofsky, D Faiman and J Gale. pp 117–128. Gordon and Beach, London.

2 Richmond A 1983 Phototrophic microalgae. *In* Biotechnology, Vol. 3. Eds. H J Rehm and G Reed. pp 109–143. Verlag Chemie, Weinheim.

Plant and Soil 89, 137–157 (1985). Ms. BR 11

Macroalgae (seaweeds): industrial resources and their utilization*

J. McLACHLAN
National Research Council of Canada, 1411 Oxford Street, Halifax, N.S. Canada B3H 3Z1

Key words Agar Alginate Applications Carrageenan Hydrocolloids Resources Seaweed extractives

Summary Seaweeds are distributed widely in all oceans of the world, but exploited in limited areas. These plants have been used for food and as soil conditioners for centuries, with industrial application of extractives a relatively recent innovation generally limited to the more developed areas of the world. There are relatively few species of industrial importance, and these occur in both the red (Rhodophyta) and brown (Phaeophyta) algal divisions.

The principal value of phycocolloids is in their gelling and stabilizing properties. Species of red algae may elaborate agar or carrageenan. Both are complex polysaccharides with innumerable structural possibilities, hence physical properties, and phycocolloids from various species can have specific applications. Carrageenans are employed mostly in the food industry, and agars in both the food and pharmaceutical sectors of the economy. Alginates, another family of complex polysaccharides, are obtained from various species of brown algae and are the third major algal gum of commerce. Alginates are used in the food and pharmaceutical industries and have other industrial applications.

Introduction

Seaweeds have been exploited for centuries by maritime societies, usually as a green vegetable, but crude extracts from these plants are also eaten directly and used in the preparation of foods[41]. Seaweeds have also been used as supplements in animal meals, as fertilizers, and as soil conditioners. As a dietary supplement seaweeds remain important, although consumption is limited almost exclusively to the Far East where, however, the monetary value is significant. This is best illustrated by the 'Nori' (Porphyra) industry in Japan, which alone turns over probably more than 1000 million dollars per annum[23]. Seaweed meal in animal rations continues to be an important resource[14], especially in northern Europe.

The use of phycocolloids, or gums from seaweeds, is widespread, and the industry is important in many areas of the world. During the past three or four decades the industry has grown spectacularly[28], paralleling the growth of the convenience-food industry, which is largely a post-war development. Two segments of the industry, carrageenan and especially furcellaran, a low-sulfate carrageenan[9], were started as a result of World War II when traditional sources of agar were unavailable to the Allies. Because of the importance of agar in

* Issued as NRCC No. 23603.

microbiological media, extensive searches were made for substitutes[18,42]. Today the primary uses of furcellaran and carrageenan are in the food industry, although the latter and other gums continue to be evaluated as substitutes for agar[2,15].

Industrially, seaweed extracts are an important part of the large gum or hydrocolloid industry that is based on plant extracts or modified extracts[32], and seaweed extracts have an estimated annual turnover value of several hundred million dollars[22]. Although these extracts are utilized most extensively in developed countries, the use of agar extractives in developing countries is increasing rapidly (J R Moss, pers. comm.).

The major groups of seaweeds include the Chlorophyta, Phaeophyta, and Rhodophyta, but only species of brown and red algae are important as resources of industrial gums. These phycocolloids are classified into three generic types: alginate from brown algae, and agar and carrageenan from red algae. All species of brown algae synthesize alginic acid. Agars and carrageenans, or other polymers with the same basic structures are the main polysaccharides in red algae[19,30], but apparently never co-exist in the same species and usually not in the same genus[19,30].

The phycocolloids of industrial interest are associated with the cell wall and the intracellular matrices. Their biological functions in the plant include maintaining the integrity of the cells and providing mechanical strength. Ion-exchange functions are also important, and it is necessary that these colloids be present as a gel, keeping the spaces swollen[36]. Furthermore, gel formation has specific ion effects that are of industrial importance.

The industrial algal heterosaccharides are copolymers of glycans substituted with anionic groups. More than 50% of the dry weight of the alga may comprise these gums[19,30], although commercial values of 20 to 30% are more usual[8,27]. The important technical uses of algal gums depend upon their physical properties, which can be modified considerably by slight changes in structure[30]. The fine-structural details of most polysaccharides, including algal extracts, remain to be determined[20], not only because of the complexity of the molecules involved but also because of the laborious and expensive techniques required for analyses[30]. However, recent advances in instrumentation are providing rewarding results[4,5,16]. It must be emphasized that these algal gums occur as families of polysaccharides with innumerable structural possibilities, and the physical properties are a statistical function of the complex mixtures[29]. The idealized polysaccharides have yet to be isolated even though the ideal structure is usually illustrated. Seaweed colloids continue to be characterized according to their component units and solubilities[20].

The commercial interest in phycocolloids results from their behavior in aqueous solutions, where they form viscous substances including gels. These gums display syneresis and, being polyelectrolytes, can be precipitated by inorganic ions[36]. Carrageenan and agar gels are thermoreversible, and both show marked hysteresis[17,35,37], whereas alginates are stable up to the boiling point of water[36]. All gels become stronger with decreasing temperature, and the gel strength increases exponentially with concentration[36].

Alginate

Uronic acids, α-L-guluronic and β-D-mannuronic linked $(1 \rightarrow 4)$, form the alginate family of polysaccharides (Fig. 1). These monomers occur in three types of blocks: one containing mostly mannuronic acid, one mostly guluronic acid, and a third of intermediate composition[16,29]. The ratio of mannuronic/guluronic acids can vary from 0.4 to > 20[29], with fucalean receptacles containing almost pure mannuronic acid residues[20]. While these ratios clearly correlate with the properties of alginates, they bear no taxonomic significance[30]. Even within the same plant there is considerable variation with, for example, the age of the tissue[40], location, and vegetative or reproductive function[29]. Alginates have a strong affinity for divalent cations, especially calcium[40], which in turn increases the gel strength[26]. Ionic binding is associated principally with the guluronic acid residues[36], and tissues with high proportions of this acid, such as the stipes of Laminaria, are thus much more rigid than the receptacles of Ascophyllum, which are rich in mannuronic acid. However, there are so many variables that comparisons become nearly meaningless[29].

Alginates are extracted from either wet or dried seaweeds, depending upon the process[13,27]. As alginates in the native state are complexed with divalent metal ions, the colloid is generally insoluble[36]. The first step in the extractive procedure is thus to exchange the alkali-metal cations for hydrogen ions by treatment with dilute mineral acid, which also removes laminaran and degrades fucoglycans[29].

Subsequent treatments consist of extraction with sodium carbonate, precipitation as calcium alginate, conversion to alginic acid, usually by washing with hydrochloric acid, and finally reacting with a base to form the desired salt. Bleaching may be done at any one of several steps. The principal water-soluble salt is sodium alginate, with lesser amounts of potassium and ammonium salts being manufactured[26]. Propyleneglycol alginate (PGA), an ester, is the only commercially available derivative of alginate; however, large quantities are utilized[26].

Alginate has, for technical reasons, been a difficult colloid to process,

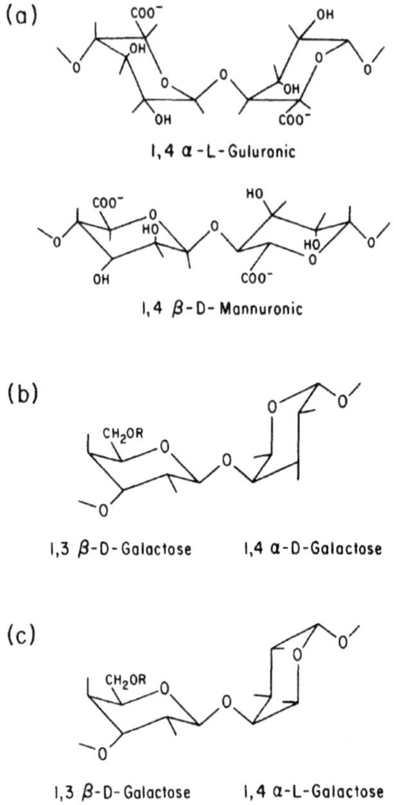

Fig. 1.(a) Uronic acid repeating units in alginic acid; (b) galactose reapeating units in carrageenan; (c) galactose repeating units in agar.

and there are many records of commercial failures. Thus, not unexpectedly, proprietary details of processing are closely guarded by the industry[26]. Nevertheless, alginates of low quality are being processed. Furthermore, the nature of the resource is also an important factor in determining the quality of the end product.

Carrageenans and agars

The predominant polysaccharides in red seaweeds are based on galactose polymers in which the repeating units are glycosidically linked alternately α-1,3 and β-1,4[19,20,29,30,40]. Carrageenans occur as D-galactose units only, whereas in agars the 4-linked unit is α-L-galactose (Fig. 1). Further, the 1,4-linked residues may also be present as 3,6-anhydrogalactose. Each of the repeating units can be modified by sulfation at various positions, and a galactan polysaccharide lacking half-sulfate ester groups has not been isolated[29]. The fine structure of the galactose units can be further modified by O-methylation and pyruvic acid as a ketal group (Table 1). These innumerable sequential arrangements all affect the physical properties of the phycocolloids.

A wide range of sulfated galactans can occur within the same species, including the same thallus[40]. Moreover, it is now well recognized that in several families, significantly different carrageenans occur in sporophytic and gametophytic stages of the life history[19,20,21]. Species of commercial interest are in the Gigartinales, where both types of polysaccharide are known, and the Gelidiaceae (Nemaliales) where only agars have been reported.

Table 1. Common substitutions of galactose units in carageenans and agars

β-Galactose	α-Galactose (L = Agar, D = Carrageenan)
D-galactose 2-sulfate	3,6-anhydrogalactose
D-galactose 4-sulfate	galactose 2-sulfate
D-galactose 6-sulfate	galactose 6-sulfate
6-O-methyl-D-galactose	3,6-anhydro-D-galactose; 2-sulfate
4,6-O-(1-Carboxyethylidine)-D-galactose	3,6-anhydrogalactose; 2-O-methyl-L-galactose
	2-O-methyl-L-galactose
	4-O-methyl-L-galactose

Carrageenans

In this family of galactans, the galactose is present as repeating D-units (Fig. 1). The carrageenans of commercial interest have a relatively high content (> 20%) of sulfate ester. Major differences among the forms of carrageenan (Table 2) are related to the position of the sulfate groups on both the α- and β-galactose units and the occurrence of 3,6-anhydro-D-galactose in the chain[17,37]. Both pyruvate and O-methyl groups, while apparently rare, have been reported for carrageenans as minor constituents[20] and seem to be of no industrial significance.

The carrageenans of commerce[17,37], are referred to as the lambda-, kappa-, and iota-families (Table 2). These polysaccharides are extracted into hot water (≃ 100°C) under slightly alkaline conditions, which aids the extraction and eliminates the 6-sulfate ester with formation of 3,6-anhydro-D-galactose. This step is followed by filtration and then concentration by evaporation. Carrageenan is then recovered by drying or precipitation in alcohol. The precipitate is pressed and finally dried and milled[37]. Because of syneresis furcellaran is recovered by

Table 2. Substitution in idealized carrageenans

Type	β-D-Galactose	α-D-Galactose
λ	2-sulfate	2,6-disulfate
κ	4-sulfate	3,6-anhydrogalactose
ι	4-sulfate	3,6-anhydrogalactose; 2-sulfate

freeze-thawing[9]. As with alginates, details of the processing are proprietary information.

A recent development in the industry has been the use of seaweeds treated with alkali, the product being referred to as alkali-modified carrageenan. The treatment with hot alkali 'locks in' the carrageenan while removing most other components. This is followed by neutralizing with dilute acid, washing, drying, and milling to a powder. The product contains mostly carrageenan, and in certain applications is employed without further processing (J R Moss, pers. comm.). This process is being used in areas where the resource is obtained. The alkali-modified powder can be further refined to pure carrageenan through the usual procedures and processing facilities.

Carrageenans, which are strongly anionic because of sulfate esters, are complexed with cationic and amphoteric materials. Water is the principal solvent, and solubility is affected by the sulfate groups and the 3,6-anhydrogalactose units. All carrageenans are soluble in hot water and hot milk, and the sodium salts are soluble in cold water[17,37]. Carrageenans react with proteins by a poorly understood mechanism, which is a function of the charged sulfate groups and the isoelectric point of the protein. This reactivity is probably the single most important commercial characteristic, as milk proteins are thus agglomerated and become stable against heat and precipitation by calcium ions.

Dilute aqueous suspensions of carrageenans are viscous, and the gelling types set from 40 to 70°C. The presence or absence of sulfate on the C-4 of the β-galactose unit determines the gelling ability[20]. The kappa-form of carrageenan is soluble at about 70°C, and forms a firm, rigid gel (with K^+ the strength is $> 1000\,g\,cm^{-2}$). Gels of the iota-form are more elastic and show reduced syneresis. Lambda-type carrageenans, which are devoid of 3,6-anhydrogalactose, do not gel; however, these carrageenans produce the highest viscosities in water.

Agars

Agar, the generic term for a family of related polysaccharides, is among the most potent gel-forming agents[35], and gelation is detected at very low concentrations. Agars (Fig. 1) are distinguished from carrageenans by the β 1,4-galactose unit being in the L-form[11,20,35].

Most agars have a low (< 5%) sulfate content. Although the agarose fraction of agar is neutral (Fig. 1), agar also consists of strongly ionized acidic polysaccharides that exist predominantly as the calcium salt[35].

The agar molecules may be variously substituted in all positions (Table 1) with innumerable sequential arrangements. These possibilities have complicated the nomenclature, and only the extreme types of agar have been characterized[10,44]. There is a continuous spectrum of agars from essentially uncharged agarose to highly charged sulfated and pyruvated types[10,20,44]. Agarose is low in both sulfate and pyruvate, and most of the 4-linked units are 3,6-anhydro-L-galactose; these polysaccharides have a high gel strength. Because of the relatively low concentrations, little is known of the sulfation pattern in agars[20]. Sulfation is, however, of great commercial importance as it interferes with gelation. Agars high in sulfate and low in 3,6-anhydrogalactose form poor gels.

Pyruvate is a common feature of agars (Table 1), with concentrations ranging from 0.04 to 2.9%[20]. Pyruvate, which occurs as a ketal, contributes to the ionic character of the molecule and is undesirable for some applications[12]. The pyruvated form is characteristically low in sulfate and high in 3,6-anhydrogalactose.

Agars are unique in that gelation occurs relatively far below the melting temperature, and many uses depend upon this characteristic. Increased gelling temperatures have been related directly to the content of 6-O-methoxyl-D-galactose[11], which may be up to 20% of the D-units (Table 1), but apparently has no effect on gel strength[29]. Agars from Gracilaria generally have a relatively high 6-O-methoxyl content, and thus high gelling temperatures. Methyl groups are regularly found in all agars[3], and in addition to the 6-O-methyl on the D-galactose, substitution (Table 1) may occur as 3,6-anhydro-2-O-methyl and 4-O-methyl-L-galactose[20,45].

Commercial preparations of agar are obtained mostly from species of Gelidium and Gracilaria (Table 9). Agar is extracted into hot water, filtered and allowed to gel. The gel is usually frozen and, on thawing, agar is recovered free from salts, pigments, and water constituents. At the lowest level of technology, freezing is done out-of-doors. In most instances the technology is not sophisticated[13,22], and in some procedures the gel is recovered by precipitation rather than freeze-thawing.

The initial stages of extraction of Gracilaria usually include treatment with alkali. This is effective in removing the 6-sulfate from the L-galactose with formation of a 3,6-anhydro-bridge. This is effective with many resources of Gracilaria, and considerably upgrades the quality of the agar[24]. In some cases these gels contain less sulfate and are stronger (with gel strengths exceeding 1000 g cm^{-2}) than those from Gelidium[11,12]. The material may be processed no further than the initial treatment with alkali.

Uses of phycocolloids

The global turnover in the hydrocolloid industry is at least 1600 million US dollars per annum[8], excluding starches, and involves at least 250 000 t[8,33] of refined plant gums. Phycocolloids contribute only about 40 000 t (Tables 3 and 9), with an annual turnover value of at least 300 million US dollars[13,22]. These gums are at the high end of the price scale for hydrocolloids (Table 4), with agar being the most valuable, followed by carrageenan, and last alginate. Most agar is processed in Asia, whereas alginate and carrageenan are processed in North America and Europe (Table 3).

Table 3. Estimated production (tonnes) of phycocolloids

Area	Agar	Carrageenan	Alginate
Asia (Korea & Japan)	3500	500	1900
Latin America	800	–	100
Europe	2000	7900	12800
North America	200	4500	6700
Others	500	–	–
	7000	12900	21500

Adapted from ref. [22].

The industrial uses of hydrocolloids are extensive (Table 5), principally to alter the properties of aqueous solutions, where they serve primarily as thickening and gelling agents. Additional uses include emulsification, stabilization, and encapsulation. The phycocolloids have remarkable uses in a variety of applications, especially in the food industry. One type of phycocolloid may compete with another and with other gums. It should, however, be appreciated that many preparations include several gums (Table 6, 7) acting synergistically.

The selection of gums for commercial utilization is based on several considerations[43], only in part on the physical and chemical properties of the polysaccharide. The cost of the gum is important, but as the cost per functional unit rather than the cost per unit mass, because some gums are more effective than others at low concentrations. Constancy of cost is also important and, because of the highly competitive nature of the industry, the costs of phycocolloids have increased rather more slowly than other costs. Stability of supply and composition

Table 4. Estimated volume and price of selected hydrocolloids and amounts used in food (statistics for USA)

Polysaccharide	Volume (t) (1980)	US$/kg (1980)	Food (t) (1974)
SEAWEED EXTRACTS			
Agar	5000	19.00–20.50	180
Alginate	10000	4.00– 4.75	2400
Carrageenan	5000	6.40	2100
SEED GUMS			
Guar gum	90000	1.60– 1.85	4700
Locust bean gum	17500	4.00– 4.65	1500
PLANT EXUDATES			
Gum arabic	67500	1.60– 3.00	3600
Karaya gum	4500	2.65– 3.10	1600
Gum tragacanth	1500	29.75–83.80	425
PLANT EXTRACT			
Pectin	10000	6.70	1700
CELLULOSE DERIVATIVES			
Carboxymethyl cellulose		3.00	12400
FERMENTATION PRODUCT			
Xantham gum		7.90–10.80	570

Adapted from ref. [33].

Table 5. Percentages of the hydrocolloid market in various applications (excluding starches)

Use	Percent
Detergents and laundry products	16
Textiles	14
Adhesives	12
Paper	10
Paint	9
Food	8
Pharmaceuticals and cosmetics	7
Other	24

Adapted from ref. [33].

has been a special problem in the agar industry[8], resulting in major yearly fluctuations[28]. Even in the carrageenan industry, a previously minor resource now dominates the industry. Perhaps the most critical factor in the industry is the possibility of replacement of one gum by another. It is not clear if any gum is entirely irreplaceable. Although the search for agar substitutes led directly to the development of the furcellaran industry[9] and stimulated the utilization of carrageenophytes, no completely satisfactory replacement for microbiological-grade agar has been found; additionally, agarose is an invaluable aid in medical research and genetic engineering[12,31]. There are, however, numerous

Table 6. Distribution of usage of selected hydrocolloids in the food and pharmaceutical industries

Market	CAR	ALG	AGR	GG	LBG	GA	PEC	XAN	CMC
FOODS									
Tinned	X	X	X						
Dairy products	X	X		X	X			X	X
Bakery products		X	X	X		X	X	X	X
Glassed foods		X			X		X		
Salad dressings		X						X	
Frozen foods		X	X	X				X	X
Noncarbonated beverages		X				X		X	X
Pet foods	X			X	X		X		
Other	X	X		X	X	X		X	
CONFECTIONARY	X		X			X	X		
SOFT DRINKS	X					X			
BREWING		X				X			
PHARMACEUTICALS	X	X	X	X	X	X	X		X
COSMETICS	X		X	X		X			X

Adapted from ref. [33].
CAR, carrageenan; ALG, alginate; AGR, agar; GG, guar gum; LBG, locust bean gum; GA, gum arabic; PEC, pectin; XAN, xanthan gum; CMC, carboxymethyl cellulose.

Table 7. Ingredients in a fabricated food represented by an 'ice cream' produced in Canada

Fabricated	Authentic
Milk solids	Cream
Sugar	Egg
Glucose	Vanilla
Whey powder	Sugar
Sodium alginate	
Sorbic acid	
Lecithin	
Mono- and disaccharides	
Polysorbate 80	
Guar gum	
Locust bean gum	
Carrageenan	
Sodium carboxymethyl cellulose	
$CaSO_4$	
Salt	
Natural and artificial flavors	
Color	

examples of substitution of one gum for another. Presently, the inclusion of agar in prepared foods in North America and Europe is minor (Table 4), although in the past it was used more extensively. Carrageenan as a binder in toothpaste was largely replaced by carboxymethyl cellulose[8], which in turn has now been replaced by xanthan gum (J R Moss, pers. comm.). Another factor in the industry is the recent introduction of the product referred to as modified carrageenan (*i.e.*

alkali-treated), which is much cheaper than refined carrageenan and can be produced with less technical sophistication. An estimated 3500 t of modified carrageenan is currently being produced, mostly for pet foods, with lesser amounts in air-freshener gels and other applications where a clear gel is not required, such as in ice cream (J R Moss, pers. comm.). However, food and drug authorities (*e.g.* FDA, WHO) may require extensive animal tests before further approval is given for inclusion in foods, thus limiting immediate expansion of the use of this form of carrageenan. Substitutions have also occurred in the alginate industry, and at present non-food usages are largely limited to textile printing (Table 8), principally of cotton. Although production of cotton textiles is increasing in developing countries, growth in the use of alginates in developed areas is slower because of synthetic fibers.

Table 8. Global applications of alginates in various industries ($\simeq$ 20 000 tonnes)

Use	Percent
Textile printing	50
Food	30
Paper	6
Welding rods	5
Pharmaceuticals	5
Other	4

Adapted from ref. [22].

The phycocolloid industries must, in order to remain viable and competitive, continually develop new applications and demands for their products. The agar industry goes up and down on 'cost bust – resource bust' cycles and the future remains unpredictable. However, as noted, there are no satisfactory replacements for microbiological-grade agars, for which the demand is increasing, and uses in foods are largely in the Orient where traditions are strong. Processors of carrageenans have been reasonably successful in creating new demands and applications, and this sector of the industry remains very dynamic. The alginate industry seems to have visibly stagnated, perhaps even declined. There have been almost no new innovations in the industry since the introduction of propyleneglycol alginate. Possibly this is why alginate of bacterial origin has not developed further, as these gums are similar to those isolated from seaweeds[33], thus offering no new possibilities to an industry that has apparently not been limited by the availability of resources.

It is very probable that microbial polysaccharides, that is semisynthetics, have a future. Certainly xanthan is competitive in the polysaccharide industry (*cf.* Tables 4, 6), as is dextran, and new polymers may soon become available, competing with existing gums and having novel functions[33]. Indeed, genetic engineering offers the potential of making traditional gums available from new sources as well as custom-designing polysaccharides.

Further demands and new developments in the colloid industry will undoubtedly come with the introduction of fabricated foods. Such foods have already been accepted (Table 7), but the major potential lies in supplying foods to areas of the world where there are critical shortages and having them accepted by those who need them. In addition to the utilization of non-traditional food sources, reduced transportation costs and the easier storage and longer life of fabricated foods are important. Certainly, the technology that has been developed with pet foods is relevant to these applications.

Agar

An estimated 7000 t of agar are used annually (Tables 3, 9)[24], both in food and as a microbiological medium; almost no agar is used for industrial purposes. Most of the world's agar is produced in the Far East (Table 3), with the major users also in this region. Microbiological

Table 9. Genera of seaweeds used in the extractive industry with estimates* of exploitation and yield of extract (both as dry tonnes)

	Resource	US$ Value	Product	US$ Value
ALGINATE	95×10^3	23×10^6	21.5×10^3	122×10^6
Ascophyllum				
Durvillea				
Ecklonia				
Laminaria				
Lessonia				
Macrocystis				
Sargassum				
CARRAGEENAN	45×10^3	18×10^6	13×10^3	71×10^6
Chondrus				
Eucheuma				
Furcellaria				
Gigartina				
Hypnea				
Iridaea				
AGAR	36×10^3	32×10^6	7×10^3	117×10^6
Gelidium				
Gracilaria				
Pterocladia				
	176×10^3	73×10^6	41.5×10^3	310×10^6

* Excluding China and the USSR.
Adapted from refs. [13,22].

applications are increasing, owing partly to new legislation regulating safety standards and partly to rapidly growing medical facilities in developing countries. For these reasons the agar industry may grow more rapidly than other sectors of the phycocolloid industry (J R Moss, pers. comm.), provided that resources can be made available.

Most of the agar available is in fact unsuitable for microbiological purposes because of relatively high gelling temperatures, but is satisfactory as a foodstuff and in food preparations. There has been a long tradition of agar use in the Far East and, unlike with the other phycocolloids, the major consumption of agar continues in this region.

Apart from microbiological purposes, agar is employed only to a minor extent in North America (Table 4) and Europe, except in the United Kingdom and Germany, where significant quantities are used in confectioneries[22]. Many previous applications have been replaced by other colloids, even though agar has numerous valuable properties that may be exploited in a wide variety of applications.

Carrageenan

In contrast with agar, most of the world's carrageenan is produced and utilized in developed regions (Table 3). However, carrageenan, like agar, is used mainly in food applications, largely in convenience foods. Small amounts of carrageenan are employed by the pharmaceutical and cosmetic industries. The major industrial use of carrageenan is in air-freshener gels, tending towards modified carrageenan.

Carrageenans are valuable gelling and stabilizing agents in foods, and they are effective at relatively low concentrations, which can offset their somewhat high cost (Table 4). There is a synergestic effect among the commercial carrageenans, and the properties of solutions can be modified by admixtures. Additionally, carrageenans are used to improve and balance the behavior of other gums (*e.g.* Tables 6 and 7), especially guar gum, locust bean gum, and carboxymethyl cellulose.

As carrageenans retain a strong negative charge over the normal pH range, they are used extensively in 'milk reactivity,' which has been one of the major advantages of this group of phycocolloids. The lambda-type is used for stabilization of cold-milk products, and the kappa-type with pasteurized products. The major uses of furcellaran are with milk products.

Alginate

Most alginate production and utilization occur in North America and Europe (Table 3). The industry is largely controlled by a single transnational firm. Alginates have been available commercially for more than 50 years and have found a wide variety of applications in the food, pharmaceutical, and industrial sectors. Alginate is the only phycocolloid that is utilized in large quantities by industry, other than in food preparations (Table 8). Competition from other gums has seriously affected the industry, and even as a coating for welding rods, alginate is being replaced by carrageenan in Brazil.

Today the industrial usage of alginates is mostly in cotton printing (Table 8), where they are superior to the gums traditionally used with reactive dye stuffs[22]. Application as a sizing/coating substance has been largely replaced by the use of other gums.

Alginates are also employed as gelling agents, stabilizers, and emulsifiers. An important property of alginate is its capacity to form gels by reaction with calcium ions, and primary food uses are in dairy products and frozen desserts, including fabricated ice cream (Tables 6, 7), where it is especially effective in preventing the formation of ice crystals. The only commercially available organic derivative, propyleneglycol alginate, has found many applications, in particular in salad dressings and sauces; this material is also used to stabilize the foam of beer.

Resources

From among the several thousand species of red and brown seaweeds, only a few within the genera shown in Table 9 are used extensively in the extractive industry. Species of Phyllophora, which yield a special type of carrageenan and are exploited in the Soviet Union, are omitted for lack of additional information. Among the brown algae the genera fall in the orders Laminariales and Fucales, species of both groups contributing significantly to the resource base. These species occur in temperate waters (Fig. 2) except for Sargassum, a warm-water genus for which there is only relatively minor demand. The resources exploited are from rather localized areas and most are obtained within relatively short distances from the processing facilities. Species of Macrocystis, which provide a significant portion of the biomass for industry, are obtained off the coast of southern California. This seaweed is harvested mechanically and processed wet. Recently there has apparently been a decline in the Californian beds of Macrocystis, attributed to the intrusion of warm water (El Niño) along this coast. In turn this has led to greater exploitation of Macrocystis from Baja California in Mexico and Lessonia from Chile, but these areas also have been affected by warm water. Resources from a distance are not only more costly to obtain, but they must be processed from dried material, which is a more expensive procedure in a facility designed for wet extraction[22]. Large standing stocks of Macrocystis are found off southern South America, but there has been only minor exploitation.

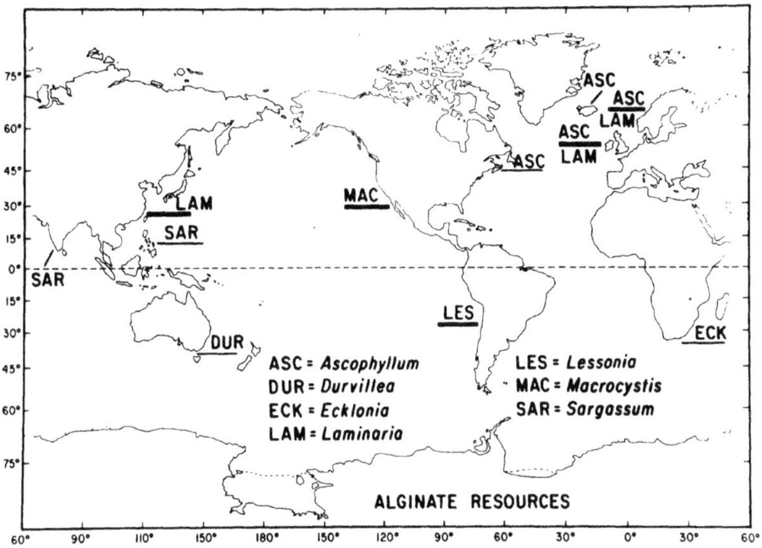

Fig. 2. World distribution of major resources harvested for alginates.

Most of the other resources in the alginate industry occur in northern Europe (Fig. 2) where *Ascophyllum nodosum* and two species of Laminaria are harvested, usually by hand, in Norway, Ireland, Scotland, and Iceland. They are dried before being processed in both the United Kingdom and Norway.

More than 200 000 dry tonnes of *Laminaria japonica* are cultivated in China[38,39]. A significant portion of this harvest is used in the Chinese alginate industry, but no other details are available. Little of this alginate has been placed on world markets, but poses a competitive threat to the existing transnational industries. Moreover, the Chinese apparently could easily and quickly double their production of Laminaria if needed (C.-K. Tseng, pers. comm.).

Special mention must be made of Durvillea in Australia. Although only 2000 to 3000 t are harvested, it comes entirely from a small island, King Island, in Bass Strait. The weed is cast ashore, collected, dried, and shipped to the United Kingdom for processing.

There seem to be no apparent global shortages of resources for alginates, and significant unexploited biomasses occur in both the North Pacific and off southern South America. The costs of exploiting these resources are at present unattractive. There are, however, no obvious exploitable resources of brown algae that offer the industry different types of alginate. The industry may thus be limited by quality rather than quantity. Cultivation, although practiced extensively in

China, is not an economically attractive alternative, since present resources are cheap[23].

The carrageenan industry (Fig. 3) initially was based on a single resource, *Chondrus crispus* (Irish moss), with minor uses of a closely related species, *Mastocarpus stellatus*; these plants were obtained from eastern Canada and northern New England. Furcellaran is obtained exclusively from *Furcellaria lumbricalis*; resources were from the Kattegat and later supplemented from eastern Canada. Irish moss is still harvested in eastern Canada, where it is an expensive resource, and also from the Atlantic coast of Europe. The industry is now, however, dominated by Eucheuma which is cultivated in the Far East, principally the Philippines. Moreover, significant quantities of several species of Iridaea are now being exploited in Chile. Species of Gigartina and *Hypnea musciformis* are utilized to a much lesser extent, the latter mostly in Brazil.

Most of the global carrageenan is processed by several large firms located in developed areas that obtain various resources from widespread but developing areas of the world. Because of blending and admixtures of gums (*cf.* Table 7), it is necessary for these firms to have a variety of resources available; frequently the blending is done at the processing stage.

The shift in the resource base to include species in addition to

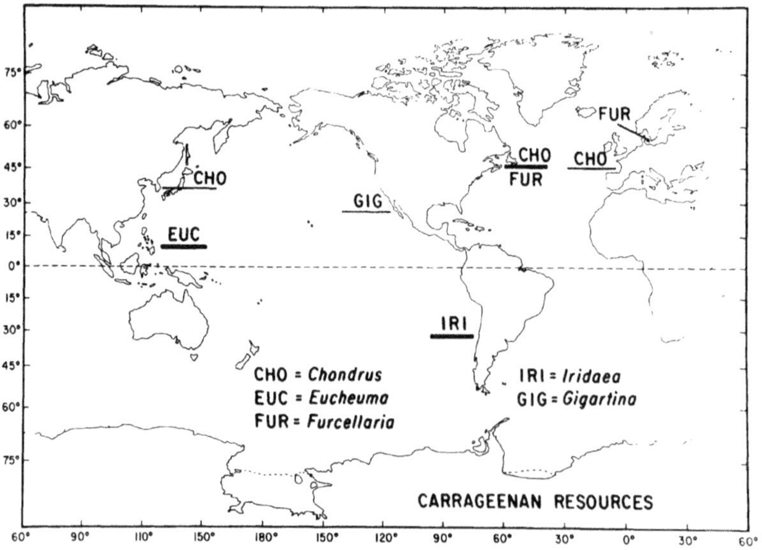

Fig. 3. World distribution of major resources harvested for carrageenans.

Chondrus crispus is interesting, since new types of carrageenans are now being used. Kappa/iota-hybrid and lambda-types are obtained from Irish moss, and *Eucheuma spinosum* (cottonii) yields an extreme κ-type and *E. striatum* an extreme ι-type[8]. Iridaea resources also yield λ-type carrageenan as well as κ/ι-hybrids. These new resources have permitted the industry to expand because of the availability of both quantity and quality. Species of Eucheuma are the only cultivated seaweeds used by the carrageenan industry. In addition to providing new resources, the resource base is essentially mobile. If present areas of cultivation become unsuitable, for biotic, abiotic or political reasons, it may be possible to shift cultivation to alternate areas. Other species also are potentially usable and, if new resources are made available to the industry, these too will probably come from cultivation, most likely in tropical areas.

The estimated value of resources of agarophytes exceeds that of the other two sectors of the seaweed industry (Table 9), and the total value of the agar is equivalent to that of alginate even though the volume is considerably less (Tables 3 and 4). Moreover, the industry is structured differently from the carrageenan and alginate industries. Most resources are obtained from developing countries, where considerable processing is also done, especially in Asia. There is also much movement of resources and finished products among countries. Japan, though a developed country, is the largest producer of agar (Table 3), importing resources from many areas of the world, and is also the largest consumer of this gum. As noted, most of the uses of agar in food in western societies have been discontinued, but in Asia particularly agar continues as both a foodstuff and an additive to foods.

The two main genera of resources for agar are Gelidium and Gracilaria (Fig. 4). Gelidium is extremely important since the extractives from its species are suitable for microbiological-grade agar, along with that from Pterocladia, which is available in only small amounts. Most resources of Gelidium are obtained from Spain, Portugal, North Africa, and Korea, and Pterocladia from the Azores. An estimated 2000 t of agar are produced annually from these sources[24].

In general, agars from species of Gracilaria are unsuitable for preparation of microbiological media because of their high gelling temperatures. These agars are used extensively where high gelling temperatures are not critical. Although agars from Gracilaria are generally characterized by forming weak gels, treatment with alkali usually results in strong gels that are commercially valuable. Major resources of Gracilaria are from Chile, Argentina and various locales in the Far East (Fig. 4). The agar industry affords a major potential for growth, perhaps more so than the other sectors of the phycocolloid industry. However,

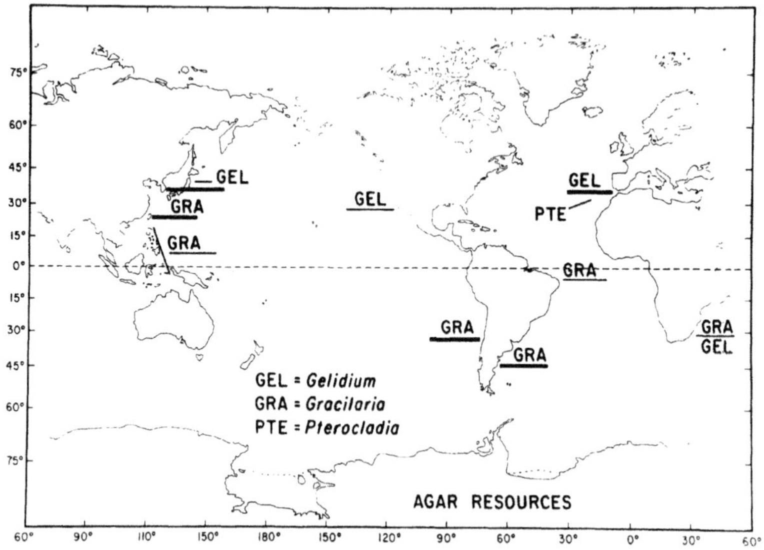

Fig. 4. World distribution of major resources harvested for agars.

because of factors already mentioned, the future of the industry is uncertain.

Further development of resources of Gelidium is problematic. Cultivation trials have not been encouraging, although appropriate management programs can possibly enhance the resource base. On the other hand, the large number of species of Gracilaria (many more than in any other genus of commercial interest) offers a potential that cannot even be predicted. Moreover, this is largely a warm-water genus[25], and some species can undoubtedly be cultivated in developing areas where mariculture can be economically viable.

There are few taxonomic problems associated with the brown algae used commercially. With red algae, both carrageenophytes and agarophytes, the problems are horrendous, not only botanically[1,6], but also industrially. For example, seaweeds for extraction may be misidentified, and this can be serious when resources are being blended. It has been pointed out[32] that prior to cultivation, Eucheuma had not gained industrial significance because of taxonomic difficulties, although even at that time the importance of Eucheuma extracts was appreciated. The present understanding of Eucheuma has improved little, but cultivation has resolved the practical taxonomic problems. Similar problems exist for Iridaea, Gigartina, and Chondrus.

With agarophytes the systematic understanding of species of Gracilaria[6] and Gelidium[34] is generally chaotic, and exploitation of species of the former genus undoubtedly has been limited because of these problems. As an example only, *G. verrucosa* is reported as the main species of commerce the world over[24]. However, this species name has been misapplied in a number of instances[7,24], and thus the species may not even be used commercially.

There are undoubtedly many species and genera of seaweeds that have potentially useful and valuable phycocolloids. The seaweed industry has limited its utilization to those species that occur in abundance and are relatively easily and cheaply obtained. There thus is a potential of undetermined magnitude for the exploitation of underutilized species. However, in order to realize this potential, a basic and urgent requirement is that a taxonomic center for species of commercial interest be established.

Acknowledgements I thank C J Bird, J P van der Meer, and W Yaphe for their comments on the manuscript.

References

1 Abbott I A 1979 The importance of taxonomy to the utilization of marine algae. Actas I. Symp. Algas Mar. Chilenas, pp 51–58.

2 Abbott I A and Chapman F A 1981 Evaluation of kappa carrageenan as a substitute for agar in microbiological media. Arch. Microbiol. 128, 355–359.

3 Araki C 1966 Some recent studies on the polysaccharides of agarophytes. Proc. Int. Seaweed Symp. 5, 3–17.

4 Bellion C, Hammer G K and Yaphe W 1981 Analysis of kappa-iota hybrid carrageenans with kappa-carrageenase, iota-carrageenase and ^{13}C N.M.R. Proc. Int. Seaweed Symp. 10, 379–384.

5 Bhattacharjee S S, Yaphe W and Hammer G K 1979 Study of agar and carrageenan by ^{13}C nuclear magnetic resonance spectroscopy. *In* Marine Algae and Pharmaceutical Science. Eds. H A Hoppe, T Levring and Y Tanaka. pp 645–655. Walter de Gruyter, Berlin.

6 Bird C J and McLachlan J 1982 Some underutilized taxonomic criteria in Gracilaria (Rhodophyta, Gigartinales). Bot. Mar. 25, 557–562.

7 Bird C J, van der Meer J P and McLachlan J 1983 A comment on *Gracilaria verrucosa* (Huds.) Papenf. (Rhodophyta: Gigartinales). J. Mar. Biol. Ass. U.K. 62, 453–459.

8 Bixler H J 1979 Manufacturing and marketing of carrageenan. Actas I. Symp. Algas Mar. Chilenas, pp 259–274.

9 Bjerre-Petersen E, Christensen J and Hemmingsen P 1973 Furcellaran. *In* Industrial Gums: Polysaccharides and their Derivatives, 2nd Edition. Ed. R L Whistler. pp 123–136. Academic Press, N.Y.

10 Duckworth M and Yaphe W 1971 The structure of agar. I. Fractionation of a complex mixture of polysaccharides. Carbohyd. Res. 16, 189–197.

11 Guiseley K 1970 The relationship between methoxyl content and gelling temperature of agarose. Carbohyd. Res. 13, 247–256.

12 Guiseley K B and Renn D W 1977 Agarose: Purification, Properties, and Biomedical Applications. FMC Corporation, Marine Colloids Division, Rockland, ME.

13 International Trade Centre UNCTAD/GATT 1981 Pilot Survey of the World Seaweed Industry and Trade. International Trade Centre UNCTAD/GATT, Geneva.

14 Jensen A 1978 Industrial utilization of seaweeds in the past, present, and future. Proc. Int. Seaweed Symp. 9, 17–34.

15 Kang K S, Veeder G T, Mirrasoul P J, Kaneko T and Cottrell I W 1982 Agar-like polysaccharides produced by a Pseudomonas species: production and basic properties. Appl. Environ. Microbiol, May 1982, pp. 1086–1091.
16 Larsen B 1981 Biosynthesis of Alginate. Proc. Int. Seaweed Symp. 10, 7–34.
17 Marine Colloids Inc. Carrageenan. Monogr. No. 1. Marine Colloids, Rockland, ME.
18 Marshall S M, Newton L and Orr A P 1949 A Study of Certain British Seaweeds and their Utilisation in the Preparation of Agar. Ministry of Health & Ministry of Supply, HMSO, London.
19 McCandless E L 1978 The importance of cell wall constituents in algal taxonomy. *In* Modern Approaches to the Taxonomy of Red and Brown Algae, Systematics Assoc. Sp. Vol. 10. Eds. D E G Irvine and J H Price. pp 63–85. Academic Press, N.Y.
20 McCandless E L 1981 Polysaccharides of the seaweeds. *In* The Biology of Seaweeds, Eds. C S Lobban and M J Wynne. pp 559–588. Univ. Calif. Press, Berkeley.
21 McCandless E L, West J A and Guiry M D 1982 Carrageenan patterns in the Phyllophoraceae. Biochem. System. Biol. 10, 275–284.
22 McHugh D J and Lanier B V 1983 The world seaweed industry and trade. ADB/FAO INFOFISH Market Rept., Vol. 6.
23 McLachlan J 1982 The seaweed industry: an appraisal of resource management and cultivation, especially as related to the extractive industry. Monogr. Biol. (N.S.). No. 2, 151–168.
24 McLachlan J and Bird C J 1983 Gracilaria in the seaweed market: prospectus. *In* Proceedings of the International Symposium on Advances and Perspectives in Aquaculture in Chile; Coquimbo, Chile, September 1983, 133–137.
25 McLachlan J and Bird C J 1984 Geographical and experimental assessment of the distribution of species of Gracilaria in relation to temperature. Helgoländer Meeresunter. 38, 319–334.
26 McNeely W H and Pettitt D J 1973 Algin. *In* Industrial Gums: Polysaccharides and Their Derivatives, 2nd Edition. Eds. R L Whistler. pp 49–81. Academic Press, N.Y.
27 Monopolies and Mergers Commission 1979 FMC Corporation, Merck & Co, Inc, Alginate Industries Limited: A Report on the Proposed Mergers, HMSO, London.
28 Moss J R 1978 Essential consideration for establishing seaweed extraction factories. *In* The Marine Plant Biomass of the Pacific Northwest Coast. Ed. R W Krauss. pp 301–314. Oregon State University Press, Corvallis.
29 Painter T J 1983 Algal polysaccharides. *In* The Polysaccharides, Vol. 2. Ed. G O Aspinall. pp 195–262. Academic Press, N.Y.
30 Percival E 1978 Do the polysaccharides of brown and red seaweeds ignore taxonomy? *In* Modern Approaches to the Taxonomy of Red and Brown Algae. Systematics Assoc. Sp. Vol. 10. Eds. D E G Irvine and J H Price. pp 47–62. Academic Press, N.Y.
31 Pluzek K 1981 Applications of agarose in biomedical techniques in relation to physical and chemical properties. Proc. Int. Seaweed Symp. 10, 711–718.
32 Sand R E and Glicksman M 1973 Seaweed extracts of potential economic importance. *In* Industrial Gums: Polysaccharides and their Derivatives, 2nd Edition. Ed. R L Whistler. pp 147–194. Academic Press, N.Y.
33 Sanford P A and Baird J 1983 Industrial utilization of polysaccharides. *In* The Polysaccharides, Vol. 2. Ed. G O Aspinall. pp 411–490. Academic Press, N.Y.
34 Santelices B 1977 A taxonomic review of Hawaiian Gelidiales (Rhodophyta). Pac. Science 31, 61–84.
35 Selby H A and Wynne W N 1973. Agar. *In* Industrial Gums: Polysaccharides and their Derivatives, 2nd Edition. Ed. R L Whistler. pp 29–47. Academic Press, N.Y.
36 Smidsrød O and Grasdalen H 1984 Polyelectrolytes from seaweeds. Proc. Int. Seaweed Symp. 11, 19–28.
37 Towle G A 1973 Carrageenan. *In* Industrial Gums: Polysaccharides and Their Derivatives, 2nd Edition. Ed. R L Whistler. pp 83–114. Academic Press, N.Y.
38 Tseng C-K 1981 Commercial cultivation. *In* The Biology of Seaweeds, Eds C S Lobban and M J Wynne. pp 680–725. University of California Press, Berkeley.

39 Tseng C-K 1981 Marine phycoculture in China. Proc. Int. Seaweed Symp. 10, 123–152.
40 Turvey J R 1978 Biochemistry of algal polysaccharides. *In* Biochemistry of Carbohydrates, Vol. II. Ed. D J Manners. pp 151–177. University Park Press, Baltimore.
41 Waaland J R 1981 Commercial utilization. *In* The Biology of Seaweeds. Eds. C S Lobban and M J Wynne. pp 726–741. University of California Press, Berkeley.
42 Walker A W Day A A 1943 Extracts from Irish moss as a substitute for agar in bacteriological culture media. Food Res. 8, 435–443.
43 Whistler R L 1973 Factors affecting gum costs and applications. *In* Industrial Gums: Polysaccharides and Their Derivatives, 2nd Edition. Ed. R L Whistler. pp 5–25. Academic Press, N.Y.
44 Yaphe W and Duckworth M 1972 The relationship between structures and biological properties of agars. Proc. Int. Seaweed Symp. 7, 15–22.
45 Zanlungo A B 1980 Polysaccharides from Chilean seaweeds. Part IX. Composition of the agar from *Gelidium lingulatum.* Bot. Mar. 23, 741–743.

Plant and Soil 89, 159–183 (1985). Ms. BR 12

Outdoor cultivation of sea vegetables

Y. LIPKIN
Department of Botany, The George S. Wise Faculty of Life Sciences, Tel-Aviv University, Tel-Aviv 69978, Israel

Key words Marine algae Sea vegetables Thalassonomy

Summary The outdoor cultivation of sea vegetables is carried out on a large scale in the Orient, mainly in Japan, China, Taiwan, Korea and the Philippines. Food crops are the most important among the sea vegetables cultivated, with Porphyra and Undaria being the more important in Japan and Laminaria in China. Eucheuma, an industrial crop containing the phycocolloid carrageenan, is cultivated in wide areas of the Philippines. The cultivation of the major food crops, which developed quickly over the past 30 years, is largely based on the results of research, especially with regard to seeding procedures, which have reached a certain level of sophistication.

The major crops of sea vegetables are cultivated attached to ropes or nets located in a suitable site and habitat. Crops of limited economic value, however, are still cultivated by the old, primitive method of planting on stones on the ocean bed and other similar means. The location and timing of farming are selected with regard to the requirements of the plants for light, temperature, water movement, exposure to air (for the intertidal species), *etc.* Cultivation of seeding material of the three food crops and seeding of ropes and nets is carried out indoors under more or less controlled conditions. When the sporelings become established they are transferred to cultivation grounds in the ocean. When the plantlets grow too densely (in Laminaria) they have to be separated and replanted at the correct distances. This is done several weeks after transplantation to the ocean, when they are large and sufficiently strong. Eucheuma and other industrial crops are propagated vegetatively, using cuttings and fragments as planting material. Where seawater is lacking in nutrients, fertilizers are applied to guarantee a higher yield. The harvest is carried out manually, except for Porphyra, for which mechanical harvesters are used. Sea vegetables are attacked by pathogens that may cause severe damage to the crops. Diseases caused by improper growing conditions are also known. Grazers may also inflict losses In all major crops the strains cultivated have been selected. In a few cases hybridization and other genetic techniques have been used to obtain domesticated varieties that can grow and yield far beyond the limits of their wild-type parents.

Despite the fact that some mechanization has been introduced into the cultivation of sea vegetables, it is still by and large a highly labor-intensive enterprise. Nevertheless, it competes well with terrestrial crops in the Orient from the economic point of view.

Interest in the cultivation of sea vegetables is widespread in the West and much experimental work aimed at its materialization has been underway during the last *ca.* twenty years.

Introduction

Outdoor cultivation of desirable marine macroalgae, the sea vegetables (this term is used in the extended meaning embracing all cultivated seaweeds, not just food crops), has been practiced in Japan at least since the beginning of the 17th century[98] and in China for about the last 200 years[87]. It developed from previous utilization of the natural resources and started as the mere provision of new suitable surfaces for settlement of useful algae. The first alga to be cultivated was Porphyra (nori in Japanese, zicai in Chinese), which is economically the most important cultivated alga to this day. Until the fifties of our century, cultivation of sea vegetables has slowly expanded geographically, additional edible algae have been included, and techniques have been slowly improved and developed.

A purely scientific achievement, the clarification of details of the life history of the European *Porphyra umbilicalis*[24], inspired studies by Japanese scientists, and the subsequent study by Kurogi on the life history of *P. tenera*, published in 1952 and 1953[42], set the Japanese nori industry on the course to modernization. A similar development took place in China, with studies by Tseng being published in 1954–56 and translated into practice in the early sixties[87]. The development affected the studies and cultivation of other food crops as well, chiefly Laminaria (cultivated in Japan since early in the 18th century[34] and in China from 1943[14]) and Undaria (cultivated in Japan since 1960[2] and in China since 1935[87]).

The last three decades have seen a tremendous development in sea vegetable cultivation, due to achievements such as breeding warmth-tolerant strains of Laminaria and developing the technique of summer seeding in the sixties and seventies[8,14], which opened a new era in the cultivation of this crop in China. The area farmed is thus expanding rapidly; a continuously growing number of species useful for a number of purposes is being incorporated; new methods and techniques are being introduced; and additional regions of the world are becoming involved. Nowadays the cultivation of sea vegetables is widespread, mostly in the countries where it was initiated – Japan and China – and to a lesser extent in Korea, the Philippines and Taiwan. Intensive experimental work is underway, aimed at its establishment in many other parts of the world.

Cultivation of industrial crops was started rather recently, but has developed relatively rapidly. In the early seventies, cultivation of the carrageenophyte Eucheuma became established in the Philippines, where it is farmed on a considerable scale[42]. In China, however, where its cultivation was begun in 1960, it has not developed much[48]. The agarophyte Gracilaria is widely cultivated in Taiwan, usually in polyculture with other marine organisms.

The production of marine algal biomass for energy conversion or for the removal of excess nutrients entering the marine environment in sewage water is still at the experimental stage *cf.*[75,93]. Practically all of the commercial cultivation of sea vegetables is carried on out-of-doors. However, some crucial phases, namely the cultivation of seeding material and the seeding process itself are done under controlled conditions indoors. Experimental work on crop cultivation indoors is underway, mainly in North America. Neither topic will be dealt with here in detail.

Despite the marked differences between the natures of the terrestrial and the marine environments, and those of the organisms involved, many of the problems and activities associated with conventional agronomy have equivalents in practical thalassonomy (lit. management of the sea, the art and science of marine crop production, 'marine agronomy'). Some are, *e.g.* the choice of crop, site selection, 'bed' preparation, sowing, weeding, fertilizing, pest and disease control, and harvesting. Equivalents of other agricultural practices, such as soil tillage, are not yet practiced in the commercial cultivation of sea vegetables, although they have been tested in experimental systems.

The differences in environment, however, naturally necessitated modifying some accepted agricultural techniques or devising entirely new ones to cope with the demands of the marine environment and conform with the nature of the marine plants. One example is the manner of planting, which is totally different with the sea vegetables. Since the plants require only physical support from their substrate, and extract their nutrients from the surrounding seawater, they may be grown on an artificial firm substrate at a level where light conditions are at their best, no matter how deep the ocean bottom is at the site. When cultivated at sea, sea vegetables are therefore always cultured attached to a substrate; when grown in ponds on land, however, they may be cultivated in the unattached state.

Here commercial cultivation out-of-doors will be emphasized, while experimental cultivation will be mentioned only briefly. The major questions concerning sea vegetable cultivation will be discussed, with the similarities and dissimilarities between it and the cultivation of terrestrial crops borne in mind.

Crops cultivated

Only a few sea vegetables are commercially cultivated, only four of them on a relatively large scale (Table 1). Three of these, Porphyra, Laminaria (kombo or konbu in Japanese, kunbu in old Chinese, haidai in modern Chinese), and Undaria (wakame in Japanese, quandai-cai in Chinese), are food crops, cultivated mainly in Japan and China[4] and to a lesser extent in Korea[60]. The fourth, the industrial crop Eucheuma, a source of the phycocolloid carrageenan, is cultivated in the Philippines[19,20,66] and on a small scale in China[48,87]. Its yields are claimed to have recently met the world demand for Eucheuma carrageenan[42,46].

Other sea vegetables are cultivated on a much smaller scale. These

Table 1. Sea vegetable crops cultivated, world-wide farming areas and estimated yields (year in parentheses)

Crop	Species cultivated (* = most important in cultivation)	Countries	Area cultivated (10^3 hectares)		Yield (10^3 MT dry weight)		Reference
PORPHYRA			75	(1978)	30	(1978)	55, 87
	* *P. tenera*	Japan, China					4, 87, 98
	* *P. yezoensis*	Japan, China					4, 87, 98
	P. kuniedai	Japan					4, 87, 98
	P. angusta	Japan					4, 98
	P. pseudolinearis	Japan					4, 87, 98
	* *P. haitanensis*	China					87
	P. seriata	Japan					87
	P. arasaki	Japan					87
	P. guangdongensis	China					87
LAMINARIA			50–60	(1979)	300	(1979)	87, 97
	* *L. japonica* (inc. var. *ochotensis*)	Japan, China, Korea					36, 87
	L. angustata	Japan					35
	L. religiosa	Japan					35
UNDARIA	* *U. pinnatifida*	Japan, China, Korea	18	(1979)	25	(1979)	55 86, 87, 98
	U. undarioides	Japan					55, 66
	U. peterseniana	Japan					55
EUCHEUMA					20	(1979)	
	* *E. denticulatum* (= *E. spinosum*)	Philippines					42, 48, 87
	* *E. striatum*	Philippines					45, 87
	E. cottonii	Philippines					45, 87
	E. gelatinae	China					86, 87

Table 1 (*continued*)

Crop	Species cultivated (* = most important in cultivation)	Countries	Area cultivated (10^3 hectares)	Yield (10^3 MT dry weight)	Reference
GRACILARIA			0.3 (1979)	12 (1979)	15
	* *G. verrucosa*	Taiwan, China, Japan			5, 15, 81, 86 87
	G. gigas	Taiwan			15, 81
	G. lichenoides	Taiwan			15, 81
	G. compressa	Taiwan			78
	G. coronopifolia	Taiwan, Hawaii			4, 21
	G. bursapastoris	Hawaii			21
GLOIOPELTIS	*G. furcata*	China, Japan		0.3 (1979)	4, 87
MONOSTROMA			1.5–7.5 (1979)	1–2 (1972)	87
	* *M. nitidum*	Japan	(includes Enteropmorpha)		66, 78
	M. latissima	Japan			78
ENTEROMORPHA			included in Monostroma		87
CAULERPA	*C. racemosa*	Taiwan			4
GELIDIUM		Japan	very little		5, 60, 66
GELIDIELLA	*G. acerosa*	India	very little		49, 58, 59

include the edible green algae Monostroma, Enteromorpha, Ulva and *Caulerpa racemosa*, the red agarophytes Gracilaria, Pterocladia and Gelidium, and the red alga Gloiopeltis, cultivated for human consumption as well as for the production of glue[4] and a sizing material for the textile industry[87]. Fresh Gracilaria is also used to feed small cultured abalone (Haliotis)[15]. These sea vegetables as well are cultivated mainly in Japan, China and Taiwan[4, 17, 52, 60, 78, 81, 87].

Selection of farm location and preparation of site for cultivation

For cultivation out-of-doors, where the possibility of controlling the environmental conditions is rather limited, the selection of the right site is crucial for the success of the crop. Different crops, naturally, have different requirements for cultivation conditions. For example, for Porphyra, Yoshida and Akiyama[98] stated that it is most important to select a site that is protected from severe currents or waves that may destroy or damage cultivation gear, a site where the tidal range during spring tide is no less than 1–1.5 m and the current speed about 20 cm s^{-1} (10 when the seawater is rich in nutrients, 30 when it is poor). Where currents are too strong or waves too large, the construction of wave-breakers or weirs may make a site suitable for nori cultivation[98]. Undaria, on the other hand, requires strong currents[55]. It grows best above a sandy seabed with scattered rocks and pebbles. *Eucheuma spinosum* requires a depth of 1–3 m, rather cool water, and a strong enough current (Neish in[21]).

Gracilaria pond sites in Taiwan are selected where a supply of fresh water is available to compensate for evaporation losses, where winds are not so strong as to dislodge the plants from their places, where tidal fluctuations are large enough to enable periodic exchange of water in the pond, where the bottom is sandy loam, and where fluctuations in pH are preferably only between 8.2 and 8.7 but no more than between 6 and 9 (ref. 80).

Setup of sea vegetable farms

The vast majority of sea vegetable farms are located in the sea. Cultivation in ponds or tanks on land is carried out to a much smaller extent.

At first, cultivation in the sea amounted to merely providing the desirable algae with substrates not naturally at their disposal. This was done in many ways; in China, by clearing rocks of other algae and sessile animals, by scraping and then spraying with lime the next day[87], at the right time of the year, which was the reproduction period of the particular alga. In Japan this was done by planting stones or concrete blocks on the seabed, by blasting reefs[36], by anchoring plastic floats at various depths[4] in the subtidal zone, or by sticking brush in the mud in the intertidal area[98]. These methods have largely given way to more efficient ones for the major crops in both countries, but for some less important crops like Gloiopeltis or Gelidium and for the major crops in Korea, they are still in use[4,87].

A variety of fixed or floating nets, ropes and rafts were next used as substrates for the algae. They were first made of natural fibers – hemp, coir or palm fibers, which require long soaking before use to leach out harmful substances – and then, in about 1960, of synthetic ones. Mariculturists in China and Japan developed a variety of specialized methods to suit the particular crop and region. Nevertheless, for some crops in some places rather simple methods are still used, such as those for Eucheuma cultivation in China. There, it is cultivated on rocky bottoms by inserting fragments of the plants into crevices in natural reefs, by spreading them on the bottom and holding them in place by stones, or, mostly in recent years, by tying fragments to coral branches and having divers arrange them on the bottom[87].

One group of cultivation supports includes different types of nets. Vertical nets and bamboo screens, common in the past for nori cultivation in Japan, are hardly used now. Horizontal nets are almost exclusively used for Porphyra cultivation in Japan and China[87,98] and give the highest yields of Eucheuma in the Philippines[21]. The nets used for Porphyra have 15 × 15 cm openings, whereas those used for Eucheuma have openings of 25 × 25 cm. The nets may be tied to poles inserted in the seabed at a fixed level in the intertidal zone (where the tidal range is no more than 1–2 m) for Porphyra, or in the subtidal zone *ca.* 50 cm above the reef-flats for Eucheuma in the Philippines. This is the so-called 'fixed-net' method. Where the tidal range is 3–6 mm, Porphyra is grown in the intertidal zone on so-called 'raft-type'[98] or 'lift-type'[87] nets. These are tied to poles with longer ropes and are equipped with floats. Thus, their level changes with the tidal fluctuations within a certain range controlled by the length of the ropes. The semi-floating net is an improvement, where the nets are equipped with short legs or a wooden frame that rests on the bottom during low tide, thus keeping the nets at a fixed lowest level which is above the low water level. Hence, during low water the nets are left exposed for a certain length of time, which may be adjusted by the length of the legs. With the rising tide these nets become submerged and then float. The 'floating culture'[98] or 'floating method'[87] is used in deep water. Here the nets are never exposed. The development of the intertidal Porphyra on these nets is not as good as on exposed nets, the young stages being the most susceptible to continuous immersion. Therefore, nets are transferred from the intertidal zone to the deep water only when the plantlets are big enough to stand continuous immersion (1–3 cm long). Monostroma is grown similarly[4].

Another group of cultivation constructions are the various arrangements of ropes and 'rafts.' Laminaria and Undaria are grown in China and Japan on such ropes[36,87], and for Eucheuma this is the most popular method in S.E. Asia[21]. The cultivation ropes may be arranged horizontally, in which case great lengths may be stretched between two poles inserted in the bottom or tied to anchors and floated at a desired depth. Ropes may also hang vertically from horizontal lines or from floats. Several such bamboo floats tied together form a 'floating raft'[14,87]. For Laminaria and Undaria, which are often grown together, horizontal lines may be kept at various depths, depending on the transparency of the water, from just below the surface to a depth of several meters. Cultivation of these two sea vegetables is more common on vertical ropes in China, whereas in Japan horizontal cultivation ropes are more popular. For Eucheuma, the ropes are tied between low poles about 50 cm above the reef-flat, as for nets.

Vertical cultivation ropes may be up to five meters long (usually they are 1–1.5 m long in China and longer in Japan) and may be located where the water is 10–15 m deep. They are kept in position by weights attached to their lower ends. Each may carry up to 30 plants. Since the plants at the upper part of the rope receive more light than those on the lower part, the cultivation ropes must be inverted, usually twice during the growing season[85]. The depth of the plants also needs to be adjusted from time to time according to their phase of development and the changing environmental conditions, especially the transparency of the water.

A variety of raft forms has been devised in China[14], but in recent years the single-line floating raft has been used in most cultivation grounds[87]. This type is better able to withstand strong currents and heavy waves.

An attempt was made to grow Gracilaria as well on floating rafts in China[44]. However, it seems that it is more successful on rafts fixed at certain levels[73]. In Japan, Gracilaria is grown on horizontal ropes[5].

Cultivation of Laminaria on sandy seabeds in Japan started by planting stones on the bottom for the settlement of spores, as had been done in rocky areas. Later, specially designed, fenestrated cylindrical concrete blocks were used. Both had some disadvantages. They gradually sank into the sand and the fronds were injured by the stones or blocks. Attempts were made to overcome these problems using concrete 'legs' and plastic pipes, and partial success was achieved in experiments[85].

In Taiwan Gracilaria and *Caulerpa racemosa* are cultivated in shallow ponds by evenly scattering algal fragments to grow unattached, usually in sites selected in areas that are not windy. Caulerpa will root itself, but winds may dislodge Gracilaria, which is sometimes covered with old fishing nets to keep it in place[81]; or, if the ponds are too big, a line or two of bamboo windbreakers are constructed to divide them and decrease the wind effect[15]. Gracilaria is often grown in polyculture with milkfish (Chanos), shrimp (Penaeus) or mangrove crab (Scylla)[15,81] to improve the profitability of the ponds.

Inoculation of farms

In Japan and China, the three major food crops, Porphyra, Laminaria and Undaria, are artificially seeded indoors by spores produced, released and settled under more or less controlled conditions. Seeding is performed in tanks containing fertile plant material induced to sporulate, into which the cultivation substrates (ropes, nets, etc.) are dipped. After the spores settle, the sporelings obtained are allowed to establish themselves before they are transplanted into the sea. Previous studies of the life histories of the various crop species and the environmental requirements of each phase enabled the sophistication of the seeding techniques for these crops. Both Porphyra on the one hand, and Laminaria and Undaria on the other, have microscopic, delicate vegetative phases in their life histories. In Porphyra this is the sporophyte, the so-called conchocelis phase; in Laminaria and Undaria it is the gametophyte. These are cultured indoors to maturation and fertility and serve as seeding material (the details of this indoor cultivation are out of the scope of this account). In addition, in some Porphyra species like *P. yezoensis* and *P. kuniedai*, the leafy, crop phase may produce neutral spores. These may be seeded on the nets either in indoor tanks or in the sea[87,98] and develop directly into the leafy phase.

Artificial induction of reproduction in Laminaria and Undaria enabled seeding of the cultivation substrates 2–3 months ahead of the natural reproduction season, and consequently, early transplantation of seeded ropes into the sea. This resulted in prolongation of the growing season and a much earlier harvest, compared with naturally seeded substrates. It also gave the crop plants an advantage over weed species[3,35,87]. In Porphyra, techniques involving the freezing of seeded nets (after partially dehydrating the young plantlets) enabled transplantation into the sea at any desired time and the replacement of depleted or infected nets[98].

With Laminaria, two transplantations are required. The spores settle on the seedling cords very densely and the sporelings, when grown, compete with and thus weaken each other. Therefore, the seeded cord transplanted to the sea is either cut into small pieces *ca.* 5 cm long, each with a few sporelings, which are inserted into the cultivation ropes at the correct intervals, or it is left in the sea until the young sporophytes reach the length of 10–15 cm. The sporophytes are then detached from the seeded cords and attached to (or usually twisted into) the cultivation ropes and transplanted again, this time over a wider area.

Natural seeding by spores produced in the sea, from either natural or cultivated populations, is still in use for some minor crops such as Gloiopeltis or for the major crops in areas where phycoculture is not much developed, such as Korea[87], where they are grown by the old methods on stones and rocks. Monostroma is also naturally seeded on nets placed in the sea, where spores are released, and subsequently transferred to suitable growing areas[78].

The important industrial crops, Eucheuma and Gracilaria, are propagated vegetatively from plant fragments. Fragments may be stuck in crevices in natural reefs or held in place on the bottom by stones, as is the practice with *Eucheuma gelatinae* or Gracilaria in China[87]. They may be held in place by short, split bamboo sticks pushed into the soft bottom[81,87] or may be tied to nets or ropes, as in the wide Eucheuma farms in the Philippines, or twisted into ropes (Gracilaria in Japan[5]).

Gracilaria and *Caulerpa racemosa* ponds in Taiwan are seeded by spreading vegetative plant fragments evenly over the pond, where they remain unattached[15].

Successful experimental cloning of Porphyra gametophytes by mechanically disintegrating thalli and culturing individual cells to mature plants makes it possible to obtain a large number of genetically identical plants[101]. Successful experimental cloning of Laminaria gametophytes and the ability to keep them for long periods in a vegetative state[26] make it possible to seed ropes and transplant Laminaria at any desired time. Both techniques will probably be used in commercial cultivation in the future.

Fertilization

Seawater usually contains all the nutrients required by marine plants for adequate growth. However, in some areas the nitrogen concentration may limit growth and affect the yields of commercial crops. This has occurred in some farmed areas in Japan and China. Nitrogenous fertilization increases the yields in these cases, and therefore such Laminaria and Porphyra farms are regularly fertilized in both countries[14,87,98]. According to Tseng[87], fertilization is imperative when the inorganic nitrogen concentration in the seawater is below 3.5 μM. When it is around 7.1 μM, the addition of nitrogen improves the quality of Porphyra product, and when it is 14.2 μM or more there is no need to fertilize. Gracilaria ponds in Taiwan also are often fertilized[15,81]. Seawater may become deficient in phosphate as well, but this does not occur in the existing marine farms.

At the present time, the most commonly used fertilizer is ammonium sulfate[87], but other fertilizers have been used as well, such as sodium nitrate[14], ammonium nitrate[4,14], urea, or fermented pig and chicken manure[15,81]. Fertilizers are usually applied by spraying larger farms with concentrated solutions (10%–15%), either from boats[87] or, in Japan, sometimes from helicopters[91]. The older method of hanging porous earthenware bottles containing fertilizers among the plants[14] to gradually dispense the nutrients is still used in some smaller farms. They have, however, largely been replaced by finely perforated plastic bags as slow dispensers of fertilizers[86,87]. For young sporophytes of Laminaria 'pulse' fertilization is employed[4,14,55,86]. This is periodic immersion of the plants in concentrated fertilizer solutions. This method is especially important just before placing the young sporophytes in the sea to give them a good start in their new environment[14,87].

Fertilization experiments are usually carried out in the laboratory or in greenhouse-kept tanks. A few, however were conducted in outdoor tanks[6,96]. Fertilization in natural populations of Macrocystis, to enhance its growth, involved the use of both nitrogen and phosphate sources that were distributed by helicopters[65]. The yield was increased compared with that of nearby beds, although the natural supply of nutrients was relatively high during the period of the study. Other experiments utilizing macronutrient-rich deep water[30] showed that this may sometimes be deficient in some micronutrients and therefore would be best utilized mixed with surface water[64]. The use of nutrient-containing plaster 'pellets' that release nutrients slowly was also tested successfully[49].

Weeding

Weeds may become a severe problem in all sea vegetable farms. The weeds appearing in different farms, however, vary with the habitat and geographical region. Weed control is based on knowledge of the environmental requirements or responses of both the crop and its particular weeds and on manipulating the farms so as to shift competition between the two in favor of the former. Herbicides are not used in commercial sea vegetable farms.

Several techniques of weed control are used for the different crops. The most serious weeds in Porphyra farms are the green Monostroma and Enteromorpha[87], which are also intertidal algae, but which are better adapted to lower levels of the intertidal zone than Porphyra. Therefore, the most common procedure in their control is to tie the fixed horizontal nets so that they are exposed for about 4–5 h daily during daytime, which is too long for the weeds[98]. This necessitates moving the nets up and down on the poles every few days, according to the changes in the tidal levels[98]. With the shift to the floating net culture technique this is no longer practical, and when weed-cover on these nets becomes too dense so that it interferes with the growth of Porphyra, the nets are exposed to the air for hours or even days[87]. This harms the weeds more than it does the Porphyra. Thus, the right exposure period results in considerably lowering the density of weeds without damaging the crop. It should be noted, however, that both green algae are utilized in Japan as food and in some cases are not regarded as weeds but are harvested as by-products[5]. Sometimes they are even encouraged to grow with the Porphyra[4]. After the last harvest of Porphyra in the spring, these green algae are left to develop and are later harvested[87]. Monostroma obtains the highest prices of any sea vegetable in Japan[4]. Other techniques aimed at decreasing competition by weeds are dense seeding and care in handling of culture nets, so as not to expose the substrate for settlement of weeds[87].

In Laminaria farms, other weeds cause problems. In China, Ectocarpus, Licmophora and Enteromorpha, among others, used to settle on ropes carrying autumn sporelings and overgrow the tiny gametophytes. Thus, they prevented the development of the resulting young sporophytes until mid-winter, when the weeds matured and disintegrated, allowing the Laminaria sporophytes to resume growth[14,87]. This delayed their maturation by a few months. The problem was solved by development of a method to obtain sporelings in the summer to produce plantlets 2–3 months ahead of the usual season and then to maintain them at low temperatures indoors for transplantation into the sea in the fall, when the sea temperatures drop enough for them to grow. Under such conditions young Laminaria sporophytes have an advantage over settling spores of weeds and therefore can compete successfully with them. This method was developed in Japan in 1966[35,36], and in

China it is now the only method used[8,87]. Obtaining the right density of sporophytes on culture substrates also helps keep the weeds low. For Laminaria the right density is '2–30 seedlings per field at 400 × magnification'[99]. With Porphyra, attempts are made to ensure that spores settle as densely as possible[87].

In Eucheuma farms in the Philippines no weeding is done[74], since Eucheuma usually outcompetes the weeds under favorable conditions.

In China the 'fixed-raft method' was developed for the cultivation of *Gracilaria verrucosa*, primarily to cope with the severe weed problem that hindered Gracilaria cultivation on floating rafts[73]. The fixed rafts, which stay at the same level, get less light during high water, which deprives the weeds of the high light they require and allows Gracilaria, which needs less light, to grow better.

In Taiwan the main weeds in Gracilaria ponds are Enteromorpha, Chaetomorpha and Ectocarpus[15]. Here fish are used to control the weeds. Milkfish (*Chanos chanos*) and Tilapia are stocked in the ponds at a density of 300–400 individuals per hectare[81] (about 1000 according to Chiang)[87]. When the weed populations decrease, the fish are removed to prevent damage to the crop[81]. According to Chiang[87], Penaeus and Scylla grown with the Graciliaria do the same thing.

A few experiments focused on chemical or biological weed control have been conducted. Cui *et al.*[16] reported the use of citric acid to counter the growth of epiphytes, especially Enteromorpha, in Porphyra farms. The use of various invertebrates to check weeds has also been reported (*e.g.* the isopod *Idotea baltica*[80]; the amphipod *Gammarus lawrencianus*[80]; the snails *Trochus niloticus*[61], Lacuna, Calliostoma and Mitrella[58]; and the opistobranch *Stylocheilus longicauda*[61]). Fish such as *Acanthurus triostegus* have been used experimentally for the same purpose[61], but they also eat the tips of crop plants.

In Hawaii, Gracilaria outplanted on reefs is kept free of epiphytes by the shifting sand in the habitat[21].

Harvesting

In most places most crops are usually harvested manually. The harvest, hand-picked from the nets, lines or ropes attached to rafts, is collected in shallow boats, with the harvesters either in the boats or wading in the water, depending on the crop, method of cultivation, season, geographical location and water depth. Where the old methods of stone planting, rock clearing, sticking of fragments in reefs, or placement of fragments attached to coral pieces on the bottom are still in use, divers pick the mature plants[5], a practice that is declining.

Only in the advanced nori industry in Japan is mechanized harvesting widely used[95]. Several types of mechanical harvesters are in use, each especially designed for a particular method of cultivation[98]. Basically, they consist of a rotary cylinder with blades fixed at the periphery, such as in old-fashioned lawn mowers. Each harvester is placed on a boat with the long axis parallel to that of the boat. The line or net on which the plants grow is lifted out of the water on one side of the boat and placed a few centimeters above the blades (and kept there by rollers) with the plants hanging downwards within reach of the blades. The turning blades cut the plants a few centimeters from their bases. The boat proceeds sidewise along the line or net, and the harvest accumulates on its bottom. The line or net is lowered back into the water on the opposite side of the boat for further growth until the next harvest. Porphyra may be harvested 3–4 times from a net, usually once a fortnight (a tidal cycle). During this period it grows 10–15(–20) cm in length, and is cut back to about 3–4 cm during each harvest[4,60]. After 3–4 harvests, the nets are replaced. Monstroma, when cultivated *per se*, is harvested in a similar manner three times per net, with the nets then being replaced[4].

For Laminaria a method of harvest by tip-cutting during the growing season was developed in northern China[14]. This serves two additional purposes: it prevents losses due to tearing parts of the thallus, and it enhances the growth of the plant. Consequently, the yield increases by 12–15%[94]. Since the intercalary meristem is situated at the base of the blades, their tips are the oldest parts. Up to a third of the blade is removed with every harvest. The time interval between harvests depends on the local conditions governing growth rates. Thinning the canopy improves light conditions and water movement. It allows more light to penetrate to the lower plants, compensating for the loss of photosynthesizing area taken with the harvest. It also allows the remaining parts of the plants to float at an angle, rather than hang vertically, so that they capture more light[87,94]. The last harvest at the end of the growing season takes place in China in June–July[86,87], when the plants may have reached a length of 3–4 m[87].

Undaria is harvested only at the end of the growing season (in China in March)[87], when it is *ca.* 1 m long[60].

Eucheuma in shallow farms in the Philippines is harvested once a month by manually breaking off portions of the plants, leaving plant masses a few centimeters in diameter to resume growth and produce the next harvest[23]. In some places, where Eucheuma is cultivated on ropes, the ropes are untied from the poles at harvest time, hauled by canoes to shaded, comfortable places to be stripped of the crop, and returned after reloading with fresh seeding material to the cultivation areas[21]. In China, bottom-cultivated Eucheuma is harvested once a year, with divers collecting the entire crop[86,87]. Gracilaria in ponds in Taiwan is harvested either by hand-picking or by nets, every 30–35 days in summer and every 45 days in winter. During the summer it is sometimes possible to harvest every ten days[81]. A third to a half of the plant material harvested is removed, and the remainder is divided into small pieces and returned to the ponds as seeding material[15].

Breeding and stock improvement

As long as natural seeding was the rule in the Porphyra, Laminaria or Undaria industries, the variability of the cultivated plants was similar to that in the natural populations. The development of artificial seeding techniques enabled the improvement of stock by selection and breeding. In Japan Porphyra strains have been selected from among the wild types on the basis of yield, quality, resistance to diseases, *etc.*[98] Many of these selected strains are now cultivated commercially. Hybridization between different types of local and foreign species has also been done in Japan[98] as well as in China[100]. However, Yoshida and Akiyama[98] stated that no commercially important strain has emerged from such hybridization. By contrast, in China hybridization between different types of *Laminaria japonica* and subsequent selection of the hybrids resulted in 1962 in the development of the Haiqing no. 1 strain[87] (Hai Ching no. 1)[14], which proved to produce higher yields but, more important, to be tolerant to higher temperatures than were previously cultivated strains. This enabled the extension of Laminaria cultivation further south, to cover the entire Chinese coast. Later, the same techniques resulted in additional commercial varieties, Haiqing no. 2 and no. 3[87]. X-ray mutagenesis has also been adopted, resulted in two commercial varieties with high yield and high iodine content, which were introduced into large areas[17,87]. This work is being continued, and more and more commercial varieties have been produced, such as the high-yielding Haiza no. 1, which was introduced into wide areas of northern China from 1980 onwards[38].

Among other sea vegetables, breeding through hybridization has not yet reached a commercial level, although Saito[76] performed interspecific hybridization in Undaria and obtained strains that could stand high temperatures better than either parent.

Stock improvement of the vegetatively propagated Eucheuma in the Philippines went along different lines. Here, selection was performed through early removal of the less desirable types by harvesting, and their replacement with vegetative portions of faster growing types[20,22,23]. The Tambalang strain is the result of such selection[21].

Strain selection of sea vegetables that are not yet cultivated on a full commercial scale has also been attempted, and in a few cases preliminary success was achieved. The fast growing, usually infertile, kappa-carrageenan-producing T4 strain of *Chondrus crispus* has been isolated and extensively studied[51]. Selection for infertile lambda-carrageenan-producing strains of Chondrus was also performed[11], as well as for strains capable of growing in warmer water than can most other ones[13]. Strains of the agarophyte *Gracilaria tikvahiae* superior to the best of the wild types were obtained through mutagenesis and selection[20].

Diseases

Diseases, especially those caused by pathogens, are not yet as great a problem among sea vegetables as they are among terrestrial crops. With more and more genetically uniform varieties introduced into cultivation, however, they are bound to become a greater problem. At present, only the more established crops, Porphyra and Laminaria, and to a lesser extent Undaria, are sometimes seriously affected by diseases. Naturally, more diseases have been encountered in crops that have been under cultivation longer. Therefore, the number of known Porphyra diseases is larger than those known for Laminaria. Fewer still are known for Undaria. Diseases affect both seeding material grown indoors and the crops cultivated out-of-doors. Most diseases encountered in sea vegetable farms are ecophysiological in nature. These are physiological disturbances caused by environmental conditions. Only a few are caused by pathogens.

Among diseases of Porphyra caused by pathogens, the most common and most serious is red rot[40,97], or red wasting disease[40,87], which attacks the crop in both Japan and China. It is caused by the fungus *Pythium porphyrae*[39], may spread quickly by zoospores, and may completely wipe out a crop. It has also been reported from natural populations of several Porphyra species in other parts of the world (caused by *P. marinum* on the Pacific and Atlantic coasts of North America[29,40] and in the Mediterranean[3]) and is therefore a threat to Porphyra cultivation everywhere. Other fungi, algae or bacteria sometimes infect Porphyra or Laminaria, causing damage to crops. Among these are Olpidiopsis, Leucothrix, benthic diatoms[98], or Pseudomonas[87] in Porphyra; and Pseudomonas[10], Micrococcus or mycoplasma-like organisms[87] in Laminaria. Some of them (such as Leuchothrix[98]) infect weakened plants and others may pass with seeding material (such as Olpidiopsis, which passes through the carpospores of Porphyra[98]) and affect the next generation. The brown alga *Streblonema aecidioides* infects Undaria in Japan, lowering the quality of the product[97]. In other infectious diseases, the pathogen is still unknown[33]. Only a small number of them can be treated by drugs (e.g. Pseudomonas, causing the fall-off disease of Laminaria summer seedlings, which may be treated with erythromycin[10], or the bacteria causing yellow spot disease of the conchocelis phase of Porphyra, where dihydrostreptomycin helps[98]. Both are treated in the indoor seeding tanks). A number of marine algae are infected by viruses[18,50]. No commercial crop has hitherto been reported among them.

Many diseases are caused by physiological disorders resulting from unfavorable environmental conditions, for example, white rot, hole rot, tumor, bud damage[98], or green disease[47] in Porphyra; green rot and white rot[87] in Laminaria; or the 'ice-ice' phenomenon in Eucheuma[23,88]. Causes of such diseases may be the lack of light (green rot of Laminaria[87]) or excess light (white rot of Laminaria[87]); a low supply of nutrients, especially nitrogen or phosphorus[88]; very low temperatures, especially in exposed Porphyra, or too high temperatures; too calm seas, resulting in stagnant water; polluted seawater or air (to exposed Porphyra); quick changes in the environment; mechanical impact of solid particles; too low or too high salinity, *etc.*; or a combination of a number of factors. The causes of other diseases, such as the detachment of sporophytes of Undaria, are unknown[4].

Treatment of the diseases depends on knowledge of their causes. Exposing Porphyra to air for longer periods might help in some cases (exposure period of 4–5 h day^{-1} is recommended as a preventive measure for a number of diseases in Japan, including red rot[98]), whereas lowering the nets or lines for longer immersion may help in others. Nitrogenous fertilization is also used successfully in some cases. Sometimes the only way to reduce damage is an early harvest. Chemical drug treatment seems to be restricted to use in tank culture of seeding material.

Pests

Several herbivores damage sea vegetable crops. The most common are fish and sea urchins. In south China herbivorous fish sometimes graze upon cultured Porphyra to an extent requiring the use of protective nets. A similar approach was taken in the Philippines to protect Eucheuma farms from grazing fish[74]. In the past urchins caused damage to oversummering fronds of *Laminaria japonica* in China. Since summer seeding was introduced, this has no longer been a serious problem[87]. In *Caulerpa racemosa* ponds in Taiwan, Cerithium snails may multiply to levels damaging the crop and should be removed[4].

In Japan, the hydrozoan *Sertularella miurensis* or the bryozoan *Membranipora serrilanella* sometimes settle on Laminaria blades, lowering product quality. The problem is more serious in the second year of Laminaria growth, since the bryozoan settles at the end of the first year. Early harvest of the Laminaria by the end of their first year, possible for plants grown from summer seeding, prevents much of the damage[37].

In tropical waters herbivorous fishes, especially siganids and some sea urchins[20,45] (*e.g.* Diadema, Tripneustes[20]), may sometimes completely destroy the Eucheuma crop in bottom plantations[19,20]. Planting on horizontal nets or lines situated *ca.* 50 cm above the bottom reduces damage by these grazers[19], so that in Eucheuma farms this is usually not a serious problem[67]. Other measures taken against fish and urchins in Euchema farms are: hanging small bags containing paradichlorobenzene around net units to deter fish, or causing heavy clouds of lime to drift among the plants to kill or drive away urchins[20].

Experiments with some commercially important algae indicated that small creatures like isopods, amphipods or small snails, when abundant, can considerably damage crops. In tank cultures of the agarophyte *Gracilaria tikvahiae* (as *G. foliifera*) and of the carrageenophyte *Agardhiella subulata* (as *Neoagardhiella baileyi*), the isopod *Idotea baltica* and the amphipod *Amphithoe valida* consumed considerable portions of the algae when their productivity was very low[63]. Idotea preferred Gracilaria over Agardhiella, whereas Amphithoe preferred Agardhiella[63]. Similarly, *Idotea baltica*, the snail *Lacuna vincta*, and the amphipod *Gammarus oceanicus* caused considerable damage to the carrageenophyte *Chondrus crispus* in flask experiments[79].

In cultures at sea of the carrageenophyte *Iridaea cordata*, the snails *Lacuna variegata* and *Lirularia lirulata* perforated blades of the alga to such an extent that considerable portions were lost to wave action[55]. Such pests are susceptible to a number of insecticides[79]. These may be used with tank cultivation, but in farms in the sea they can harm the environment. Fish readily consumed crustacean pests, but snails had to be crushed prior to feeding them to fish[79]. The latter technique can hardly be considered feasible for control of snails on commercial farms.

Experimental outdoor cultivation

There is very great interest in the cultivation of marine algae in many parts of the world, which is reflected in intensive experimental work. Much of this work was done in North America and was recently reviewed by Mathieson[51] (including studies in the U.S. possessions and Trust Territories in the Pacific). Most of it will therefore not be mentioned here. Much of the experimental work has been carried out indoors. This, as well, will mostly not be referred to here. The experiments have in some cases reached the pilot plant level, with some of the projects being proven successful (*cf.*[51]). Nevertheless, in the West none of them has yet become a fully commercial enterprise.

In addition to North America, investigations are underway also in Central and South America (*e.g.* the West Indies[83], Brazil[96]), Europe (*e.g.* England[39], France[6], Italy[31,82]), Asia (*e.g.* Israel[28], India[9,49,68,69,72], Indonesia[84]), Africa (*e.g.* Djibouti[7], Tanzania[54]), and the Pacific Islands (*e.g.* Hawaii[21,53,74], Micronesia[61]).

A relatively large number of useful algae have been investigated. They include members of at least 25 genera, for a number of which several species are involved. Of these, the majority have not been commercially cultivated before, although natural populations of most have been utilized. These genera are the following: Macrocystis[25,30,62,65,91,93], Gracilaria[21,28,43,61,72,75,82,83,96], Chondrus[6,58], Iridaea[1,55–58,89], Gigartina[58,89,90], Eucheuma[71,84], Hypnea[28,43,53,96], Palmaria[53,58,89], Porphyra[58,89], Laminaria[27,39], Gelidium[91], Alaria[39], Saccorhiza[39], Sargassum[9,91], Gelidiella[49,68,69], Acanthophora[96], Pterocladia[28,96], Cymathere[27], Agardhiella (as Neoagardhiella)[75,89], Plocamium[89], Schizymenia[89], Callophyllis[89], Farlowia[89], and Prionitis[89]. While most of the sea vegetables currently cultivated on a commercial scale are food crops, almost all algae listed above are industrially important. In addition to the conventional purposes for which sea vegetables are cultivated commercially, *i.e.* for food and for the extraction of phycocolloids, investigations aimed at the production of biomass for energy conversion[39,43,61,75] or the purification of waste waters by macroalgae have been carried out[31,43,61,75,82].

The cultivation methods involved in the experimental work include those used in commercial cultivation[1,7,27,39] *cf.*[51,55–58,68,69,72,87,91]. However, a variety of other methods were developed, most of them aimed at a greater control of cultivation conditions. Consequently, large plastic cylinders suspended in the sea were used[68], as well as small to very large and complex constructions located offshore, where water depth is several hundreds of meters[30,93], and a variety of raceways *e.g.*[21,75] and tanks[1,6,43,89,90,96] located on land.

A wide variety of subjects have been studied. Some, like those dealing with fertilization, seeding, or stock improvement, have been referred to above. Others, such as control of the different stages of the life history – induction of reproduction, germination and growth – and questions concerning the storage of 'seed' material[71] or genetic manipulation to obtain 'domesticated' plants[12,62], have not. Most of the experiments have concerned study of the influence of various factors on the growth rates and/or yields of the different sea vegetables studied. The usual growth rates observed were around 3.5–4% daily. However, considerably higher growth rates were shown by some algae under favorable conditions (5.8% day^{-1} by Macrocystis[90], 7.9% day^{-1} by Sargassum[90], 7.6% and 8.3% day^{-1} by two species of Gracilaria[61]). Some of the higher yields obtained were: 12–17 g dw m^{-2} day^{-1} for Gracilaria[43], 16–22 g dw m^{-2} day^{-1} for Iridaea[58], or 17 g dw m^{-2} day^{-1} for Agardhiella[75] (as Neoagardhiella), yearly average values. These values compare well with yields of microalgae or fast-growing terrestrial crops[43]. During the growing season, higher values of more than 40 g dw m^{-2} day^{-1} were obtained[75].

Prospects of sea vegetable thalassonomy

The cultivation of sea vegetable food crops, which markedly developed during the past few decades, seems to have presently closely edged up to its widest potential, with only a limited further expansion to be expected, based on developments in the Far East market. This will be true unless an as yet unforeseen major change takes place in the eating habits of the populations in the developed countries of the world, caused either by a shift of preferences or by necessity. By contrast, a considerable development and expansion of the cultivation of sea vegetable industrial crops is much more likely, as the demand for their products is world-wide rather than being restricted to only a small number of countries.

This future development of thalassonomy will no doubt depend on breeding of truly domesticated varieties of superior yields, quality of product, and success under prescribed cultivation conditions. The best of wild types selected from the natural populations will no longer be satisfactory.

The rapid accumulation of knowledge concerning the environmental requirements of the industrial sea vegetables, the widening understanding of the chemical structures of the industrially important substances produced by them and its bearing on their physical properties, and the experimental effort put into conventional and less conventional genetic research will provide the tools which future breeders of such varieties may use for creation of the appropriate genotypes and for a soundly based selection. This achieved, obtaining the varieties on which to base successful thalassonomy will not be too far off. This alone, however, will not guarantee the success of sea vegetable thalassonomy, as there is still much to be learned about cultivation techniques as well.

Although sea vegetable thalassonomy started on the basis of trial and error, much of its success in the past, especially during the last few decades, is the outcome of achievements of scientific research. So also are the developments mentioned, expected in the future, that will turn it into a modern, sophisticated enterprise. Yet it should be borne in mind that, along with the developments in this field, new problems are bound to arise and thus, intensive scientific research will have to accompany the developing thalassonomy for many years after it becomes such an enterprise.

References

1 Adams R W and Austin A 1979 Potential yields of *Iridaea cordata* (Florideophyceae) in natural and artificial populations in the northeast Pacific. Proc. Int. Seaweed Symp. 9, 499–507.

2 Akiyama K and Kurogi M 1982 Cultivation of *Undaria pinnatifida* (Harvey) Suringar, the decrease in crops from natural plants following crop increase from cultivation. Bull. Tohoku Reg. Fish. Res. Lab. 44, 91–100.

3 Aleen A A 1980 *Pythium marinum* Sparrow (Phycomycetes) infesting *Porphyra leucostica* Thuret in the Mediterranean Sea. Bot. Mar. 23, 405–407.

4 Bardach J E, Ryther J H and McLarney W O 1972 Aquaculture. The Farming and Husbandry of Freshwater and Marine Organisms. Wiley-Interscience. New York.

5 Bonotto S 1976 Cultivation of plants. Multicellular plants. *In* Marine Ecology, Vol. 3, pt. 1, Ed. O Kinne. pp 467–501. J. Wiley and Sons, London.

6 Braud J P and Delepine R 1981 Growth response of *Chondrus crispus* (Rhodophyta, Gigartinales) to light and temperature in laboratory and outdoor tank culture. Proc. Int. Seaweed Symp. 10, 553–558.

7 Braud J P and Perez R 1979 Farming on pilot scale of *Eucheuma spinosum* (Florideophyceae) in Djibouti waters. Proc. Int. Seaweed Symp. 9, 533–539.

8 Cai P X 1983 Summering of Laminaria sporophyte in south China and the cultivation of summer sporelings. Abstr. Int. Seaweed Symp. 11, 27.

9 Chauhan V D and Taqui Khan M M 1983 Marine algae Sargassum, their utilization and cultivation. Abstr. Int. Seaweed Symp. 11, 31.

10 Chen D, Liu X Y, Liu X Z, Yu Y D, Yang Z H and Qui S H 1983 Studies on alginic acid-decomposing bacteria. III. The cause of rot disease and detaching of Laminaria sporophytes in sporeling culture stations and their preventive measures. Abstr. Int. Seaweed Symp. 11, 33.

11 Chen L C M, Morgan K and Simpson F J 1982 Selection of tetrasporophytes of Irish moss. Bot. Mar. 25, 35–36.

12 Cheney D P 1982 Genetic modification in seaweeds: applications to commercial utilization and cultivation. *In* Biotechnology in the Marine Sciences, Proc. MIT Sea Grant Conf., pp 1–15, Wiley Press, New York.

13 Cheney D, Mathieson A and Schubert D 1981 The application of genetic improvement techniques to seaweed cultivation: I. Strain selection in the carrageenophyte *Chondrus crispus*. Proc. Int. Seaweed Symp. 10, 559–567.

14 Cheng T H 1969 Production of kelp – a major aspect of China's exploitation of the sea. Econ. Bot. 23, 215–236.

15 Chiang Y M 1981 Cultivation of *Gracilaria* (Rhodophycophyta, Gigartinales) in Taiwan. Proc. Int. Seaweed Symp. 10, 569–574.

16 Cui G F, Zheng Q S, Gu S B and Wang X K 1983 A preliminary account on preventing weed algae from growing on the nets of *Porphyra yezoensis*. Abstr. Int. Seaweed Symp. 11, 52.

17 Dawes C J 1981 Marine Botany, J. Wiley and Sons, New York.
18 Dodds J A 1979 Viruses of marine algae. Experientia 35, 440–442.
19 Doty M S 1973 Farming the red seaweed, Eucheuma, for carrageenans. Micronesica 9, 59–73.
20 Doty M S 1979 Status of marine agronomy, with special reference to the tropics. Proc. Int. Seaweed Symp. 9, 35–58.
21 Doty M S 1980 Outplanting Eucheuma species and Gracilaria species in the tropics. *In* Pacific Seaweed Aquaculture, Proc. Symp. Useful Algae. Eds. I A Abbott, M S Foster and L F Eklund. pp 19–22. California Sea Grant College Program, Inst. Mar. Resources, Univ. Calif., La Jolla.
22 Doty M S and Alvarez V B 1973 Seaweed farms: a new approach for U.S. industry. Proc. 9th Ann. Conf. Mar. Tech. Soc., pp 701–708.
23 Doty M S and Alvarez V B 1975 Status, problems, advances and economics of Eucheuma farms. Mar. Tech. Soc. J. 9(4), 30–35.
24 Drew K M 1949 Conchocelis phase in the life history of *Porphyra umbilicalis* (L.) Kutz. Nature 164, 748–749.
25 Druehl L D 1983 Morphological and physiological responses of *Macrocystis pyrifera* to nitrate enrichment. Abstr. Int. Seaweed Symp. 11, 60.
26 Druehl L D and Boal R 1981 Manipulation of the laminarialean life-cycle and its consequences for kombu mariculture. Proc. Int. Seaweed Symp. 10, 575–580.
27 Druehl L D and Lloyd K E 1983 Long line cultivation of three species of Laminariaceae. Abstr. Int. Seaweed Symp. 11, 61.
28 Friedlander M and Lipkin Y 1982 Rearing of agarophytes and carrageenophytes under field conditions in the eastern Mediterranean. Bot. Mar. 25, 101–105.
29 Fuller M S, Lewis B and Cook P 1966 Occurrence of Pythium sp. on the marine alga Porphyra. Mycologia 58, 313–318.
30 Gerard V and North W 1981 Kelp growth on an ocean farm in relation to fertilizing. Proc. Int. Seaweed Symp. 10, 581–586.
31 Giaccone G, Princi M, Feoli E, Lokar Cassini L, Rizzi Longo L and Tortul V 1979 Valutazione delle risorse vegetali lagunari del Basso Tirreno e sperimentazione di coltivazione controllata dell'alga rossa *Gracilaria verrucosa* e di altre agarofite in Sicilia. Atti Conv. Sci.
32 Guist G G Jr, Dawes C J and Castle J R 1982 Mariculture of the red seaweed, *Hypnea musciformis.* Aquaculture 28, 375–384.
33 Guo X K 1983 A kind of Laminaria disease in Dalian. Abstr. Int. Seaweed Symp. 11, 298.
34 Hasegawa Y 1966 Cultivation of Laminaria in Japan. 11th Pacif. Sci. Congr., Tokyo. Sect. VII Fish., Symp. 36. Algae in the Pacific. Aug. 27 1966, p 5.
35 Hasegawa Y 1972 Forced cultivation of Laminaria. Proc. Int. Seaweed Symp. 7, 391–393.
36 Hasegawa Y 1976 Progress of Laminaria cultivation in Japan. J. Fish. Res. Board Can. 33, 1002–1006.
37 Hasegawa Y 1978 Immediate problems facing Laminaria cultivation. Bull. Hokkaido Reg. Fish. Res. Lab. 43, 89–92.
38 Jiang B Y, Qu W Y and Tang Z J 1983 The breeding and utilization of a new variety of *Laminaria japonica* Aresch. Abstr. Int. Seaweed Symp. 11, 104.
39 Kain J M and Holt T J 1983 Cultivation of Laminariales species on ropes in the Irish Sea. Abstr. Int. Seaweed Symp. 11, 106.
40 Kazama F Y 1979 Pythium 'red rot disease' of Porphyra. Experientia 35, 443–444.
41 Kurogi M 1980 Biological research on seaweed. *In* Aquaculture in Shallow Seas: Progress in Shallow Sea Culture, Ed. T Imai, pp 3–56. A. A. Balkema, Rotterdam.
42 Laite P and Ricohermoso M 1981 Revolutionary impact of Eucheuma cultivation in the South China Sea on the carrageenan industry. Proc. Int. Seaweed Symp. 10, 595–600.
43 Lapointe B E, Williams L D, Goldman J C and Ryther J H 1976 The mass outdoor culture of macroscopic marine algae. Aquaculture 8, 9–21.
44 Li R Z, Chong R Y and Meg Z C 1983 Preliminary study on raft cultivation of *Gracilaria verrucosa* and *Gracilaria sjoestedtii.* Abstr. Int. Seaweed Symp. 11, 130.

45 Lim J R H 1983 Effects of the environment on the growth of Eucheuma in the Danajon Reef Farm, northern Bohol, Philippines. Abstr. Int. Seaweed Symp. 11, 137.
46 Lim J R and Porse H 1981 Breakthrough in the commercial culture of *Eucheuma spinosum* in northern Bohol, Philippines. Proc. Int. Seaweed Symp. 10, 601–606.
47 Lin D H and Lin Y S 1983 Studies on the green disease of *Porphyra haitanensis* Chang et Zheng. Abstr. Int. Seaweed Symp. 11, 139.
48 Liu S J and Zhuang P 1983 The commercial cultivation of Eucheuma in China. Abstr. Int. Seaweed Symp. 11, 147.
49 Mairh O P, Thomas P C, Ramavat B K and Sreenivasa Rao P 1979 Fertilizer pellets and their application in the field cultivation of *Gelidiella acerosa* (Forssk.) Feld. et Hamel. Abstr. Int. Symp. Marine Algae of the Indian Ocean Region, Central Salt and Marine Chemicals Research Institute, Bhavnagar, p. 25.
50 Martin E L and Benson R L 1982 Algal viruses, pathogenic bacteria and fungi: introduction and bibliography. *In* Selected Papers in Phycology II, Eds. J R Rosowski and B C Parker, pp 792–798.
51 Mathieson A C 1981 Seaweed cultivation: a review. Proc. 6th U.S.–Japan Meeting on Aquacult. U.S. Dept. Commer., Rep. NMFS Circ. 442, 25–66.
52 Mathieson A C and North W J 1982 Algal aquaculture: introduction and bibliography. *In* Selected Papers in Phycology II. Eds. J R Rosowski and B C Parker. pp 773–778. Phycological Society of America, Lawrence, Kansas.
53 Mshigeni K E 1976 Field cultivation of Hypnea (Florideophyceae) spores for carrageenan: prospects and problems. Bot. Mar. 19, 227–230.
54 Mshigeni K E 1983 The red algal genus Eucheuma (Gigartinales, Solieriaceae) in east Africa: an underexploited marine source. Abstr. Int. Seaweed Symp. 11, 182.
55 Mumford T F Jr 1977 Growth of Pacific northwest marine algae on artificial substrates – potential and practice. *In* The Marine Plant Biomass of the Pacific Northwest Coast. Ed. R W Krauss. pp 139–161. Oregon State University Press, Corvallis.
56 Mumford T F Jr 1977 Some biological, legal, social, and economic aspects of the culture of the red alga *Iridaea cordata* on nets in Puget Sound. Proc. Conf. The Use, Study and Management of Puget Sound, Washington Sea Grant Publ., pp 221–229.
57 Mumford T F Jr 1979 Field and laboratory experiments with *Iridaea cordata* (Florideophyceae) grown on nylon netting. Proc. Int. Seaweed Symp. 9, 515–523.
58 Mumford T F Jr 1980 Seaweed culture in Washington and British Columbia: potential and practice. Proc. Int. Symp. Coastal Pacific Marine Life. Western Washington Univ., Bellingham, pp 13–30.
59 Mumford T M Jr and Waaland J R 1980 Progress and prospects for field cultivation of *Iridaea cordata* and *Gigartina exasperata*. *In* Pacific Seaweed Aquaculture, Proc. Symp. Useful Algae. Eds. I A Abbott, M S Foster and L F Eklund. pp 92–105, California Sea Grant College Program, Inst. Mar. Resources, Univ. Calif., La Jolla.
60 Naylor J 1976 Production, trade and utilization of seaweeds and seaweed products. FAO Fish. Tech. Pap. 159.
61 Nelson S G, Tsutsui R N and Best B R 1980 A preliminary evaluation of the mariculture potential of Gracilaria (Rhodophyta) in Micronesia: growth and ammonium uptake. *In* Pacific Seaweed Aquaculture, Proc. Symp. Useful Algae. Eds. I A Abbott, M S Foster and L F Eklund, pp 72–79, California Sea Grant College Program, Inst. Mar. Resources, Univ. Calif., La Jolla.
62 Neushul M 1977 The domestication of the giant kelp, Macrocystis, as a marine plant biomass producer. *In* The Marine Plant Biomass of the Pacific Northwest Coast. Ed. R W Krauss. pp 163–181, Oregon State University Press, Corvallis.
63 Nicotri M E 1977 The impact of crustacean herbivores on cultured seaweed populations. Aquaculture 12, 127–136.
64 North W J 1977 Algal nutrition in the sea management possibilities. *In* The Marine Plant Biomass of the Pacific Northwest Coast. Ed. R W Kruass. pp 215–230. Oregon State University Press, Corvallis.
65 North W, Gerard V and McPeak R 1981 Experimental fertilizing of coastal Macrocystis beds. Proc. Int. Seaweed Symp. 10, 613–618.

66 Ohno M 1979 Ecology of cultivated sea weeds of warm water in Japan. Abstr. Int. Symp. Marine Algae of the Indian Ocean Region, Central Salt and Marine Chemicals Research Institute, Bhavnagar, p. 24.
67 Parker H S 1974 The culture of the red algal genus Eucheuma in the Philippines. Aquaculture 3, 425–439.
68 Patel J B, Gopal B V, Nagulan V R, Subbaramaiah K and Thomas P C 1979 Experimental field cultivation of *Gelidiella acerosa* at Ervadi, India. Abstr. Int. Symp. Marine Algae of the Indian Ocean Region, Central Salt and Marine Chemicals Research Institute, Bhavnagar, pp 24–25.
69 Patel N B, Gopal B V, Nagulan V R, Subbaramaiah K, Thomas P C and Mehta D J 1979 Experimental field cultivation of *Gelidiella acerosa* at Krusadai Island. Int. Symp. Marine Algae of the Indian Ocean Region. Late Abstr. No. 2.10, mimeographed.
70 Patwary M U and van der Meer J P 1983 Improvement of *Gracilaria tikvahiae* (Rhodophyceae) by genetic modification of thallus morphology. Aquaculture 33, 207–214.
71 Polne M, Neushul M and Gibor A 1981 Studies in domestication of *Eucheuma uncinatum*. Proc. Int. Seaweed Symp. 10, 619–624.
72 Raju P V and Thomas P C 1971 Experimental field cultivation of *Gracilaria edulis* (Gmel.) Silva. Bot. Mar. 14, 71–75.
73 Ren G Z, Wang J C and Chen M Q 1983 Studies on the fixed raft method of Gracilaria cultivation. Abstr. Int. Seaweed Symp. 11, 211.
74 Richohermoso M A and Deveau L E 1979 Review of commercial propagation of Eucheuma (Florideophyceae) clones in the Philippines. Proc. Int. Seaweed Symp. 9, 525–531.
75 Ryther J H, DeBoer J A and Lapointe B E 1979 Cultivation of seaweeds for hydrocolloids, waste treatment and biomass for energy conversion. Proc. Int. Seaweed Symp. 9, 1–16.
76 Saito Y 1972 On the effect of environmental factors on morphological characteristics of *Undaria pinnatifida* and the breeding of hybrids in the genus Undaria. *In* Contributions to the Systematics of Benthic Marine Algae of the North Pacific. Eds. I A Abbott and M Kurogi. pp 117–132, pl 1, Japanese Society of Phycology, Kobe.
77 Saunders R G and Lindsay J G 1979 Growth and enhancement of the agarophyte Gracilaria (Florideophyceae). Proc. Int. Seaweed Symp. 9, 249–255.
78 Segi T and Kida W 1961 On the relation between distribution of early germlings of Monostroma and tidal current in the culture ground. Bot. Mar. 2, 223–230.
79 Shacklock P F and Croft G B 1981 Effect of grazers on *Chondrus crispus* in culture. Aquaculture 22, 331–342.
80 Shacklock P F and Doyle R W 1983 Control of epiphytes in seaweed cultures using grazers. Aquaculture 31, 141–151.
81 Shang Y C 1976 Economic aspects of Gracilaria culture in Taiwan. Aquaculture 8, 1–7.
82 Simonetti G, Giaccone G and Pighnatti S 1970 The seaweed *Gracilaria confervoides*, an important object for autecologic and cultivation research in the northern Adriatic Sea. Helgolander Wiss. Meereunters. 20, 89–96.
83 Smith A H, Nichols K and McLachlan J 1983 Cultivation of seamoss (Gracilaria) in St. Lucia, West Indies. Abstr. Int. Seaweed Symp. 11, 229.
84 Soegiarto A 1979 Indonesian seaweed resources: their utilization and management. Proc. Int. Seaweed Symp. 9, 463–471.
85 Torii S and Kawashima S 1979 Experiments on propagation of Laminaria (Phaeophyceae). I. Propagation of *L. japonica* by plastic pipe (netron) in sandy substrate. Proc. Int. Seaweed Symp. 9, 473–478.
86 Tseng C K 1981 Marine phycoculture in China. Proc. Int. Seaweed Symp. 10, 123–152.
87 Tseng C K 1981 Commercial cultivation. *In* The Biology of Seaweeds, Botanical Monographs. Vol 17. Eds. C S Lobban and M J Wynne. pp 680–725, Univ. Calif. Press, Berkeley.
88 Uyenco F R, Saniel L S and Jacinto G S 1981 The 'ice-ice' problem in seaweed farming. Proc. Int. Seaweed Symp. 10, 625–630.
89 Waaland R J 1977 Growth of Pacific northwest marine algae in semi-closed culture. *In* The Marine Plant Biomass of the Pacific Northwest Coast. Ed. R W Krauss, pp 117–137, Oregon State University Press, Corvallis.

90 Waaland J R 1979 Growth and strain selection in *Gigartina exasperata* (Florideophyceae). Proc. Int. Seaweed Symp. 9, 241–247.
91 Wheeler W N, Neushul M and Harger B W W 1981 Development of a coastal marine farm and its associated problems. Proc. Int. Seaweed Symp. 10, 631–636.
92 Wheeler W N, Neushul M and Woessner J W 1979 Marine agriculture: progress and problems. Experientia, 35, 433–435.
93 Wilcox H A 1977 The U.S. Navy's Ocean Food and Energy Farm Project. *In* Ocean Energy Resources. Ed. N T Monney. pp 83–104, Am. Soc. Mechanical Engin., New York.
94 Wu C Y, Zheng S Q and Tseng C K 1981 Tip-cutting of fronds as a means of increasing production in Laminaria aquaculture. Proc. Int. Seaweed Symp. 10, 637–642.
95 Yamamoto Nori Manufacturing Co. 1978 Asakusa-nori. 16 mm film. Gakken Co. Ltd., Tokyo, 25 min.
96 Yoneshigue-Braga Y and Baeta Neves M H C 1981 Preliminary studies on mass culture of *Gracilaria* sp. using different nutrient media. Proc. Int. Seaweed Symp. 10, 643–648.
97 Yoshida T and Akiyama K 1979 Streblonema (Phaeophyceae) infection in the frond of cultivated Undaria (Phaeophyceae). Proc. Int. Seaweed Symp. 9, 219–223.
98 Yoshida T and Akiyama K 1980 Techniques of seaweed culture. *In* Aquaculture in Shallow Seas: Progress in Shallow Sea Culture. Ed. T Imai. pp 66–96, A. A. Balkema, Rotterdam.
99 Zhang D M, Miao G R and Lin Q S 1983 A summary of cultivation techniques in summer Laminaria seedlings. Abstr. Int. Seaweed Symp. 11, 280.
100 Zhang Y J, Yang Y X, Wang S R and Wang Q Y 1983 A preliminary study of the breeding and genetics of *Porphyra yezoensis* Ueda. Abstr. Int. Seaweed Symp. 11, 286.
101 Zhao H D and Zhang X C 1983 Isolation and cultivation of the vegetative cells *Porphyra yezoensis* Ueda and its potential use in commercial production. Abstr. Int. Seaweed Symp. 11, 304.

Section 3

Crop production with saline water

Plant and Soil 89, 187–198 (1985). Ms. BR 13

Salt-tolerant crops: origins, development, and prospects of the concept

EMANUEL EPSTEIN
Department of Land, Air and Water Resources, University of California, Davis, CA 95616, USA

Key words Breeding Genotypes Halophytes Saline water Salt-affected soils Selection Stress

Summary The genetic approach to the problems posed by salt-affected soils and water, *i.e.*, breeding crops resistant to salinity stress, is traced to two principal origins: the European ecological interest in halophytes, and the exigencies of growing crops in the arid and semi-arid lands of the American West. The point is made that breeding for resistance to salinity stress cannot be divorced from breeding for various other desirable traits of mineral plant nutrition and metabolism. A survey is conducted of the existing body of information on breeding for desiderata of mineral nutrition in general and salt tolerance in particular. The prospects of breeding crops for salt tolerance are discussed, with emphasis on a) its relation to breeding for resistance to other mineral stresses; b) field trials; c) collaboration between plant physiologists and geneticist-breeders; and d) extensive exploration of germplasm.

Introduction: the origins

The organizers of this symposium have asked me to present not so much an account of recent advances as a review of the origin and evolution of the concept that salt-tolerant crops can and should be developed, and some preview of what might reasonably be expected. There are two main intellectual traditions that have joined and contributed to the worldwide enterprise that this Biosalinity Workshop was a part of. The first is the European interest in the ecology and physiology of halophytes. Europe, being an amply watered continent (except for some of its southern regions), and the British Isles have well leached soils, so that there was little agricultural impetus for an interest in saline conditions. But around the turn of the century Schimper[42] in Germany and Tansley[46] in Britain became impressed by the different types of vegetation in various environments and soils. Plants in saline environments did not escape their attention, and thus the foundation was laid for the study of halophytes, Schimper stressing physiological approaches and Tansley, genetic differentiation and competition. Neither these nor other investigators at that time confined their attention to halophytes. They were interested in all geographical factors in plant distribution, including edaphic ones such as salinity.

American interest in the interplay between plants and saline soils had an altogether different origin. The early settlers on the eastern coast of the continent found climatic and soil conditions not radically different from those they were used to in Europe. That changed when the inexorable expansion brought settlers into the arid west, and places with names such as Salt River Valley, Bitter Lake, and Alkaline Flat appeared upon the maps.

The saline and alkali soils of the West were troublesome because the early settlers were farmers for the most part, not adventurers and gold miners as the mythology has it. To farm in the arid west was a novel challenge, and indeed a challenge that even knowledgeable persons did not believe possible to meet. William H. Brewer was a member of the first California State Geological Survey. He recorded his observations in letters subsequently published under the title, Up and Down California in 1860–1864[2]. Here is what he had to say about the San Joaquin Valley, today the foremost agricultural region of California (pp 202–204):

'The San Joaquin (pronounced San Waugh-keen') plain lies between the Mount Diablo Range and the Sierra Nevada – a great plain here, as much as forty to fifty miles broad, desolate, without trees save along the river, without water during nine or ten months of the year, and practically a desert. The soil is fertile enough, but destitute of water, save the marshes near the river and near the Tulare Lake . . . Barren, very barren, few trees – often one will have a prospect of a dozen or even twice that number of square miles without a tree or shrub large enough to be seen, the ground either entirely bare or with a very scanty vegetation of stunted grass and low weeds . . . The streams that form in these hills in the spring all sink when they enter the plain, and as summer advances they dry up farther and farther until they all disappear. Such is an immense region, such it must ever remain, supporting a scanty population.'

Competent observer that he was, Brewer failed to visualize the potential for irrigation that lay in the snowpack of the Sierra Nevada to the east.

E. W. Hilgard was Professor of Agriculture at the University of California from 1875 to 1906[23]. He studied another aspect of arid lands: the salinity and alkalinity of their soils[21]. He knew that by reclamation procedures including drainage and the use of amendments, such soils could be rendered suitable for agriculture, and indeed he was the foremost developer and proponent of such procedures.

Hilgard was very broadly informed and had wide-ranging interests and vision. He was aware of the reclamation of polder lands from the sea in northern Europe and expressed the hope that the lands of the Near East could once again become fertile. He would have relished seeing what has since been accomplished in reclamation in many parts of the world, our host country, Israel, prominent among them.

Finally, he had what now seem prophetic words to say about plant life and agricultural crops on such soils. He commented favorably on the use of Australian saltbushes, *Atriplex* spp., as fodder for stock in California (21, p. 469). As for crops, two of his collaborators, J. B. Davy and R. H. Loughridge, conducted a survey of the responses to salinity and alkalinity of about a hundred field crops, vegetables, and fruit trees (pp 466–484). Hilgard's conclusion from this survey was not overly optimistic: 'it may be said generally that the search for widely acceptable kinds has not been very successful' (p. 468). This is among the earliest scientific estimates of the feasibility of crop production on salt-affected soils. 'It will require more extended experience and experiments before any of these plants will be definitely adopted for propagation by farmers and stockmen' (p. 468).

During and after Hilgard's tenure at Berkeley the enterprise of irrigation agriculture expanded in California, and many scientists kept struggling with the problems posed by salt-affected soils, poor quality, saline irrigation water, and inadequate drainage. These problems were considered to be primarily in the domain of the soil scientist, the irrigation specialist, and the grower. When, in 1937, the United States Regional Salinity Laboratory, now the United States Salinity Laboratory, was established in Riverside, California, this focus was maintained, but crop evaluation of the kind already initiated in Hilgard's day was also pursued.

Development: the genetic approach

The key to existence in arid and semi-arid environments is water of good quality. The ever-growing demand for water for urban and industrial uses and for agriculture was the impetus for a conference that met in 1961 at Woods Hole, Massachusetts, under the auspices of the National Academy of Sciences-National Research Council to consider the problem of desalination. If efficient, cheap ways could be devised for getting nonsaline from saline water, many of the problems besetting arid regions could be overcome. The participants in this conference were for the most part chemists, physicists, and engineers. But James Bonner of the California Institute of Technology reminded the organizers that living cells very effectively concentrate ions from their aqueous media and maintain steep diffusion gradients across their membranes. Perhaps an opportunity could be made for someone to present evidence on this biological salt transport and 'desalination.' He was kind enough to recommend me for the job. He was aware of my interest in ion transport in plants in general and in the selectivity between sodium and potassium in saline environments in particular[9,13].

I chose to use the opportunity to assemble what evidence I could find on the genetic control of ion transport in plants, with a view to determining the feasibility of projects for selecting and breeding for salt tolerance: 'It is axiomatic in present-day physiology and biochemistry that specific physiological and biochemical performances depend ultimately on the synthesis of appropriate enzymes, and that the synthesis of enzymes in turn is gene-controlled. This should apply to those physiological performances, notably selective ion transport, involved in the ability of plants to thrive in saline substrates'[10]. Because my colleagues in the plant sciences were unlikely to read the proceedings of a desalination research conference, I was fortunate to be able to expand and update that paper for publication in the Annual Review of Plant Physiology. This I did in collaboration with R. L. Jefferies, then a visiting scientist and now at the University of Toronto[14]. 'The large varietal differences in salt transport and toleration that have been encountered make it likely that much is to be gained by an energetic pursuit of the possibilities of breeding for salt tolerance.' We noted that hardly anything had yet been done along this line.

Up to that time there existed only a few reviews on genotypic differences among plants on any aspect of mineral nutrition and metabolism, none with salt tolerance as their focus or impetus. There were, however, some prescient individual research papers. A notable one was by Lyon[29], published in 1941. He used cultivars of two tomato species, *Lycopersion esculentum* and *L. pimpinellifolium* and the F_1 generation from their cross. He observed differences in some characters in response to salinization of the nutrient solution, none in others. He wrote that to 'accompany proper soil management practices and more detailed studies of salt antagonisms and toxicity effects, it may be possible and desirable to select and breed plants for tolerance to saline conditions. Such a program should eventually involve not only the production of strains with the characteristic of general tolerance to high salt concentrations but the actual selection and breeding of strains for specific concentrations of two or more ions.'

It is from these diverse and scattered beginnings that the enterprise on which we are now engaged has grown. Our present thinking, as reflected by the contributions to this conference, is to the effect that a genetic dimension of selection and breeding has been added to the traditional approach of reclamation and drainage, essential as these measures will continue to be.

Selection and breeding for salt tolerance should not, indeed cannot, be separated from selection and breeding for other aspects of mineral metabolism and nutrition of plants. On the contrary, from the beginning it has been considered in that broader context. There are four reasons for continuing that connection. (a) The first is that salinity-tolerance breeders will have their scientific intellectual horizons broadened and their competence enhanced by contact with geneticist-breeders concerned with other aspects of mineral plant metabolism. (b) A more specific reason for such collaboration is that plants on salt-affected soils often are exposed to a number of adverse conditions which, strictly speaking, do not fall in the category of salinity or sodicity. Iron deficiency chlorosis is a prime example[3]. It has been shown that this and other difficulties can be tackled by means of selection and breeding. (c) In breeding for a given trait it is not uncommon to introduce inadvertently genes that have unexpected and, more often than not, deleterious effects on the organism. Salinity-tolerance breeders may therefore find themselves dealing willy-nilly with matters other than salt tolerance *per se.* (d) Finally, comparative physiological and biochemical studies of genotypes contrasting in salt tolerance (or sensitivity) are of both intrinsic and potentially practical importance, and such studies are likely to be more fruitful if done in the wide framework of mineral plant metabolism than within the confines of the more narrow context of salt tolerance alone.

With this in mind, I have compiled in Table 1 reviews, chapters, and volumes dealing with the genetic control of mineral nutrition and metabolism in plants. Additional references from Eastern Europe and the U.S.S.R. not available to me are given by Sarić in the volumes edited by him (Sarić[41a]) and Sarić and Loughman[41b]. (The two volumes are almost identical – the proceedings of the same symposium held in 1982 in Beograd.) No original research papers are included, nor are reviews dealing specifically with salt tolerance. That topic is amply covered in the proceedings of several conferences that preceded the present one. The principal ones are listed in Table 2. Further reviews appearing in other publications are listed in Table 3.

The still broader matters of plant resources for arid (and hence salinity-prone) lands and the direct use of halophytes have been discussed at various conferences; see for example Goodin and Northington[19], Manassah and Briskey[31], and contributions to the present volume. Another conference on plant resources was held at the Royal Botanic Gardens at Kew later in 1984. It would, however, be beyond the scope of the present paper to do more than make mention of the many activities, including publishing ventures, dealing comprehensively with the manifold problems of arid lands. Paylore[34] recently has forcefully drawn attention to the need to devise means not just for generating

Table 1. Reviews, chapters, and volumes dealing with the genetic control of the mineral nutrition and ion relations of plants[a]

Author(s) or Editor	Year	Title	Reference
Kruckeberg, A. R.	1959	Ecological and genetic aspects of metallic ion uptake by plants and their possible relation to wood preservation	27
Myers, W. M.	1960	Genetic control of physiological processes: consideration of differential ion uptake by plants	33
Millikan, C. R.	1961	Plant varieties and species in relation to the occurrence of deficiencies and excesses of certain nutrient elements	32
Epstein, E.	1963	Selective ion transport in plants and its genetic control	10
Gerloff, G. C.	1963	Comparative mineral nutrition of plants	17
Vose, P.B.	1963	Varietal differences in plant nutrition	47
Epstein, E. and R.L. Jefferies	1964	The genetic basis of selective ion transport in plants	14
Epstein, E.	1972	Physiological genetics of plant nutrition	11
Klimashevsky, E.L., Ed.	1974	Variety and nutrition	25
Läuchli, A.	1976	Genotypic variation in transport	28
Wright, M. J., Ed.	1977	Plant adaptation to mineral stress in problem soils	51
Jung, G. A., Ed.	1978	Crop tolerance to suboptimal land conditions	24
Brown, J. C.	1979	Genetic improvement and nutrient uptake in plants	3
Clark, R. B. and J. C. Brown	1980	Role of the plant in mineral nutrition as related to breeding and genetics	5
Wegrzyn, V. A., R.R. Hill, Jr. and D. E. Baker	1980	Soil fertility – crop genotype assocations and interactions	49
Duvick, D. N., R.A. Kleese and N. M. Frey	1981	Breeding for tolerance of nutrient imbalances and constraints to growth in acid, alkaline and saline soils	8
Klimashevsky, E. L.	1981	Physiological genetics of plant mineral nutrition	26
Sarić, M. R.	1981	Genetic specificity in relation to plant nutrition	40
Clark, R. B.	1982	Plant response to mineral element toxicity and deficiency	4
Devine, T. E.	1982	Genetic fitting of crops to problem soils	6
Sarić, M. R.	1982	Genetic specificity of mineral nutrition of plants	41a
Sarić, M. R. and B. C. Loughman	1983	Genetic aspects of plant nutrition	41b

Table 1 (*continued*)

Author(s) or Editor	Year	Title	Reference
Epstein, E. and A. Läuchli	1983	Mineral deficiencies and excesses	15
Gerloff, G. C. and W. H. Gabelman	1983	Genetic basis of inorganic plant nutrition	18
Vose, P. B.	1984	Effects of genetic factors on nutritional requirements of plants	48

[a] Publications dealing exclusively with particular types of inorganic ions or compounds such as heavy metals or salt are not included, nor are in-house items and reports for government agencies which are of such limited distribution as to be virtually unavailable. Publications dealing with salt tolerance are listed in Tables 2 and 3.

Table 2. Proceedings and volumes dealing with the development of salt-tolerant crops[a]

Editor(s)	Year	Title	Reference
Hollaender, A., J. C. Aller, E. Epstein, A. San Pietro and O. R. Zaborsky	1979	The biosaline concept: An approach to the utilization of underexploited resources	22
Rains, D.W., R. C. Valentine and A. Hollaender	1980	Genetic engineering of osmoregulation: impact on plant productivity for food, chemicals, and energy	37
Qureshi, R. H., S. Muhammed, and M. Aslam	1981	Proceedings of the workshop/ seminar on membrane biophysics and development of salt tolerance in plants	35
Rains, D. W.	1981	A conference on biosalinity: The problem of salinity in agriculture	36
San Pietro, A.	1982	Biosaline research: A look to the future	39
Staples, R. C. and G. H. Toenniessen	1984	Salinity tolerance in plants: strategies for crop improvement	45

[a] Proceedings and volumes of limited distribution or mainly local scope and those dealing with particular crops are not included.

information but for disseminating it in useful ways. What was said there applies to our more specialized subject as well.

Take-home lessons and prospects

What can we gain from a survey such as this? First, the task in which we are engaged is to learn how to grow food, fodder, and fiber in those large parts of the world where soils and water are affected by salt. The development of plants tolerant of these conditions is a new challenge to plant breeders, and it is useful therefore to provide for them a survey

Table 3. Reviews on the development of salt-tolerant crops not appearing in the volumes listed in Table 2[a]

Author(s)	Year	Title	Reference
Gupta, V. S.	1978	Genetic and breeding aspects of salt tolerance in plants	20
Shannon, M. C.	1979	In quest of rapid screening techniques for plant salt tolerance	43
Epstein, E., J. D. Norlyn, D. W. Rush, R. W. Kingsbury, D. B. Kelley, G. A. Cunningham and A. F. Wrona	1980	Saline culture of crops: a genetic approach	16
Epstein, E.	1983	Crops tolerant of salinity and other mineral stresses	12
Downton, W. J. S.	1984	Salt tolerance of food crops: prospectives for improvements	7

[a] Specialized reviews and those devoted mainly to particular methodologies or crops are not included.

of what has been thought and done about it so far. The present review should give them ready access to the available literature on genetic aspects of mineral metabolism and salt relations of plants. (When, for a 1976 workshop[51], I compiled a list like those of Tables 1–3, it contained only eight items, all individual review papers. Tables 1–3 contain 35 items, 10 of them multi-authored volumes. Clearly we have come a long way in a short time.)

The other important benefit we may get from such a survey is an assessment of the needs and the opportunities that we are facing in this endeavour. Such an assessment is of course a matter of judgment, and I do not expect unanimous agreement with the points I shall make.

(a) Salt-affected soils and saline water represent mineral stresses. In programs devoted to the development of plants resistant to these stresses we should view our task in the broad context of breeding for resistance to all manner of other mineral stresses, for many of them have a bearing on and occur in conjunction with salinity stress.

(b) We have for the most part explored feasibilities, conducted experimental studies, and done screening for salt tolerance on a modest scale, often under greenhouse conditions. We shall have to expand the scale of our operations and include more work under field conditions, on salt-affected soils, and even experimentally salinized ones, using irrigation water of the kind available in the areas for which the crops are being developed.

(c) With regard to the point just made about field experimentation, the refractory problem of spatial and temporal variability of salinity and sodicity must be given close attention. Experience so far is too limited to permit us to devise breeding strategies for these patchy situations. It is thus not easy to agree that selection should be done on nonsaline soils[38]. What is needed are careful analyses of field conditions on an extensive scale, statistical evaluations of the results, and the development of breeding strategies based on the results of these analyses and evaluations.

(d) We shall have to convince more plant breeders to include salt tolerance among their breeding objectives. Plant breeders have been highly successful in breeding for numerous traits such as yield, quality, and disease resistance[44]. There is no reason why breeding for salinity tolerance should be an exception. There were only three geneticist-breeders at this workshop. I for one hope that there will be many more at our next meeting.

(e) We shall have to search more extensively than we have done for sources of suitable germplasm. There are, to be sure, limits to genetic variability and limits to the degree of any stress to which plants may become adapted[1]. On the other hand, physiological traits are under genetic control, and much genetic variability is available to the breeder[11,30]. Salt tolerance is such a trait. The quest for salt-tolerance germplasm is a very recent development. There is every indication that salt tolerance exists, and in forms that breeders can exploit. Wide crosses are likely to play an increasingly large role.

(f) We should not be deluded by facile claims that genetic engineering, specifically recombinant DNA techniques, will solve our problems quickly and cheaply, but we must be alert to these and a whole range of relatively new technologies and the potentials they offer. Selection at the cell level, somaclonal variation, protoplasmic fusion, and also such older methods as mutation breeding and still other techniques may all contribute to the development of salt-tolerant crops.

(g) We shall continue doing what we have already done with some measure of success, *i.e.*, study the mechanisms of salt tolerance as a guide to further progress. We should, however, resist the temptation to make large claims for the utility of such studies for the enterprise of selection and breeding for this trait; *cf.* Woolhouse[50]. To make such mechanistic studies as useful as possible for this purpose, plant physiologists and biochemists will have to learn to make their studies meaningful to plant breeders. A much closer symbiosis than now exists between the two groups will have to evolve.

(h) Finally, we shall have to make decisions about the kind of salt-tolerant crops to aim for. I have argued elsewhere[12] for development of halophytic crops, the consumable part of which is botanically fruit (grains, tomatoes, *etc.*). Most halophytes, in contrast to more or less salt-tolerant non-halophytes, use salt itself as the major osmoticum; I enumerated the reasons why that is an energetically advantageous mechanism[12]. But as a result of salt absorption, the foliage of these halophytes has a high salt content – a disadvantage if the foliage is to be used as feed or food. Fruit, on the other hand, receives most of its solutes via the phloem, which tends to screen out much of the salt. Finally, in halophytes the adaptation to saline conditions is mainly at the cellular level rather than at higher levels of organization (tissue, organ). Therefore, the new technologies mentioned above (f) lend themselves most readily to application in the development of crops with truly halophytic adaptation to salinity[12]. All these considerations lead me to favor the development of crops with truly halophytic response to salinity, the consumable part of the plant being botanically fruit (grains, seeds, berries, pomes, *etc.*). This emphasis on fruit does not, however, apply to crops grown for fiber, chemical feedstocks, medicinals, or biomass for generating energy.

Acknowledgments I wrote this review while a guest in the Laboratory for Nitrogen Fixation Research, H J Evans, Director, and the Department of Botany and Plant Pathology, T C Moore, Chairman, Oregon State University, Corvallis, Oregon 97331, USA. My research on salt tolerance in plants is being supported by National Science Foundation Grants PCM-8304506 and (with A Läuchli) PCM-8404442. I am indebted to a number of colleagues for reprints, information on items in press, and discussions.

References

1 Bradshaw A D 1984 Adaptation of plants to soils containing heavy metals – a test for conceit. *In* Origins and Development of Adaptation. Ciba Foundation Symposium 102, pp 4–19. Pitman, London.
2 Brewer W H 1930 Up and Down California in 1860–1864. Yale University Press, New Haven. Available in a third edition. Ed. F P Farquhar. University of California Press, Berkeley, 1966.
3 Brown J C 1979 Genetic improvement and nutrient uptake in plants. BioSci. 29, 289–292.
4 Clark R B 1982 Plant response to mineral element toxicity and deficiency. *In* Breeding Plants for Less Favorable Environments. Eds. M N Christiansen and C F Lewis. pp 71–142. John Wiley and Sons, New York.
5 Clark R B and Brown J C 1980 Role of the plant in mineral nutrition as related to breeding and genetics. *In* Moving up the Yield Curve: Advances and Obstacles. Eds. L S Murphy, E C Doll and L F Welch. pp 45–70. ASA Special Publication Number 39, American Society of Agronomy, Madison, Wisconsin.
6 Devine T E 1982 Genetic fitting of crops to problem soils. *In* Breeding Plants for Less Favorable Environments. Eds. M N Christiansen and C F Lewis. pp 143–173. John Wiley and Sons, New York.
7 Downton W J S 1984 Salt tolerance of food crops: prospectives for improvements. CRC Critical Reviews in Plant Sciences 1, 183–201. CRC Press, Boca Raton, Florida.
8 Duvick D N, Kleese R A and Frey N M 1981 Breeding for tolerance of nutrient imbalances and constraints to growth in acid, alkaline and saline soils. J. Plant Nutr. 4, 111–129.

9 Epstein E 1960 Spaces, barriers, and ion carriers: ion absorption by plants. Am. J. Bot. 47, 393–399.
10 Epstein E 1963 Selective ion transport in plants and its genetic control. Desalination Research Conference Proceedings. NAS-NRC Publication 942, 284–298.
11 Epstein E 1972 Physiological genetics of plant nutrition. *In* Mineral Nutrition of Plants: Principles and Perspectives. pp 325–344. John Wiley and Sons, New York.
12 Epstein E 1983 Crops tolerant of salinity and other mineral stresses. *In* Better Crops for Food. Ciba Foundation Symposium 97, 61–82. Pitman, London.
13 Epstein E and Hagen C E 1952 A kinetic study of the absorption of alkali cations by barley roots. Plant Physiol. 27, 457–474.
14 Epstein E and Jefferies R L 1964 The genetic basis of selective ion transport in plants. Annu. Rev. Plant Physiol. 15, 169–184.
15 Epstein E and Läuchli A 1983 Mineral deficiencies and excesses. *In* Challenging Problems in Plant Health. Eds. T Kommedahl and P H Williams. pp 196–205. American Phytopathological Society, St. Paul, Minnesota.
16 Epstein E, Norlyn J D, Rush D W, Kingsbury R W, Kelley D B, Cunningham G A and Wrona A F 1980 Saline culture of crops: a genetic approach. Science 210, 399–404.
17 Gerloff G C 1963 Comparative mineral nutrition of plants. Annu. Rev. Plant Physiol. 14, 107–124.
18 Gerloff G C and Gabelman W H 1983 Genetic basis of inorganic plant nutrition. *In* Encylopedia of Plant Physiology, New Series, Vol. 15. Inorganic Plant Nutrition. Eds. A Läuchli and R L Bieleski. pp 453–480. Springer-Verlag, Berlin.
19 Goodin J R and Northington D K, Eds 1979 Arid Land Plant Resources. International Center for Arid and Semi-Arid Land Studies, Texas Tech University, Lubbock.
20 Gupta V S 1978 Genetic and breeding aspects of salt tolerance in plants. Curr. Agric. 2, 1–6.
21 Hilgard E W 1906 Soils. Macmillan, New York.
22 Hollaender A, Aller J C, Epstein E, San Pietro A and Zaborsky O R, Eds 1979 The Biosaline Concept: An Approach to the Utilization of Underexploited Resources. Plenum Press, New York.
23 Jenny H 1961 E W Hilgard and the Birth of Modern Soil Science. Collana della Rivista 'Agrochimica' 3. Pisa.
24 Jung G A, Ed 1978 Crop Tolerance to Suboptimal Land Conditions. ASA Special Publication Number 32, American Society of Agronomy, Madison, Wisconsin.
25 Klimashevsky E L, Ed 1974 Variety and Nutrition. Academy of Sciences of the USSR. Siberian Branch. Sib. Inst. Plant Physiol. and Biochem. Irkutsk, Siberia, USSR. (*In Russian*)
26 Klimashevsky E L 1981 Physiological Genetics of Plant Mineral Nutrition. Kolos Publishing House, Moscow. (*In Russian*)
27 Kruckeberg A R 1959 Ecological and genetic aspects of metallic ion uptake by plants and their possible relation to wood preservation. *In* Marine Boring and Fouling Organisms. Ed. D L Ray. pp 526–536. University of Washington Press, Seattle.
28 Läuchli A 1976 Genotypic variation in transport. *In* Encyclopedia of Plant Physiology, New Series, Vol. 2, Part B. Transport in Plants II B. Eds. U Lüttge and M G Pitman. pp 372–393. Springer-Verlag, Berlin.
29 Lyon C B 1941 Responses of two species of tomatoes and the F_1 generation to sodium sulphate in the nutrient medium. Bot. Gaz. 103, 107–122.
30 Mahon J D 1983 Limitations to the use of physiological variability in plant breeding. Can. J. Plant Sci. 63, 11–21.
31 Manassah J T and Briskey E J, Eds 1981 Advances in Food Producing Systems for Arid and Semiarid Lands. Academic Press, New York.
32 Millikan C R 1961 Plant varieties and species in relation to the occurrence of deficiencies and excesses of certain nutrient elements. J. Aust. Inst. Agric. Sci. 27, 220–233.
33 Myers W M 1960 Genetic control of physiological processes: consideration of differential ion uptake by plants. *In* A Symposium on Radioisotopes in the Biosphere. Eds. R S Caldecott and L A Snyder. pp 201–226. University of Minnesota, Minneapolis.

34 Paylore P 1982 Arid lands information transfer: who knows what? Annals Arid Zone 21, 65–72.
35 Qureshi R H, Muhammed S and Aslam M, Eds 1981 Proceedings of the Workshop/Seminar on Membrane Biophysics and Development of Salt Tolerance in Plants. University of Agriculture Press, Faisalabad.
36 Rains D W, Ed 1981 A Conference on Biosalinity: the Problem of Salinity in Agriculture. University of California, Davis.
37 Rains D W, Valentine R C and Hollaender A, Eds 1980 Genetic Engineering of Osmoregulation: Impact on Plant Productivity for Food, Chemicals, and Energy. Plenum Press, New York.
38 Richards R A 1983 Should selection for yield in saline regions be made on saline or non-saline soils? Euphytica 32, 431–438.
39 San Pietro A, Ed 1982 Biosaline Research: A Look to the Future. Plenum Press, New York.
40 Sarić M R 1981 Genetic specificity in relation to plant mineral nutrition. J. Plant Nutr. 3, 743–766.
41a Sarić M R, Ed 1982 Genetic Specificity of Mineral Nutrition of Plants. Serbian Academy of Sciences and Arts, Beograd.
41b Sarić M R and Loughman B C, Eds. 1983 Genetic Aspects of Plant Nutrition. Martinus Nijhoff Publishers, The Hague.
42 Schimper A F W 1898 Pflanzengeographie auf Physiologischer Grundlage. Verlag Gustav Fischer, Jena.
43 Shannon M C 1979 In quest of rapid screening techniques for plant salt tolerance. Hort Sci. 14, 587–589.
44 Sprague G F, Alexander D R and Dudley J W 1980 Plant breeding and genetic engineering: a perspective. BioSci. 30, 17–21.
45 Staples R C and Toenniessen G H, Eds 1984 Salinity Tolerance in Plants: Strategies for Crop Improvement. John Wiley and Sons, New York.
46 Tansley A G 1911 Types of British Vegetation. Cambridge University Press, Cambridge.
47 Vose P B 1963 Varietal differences in plant nutrition. Herb. Abstracts 33, 1–13.
48 Vose P B 1984 Effects of genetic factors on nutritional requirements of plants. *In* Crop Breeding: A Contemporary Basis. Eds. P B Vose and S G Blixt. pp 67–114. Pergamon Press, Oxford.
49 Wegrzyn V A, Hill R R Jr and Baker D E 1980 Soil fertility – crop genotype associations and interactions. J. Plant Nutr. 2, 607–627.
50 Woolhouse H W 1981 Crop physiology in relation to agricultural production: the genetic link. *In* Physiological Processes Limiting Plant Productivity. Ed. C B Johnson. pp 1–21. Butterworths, London.
51 Wright M J, Ed 1977 Plant Adaptation to Mineral Stress in Problem Soils. A Special Publication of Cornell University Agricultural Experiment Station, Cornell University, Ithaca, New York.

Plant and Soil 89, 199–226 (1985). Ms. BR 14

Genetics of salt tolerance in higher plants: theoretical and practical considerations

MOSHE TAL
Department of Biology, Ben-Gurion University of the Negev, Beer-Sheva, Israel

Key words Ecology Genetics Physiology Salt tolerance Tomato species

Summary An interdisciplinary approach to breeding for stress tolerance in plants has gained considerable recognition in the past few years. Accordingly, this article presents a synthesis of the genetic, physiological, and ecological aspects of salt tolerance in plants. An understanding of these aspects and the interrelationships between them is essential for an efficient breeding program.

A significant part of the presentation concentrates on the basic problems associated with the genetics of tolerance to stresses and of quantitative characters in general, since many of the unsolved problems relevant to the genetics of salt tolerance are still general. Significant progress in the breeding of quantitative as well as qualitative traits in multicellular organisms depends on an understanding of the genetic and epigenetic dimensions of gene action. The discussion therefore includes an overview of (1) the limited existing knowledge on the genetic control of salt tolerance and (2) the physiological mechanisms and molecular targets central to the control of salt resistance as expressed by the amount and stability of yield.

An additional subject emphasized here concerns the main strategies of adaptation of wild species to their natural habitats. An understanding of them is essential to (1) enable distinction between traits that can increase agricultural yield and traits that are favorable only for survival under natural conditions (such a distinction is essential, especially when wild species are used as a gene source), and (2) predict the best combinations of characters for efficient agricultural production in stressful environments.

General introduction

An understanding of the genetics of salt and stress tolerance in plants is essential for the efficient breeding of crop plants tolerant to environmental stresses, which has been proposed as a solution to the problem of the mounting world food demand[4,7,83,133,134]. Although the human population is increasing rapidly, food production has reached a plateau or is even decreasing[133,134]. This has resulted mainly from the great increase in the price of energy-intensive inputs[9,23,98] and from soil erosion and salination[4,9,69,78,83,133]. An additional factor is a decrease in the funding of agricultural research[4,7,133]. Although most of the present and future problem of starvation may be due to the patterns of food distribution and to political-economic policies[1], there is always a need for technological-biological solutions to help us to overcome the physiological limitations that restrict plant productivity. Such solutions may include modification of the environment (the technological approach) or of the physiology of the plant (the biological approach).

The two approaches can complement one another. The technological approach may include: 1) increased application of mineral fertilizers, 2) conservation of existing agricultural lands, 3) reclamation of lands that were once arable or of marginal lands that were never arable, and 4) maximum exploitation of local advantages[9,38,78,83,98,133,134]. However, since this approach usually requires high immediate and long-term investments, the difference between the profits and costs of high-yield crops that require high investment may be narrow. In addition, large human populations live in underdeveloped countries that cannot afford the energy-intensive inputs required for such crops[82]. These considerations contrast with the benefits of the biological approach, by which crops will be bred with somewhat lower rates of growth and reproduction but with lower nutrient requirements and more stable yields under the prevailing or more severe agricultural conditions[7,15,134].

Genetic manipulation of crops can be done by conventional breeding methods or more modern technologies[3,7,20,99,108,131]. According to Evans *et al.*[27], the application and development of cell culture techniques are opening the door to a second green revolution. Borlaug[7] agrees that the potential of the modern techniques is high but believes that, because of the complexity of the plant genome, many years will elapse before these techniques will be used successfully for breeding better crop varieties. He recommended, therefore, that most funding for agricultural research should still be concentrated on the improvement and application of conventional methods. Which of the alternatives should be more encouraged is not easily decided, since the potential benefits of the new methods will be largely a function of the investment in them.

The existence of genetic variability is a prerequisite for any breeding work. Since improvement of most crop species took place outside their native areas or under favorable conditions, their gene pools have become limited[90]. These gene pools can be enriched by the selection of mutants in cell culture[99], by conventional mutagenesis, or by the introduction of genes from wild species. The potential, achievements, and problems associated with the application of cell culture methodologies for obtaining stress-tolerant plants were discussed in recent publications[109,115]. There is extensive evidence for a vast genetic potential in the plant kingdom for adaptation to adverse environmental conditions[3,24,25,78,82,90]. The importance of determining the extent of the potential in the gene pools of wild species and of preserving them for use in plant breeding, as well as the problems associated with the transfer of genes from such species to crop plants, were discussed by Feldman[28].

In the breeding of multicellular organisms, genetics cannot be separated from epigenetics, which includes the biochemical and physiological aspects of gene action. Both types of factors, together with the environment, determine the final phenotype. Because of our virtual ignorance of the genetics of salt tolerance, many of the unsolved problems related to this subject are still of a general nature, common to other stresses and to quantitative characteristics in general. The specific discussion of the genetics of salt tolerance will be preceded, therefore, by two sections that include a discussion of basic questions associated with the genetics of complicated or quantitatively inherited characteristics, and of the general ecological and physiological principles related to survival and yield under stress conditions.

Complicated characteristics

Characters are defined as complicated or quantitative when they appear to be continuous, so that the relative contributions of individual genes to the final phenotype are not known. One of the concepts that prevailed in the past was that the inheritance of quantitative traits depends on many genes with relatively small effects, and that polygenes and major genes represent separate classes of genes[64]. However, the dominant view today is that qualitative and quantitative characteristics are the two extremes of a continuous spectrum, and consequently the distinction between them is artificial. The position of a character in this spectrum depends on the intensity of the interactions among various genes and their products, and between genes and the environment. The criticism of previous concepts of the genetics of quantitative characters was expressed by Lewontin[58] as follows: 'Quantitative genetics, upon which modern plant and animal breeding is based, is an attempt to produce knowledge by systematization of ignorance.'

It has often been demonstrated that polygenic traits can actually be determined by a small number of major genes or linked gene complexes that segregate as units[5,11,62,65,128]. Polygenes may thus simply reflect minor pleotropic effects of the major genes. The expression of the phenotype as qualitative or quantitative may vary, depending on the genetic background and/or the environment. Some examples are the inheritance of nitrate reductase in the common wheat[19,39] and of tolerance of plants to insects, diseases[55], and heavy metals[62].

Although statistical techniques have provided the breeder with powerful tools to analyze quantitative traits, they have not provided any clue on the mechanism of action of the genes involved[88]. A similar idea was expressed by Lewontin[58], who stated that 'observations on the phenotypic level may be quite sufficient to predict the outcomes of breeding schemes quite directly, without the necessity of knowing the detailed structure of the genes and its relation to the phenotype . . . But if we wish to look beyond the immediate rate of response to selection in the present set of environments we really do require knowledge of two general sorts.' This knowledge includes details of the structure of the relevant part of the genome (the genetic dimension) and of its relation to the phenotype (the epigenetic dimension), which can be obtained by molecular, developmental, and physiological genetics.

According to Rice and Carlson[88] it is not necessary to invoke the action of any new molecular mechanism for polygenic action, since known mechanisms can account for it. In their opinion, the ideal experimental system for analysis of the genetic basis of quantitative inheritance should satisfy the definition of polygenic inheritance, be amenable to the reductionist approach of molecular genetics, and allow the use of biochemical and genetic technology.

The molecular structure of the genome in eukaryotes in general and in plants in particular is only starting to be clarified. According to present knowledge[34], the genome is composed of a relatively small fraction of unique nucleotide sequences which constitute the structural genes and of a larger fraction of moderately and highly repetitive DNA. At present, most structural genes are identified after mutation by a change in the electrophoretic mobility of their protein products or by an alteration in the phenotype. Only a few have been identified directly on the DNA level[72]. Repetitive DNA sequences may control gene action during differentiation and in the adult organism[16,74].

A prerequisite for understanding the inheritance of 'complex' traits[126] is the identification of loci or gene complexes that control quantitative variation and are inherited as units (major genes). The methods that can be used for their identification include[50]: (1) the detection of linkage between gene markers already mapped on the chromosomes and the major genes to be identified; (2) statistical analyses of frequency distributions of segregating progeny; and (3) genotypic assay. A more modern approach is the isolation of the gene product. Scientists working on tolerance to heavy metals, for example, hope to isolate the mRNA of the metal-binding protein that seems to play a central role in metal detoxification[136]. Such an mRNA may then be used to prepare cDNA, which, when cloned, will provide a probe for measuring the frequency of genes involved in the production of this protein.

In plant improvement, there is much interest in the epigenetic dimensions of complicated traits in relation to the biochemical and physiological processes that control yield. The increase in yield due to the breeding of crop plants was achieved by direct selection for the production of the appropriate organs under progressively improving agrotechnical conditions[41]. Changes in individual physiological traits were not directly selected for. Although breeding can make progress without advances in physiological and biochemical knowledge[31,135], more rapid improvements in the ability of plants to cope with adverse environments depend on the adoption of new selection criteria and on an understanding of the physiological processes and genetic controls that limit crop production[4,9,20,40,83,113,134]. The value of a physiological process as a basis for selection and breeding depends on the satisfaction of four basic criteria[63]: the existence of genetic variability with respect to that process, an understanding of the genetic control of the process, its potential agronomical benefit, and its measurability in large-scale trials, *i.e.* its being detectable in a rapid, economical and reliable fashion. The agricultural benefit and the genetic control of a physiological process are difficult to evaluate together because of the hierarchical organization of such processes. According to Mahon[63], the hierarchy presented in Fig. 1 implies that processes 'near the output of the system will be more easily related to agronomically desirable traits, because there are fewer interactions with other processes, while those near the bottom will be more directly linked to initial gene products and more easily related to simple genetic control.' Information on the genetic control of physiological processes may aid in planning breeding programs and may also help to understand the stability or instability of the genotypic performance. The agricultural benefit of a physiological process can be demonstrated unequivocally only by testing it[63]. According to Fischer[31], evaluating the importance of a physiological trait for increasing yield under conditions of stress can take either the pragmatic, 'black box' approach, which attempts to

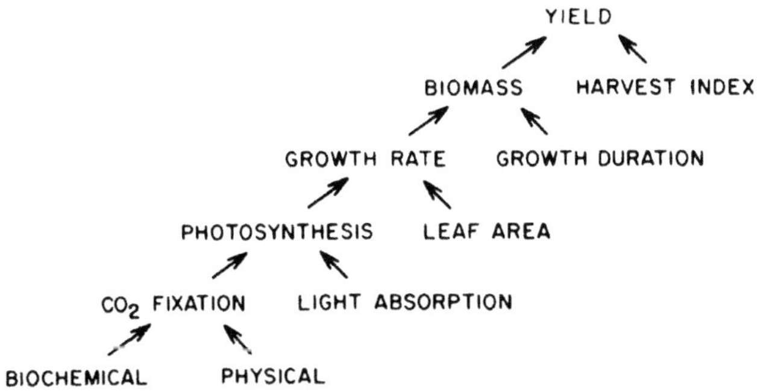

Fig. 1. Scheme of the possible hierarchy of a number of physiological and yield characters (reproduced, with permission, from ref.[63]).

proceed from yield differences to the underlying processes; or the theoretical, 'ideotype' strategy, which attempts to predict differences in yield from an understanding of the physiology.

Since there are few examples of successful application of crop physiology in breeding programs, 'one might be tempted to conclude that yield determination is so complex and there are so many different paths to higher yield that it is never going to be possible to make a worthwhile physiological formulation of guidelines for breeders'[30]. However, the gap between physiological research and its use in breeding programs may be narrowed if the research becomes more directly concerned with the question of genotypic control of yield, if techniques for less-precise but more rapid measurements of key processes are applied so that more cultivars can be studied, and if cooperation between physiologists and breeders is increased[30].

Stress tolerance: general ecological and physiological principles

Environmental stresses in general constitute the primary limit on increasing the production of many if not all crops[134]. Information on the adaptive strategies of wild species in their natural habitats is important for distinguishing between characteristics that may contribute to agricultural yield and those favorable for survival only under natural conditions. Such information may also help in predicting the combinations of characteristics that will allow effective agricultural exploitation of stressful environments. Indeed, knowledge of the strategy of stress-tolerant species in nature helped to explain the failure to breed crop varieties that were productive on infertile soils[14].

Phenomena that limit plant production or destroy biomass were defined by Grime[43] as 'stress' and 'disturbance,' respectively. He classified wild species according to their major strategies of adaptation to these conditions as follows:

1. Stress-tolerant species include those that usually occupy unproductive habitats having severe stresses and mild, infrequent disturbances. They are characterized by: (1) a slow growth rate, which enables the accumulation of reserves because the rate of resource uptake exceeds the rate of their use for growth and is slow enough not to exhaust the limited resources of the environment, (2) low morphological flexibility, which results from a relatively low proportion of highly responsive differentiating tissue, (3) long-lived absorbing surfaces (roots and leaves), which results in a decrease in nutrient cycling between the plant and its environment, (4) high physiological flexibility based on multiple mechanisms for stress tolerance, and (5) low metabolic turnover of carbohydrates, minerals, and water. Since plant adaptation to

infertile soils has many of the same basic characteristics as adaptation to other stresses[13], the interactions between the various characteristics of nutrient-stress tolerators may be representative of stress-tolerant species in general (Fig. 2).

2. Ruderal species are annuals living in productive habitats characterized by low stresses and frequent intense disturbances. Their adaptation includes fast growth and a short life cycle. Absorbed nutrients and photosynthates are allocated mainly to seed production.

3. Competitive species exist in productive but relatively stable habitats where stresses and disturbances are mild and infrequent. They are characterized by a high capacity for uptake of external resources and a high rate of growth and biomass production. The fast growth and short lifetime of absorbing surfaces enable them to adjust quickly to changes in the environment. Several characteristics typify adaptation to productive habitats (Fig. 3).

Based on many studies and observations, it is commonly accepted today that the amount of DNA in the nucleus is almost always correlated positively with the volume of the nucleus and the cell and negatively with the rate of metabolism and growth[12]. A positive correlation seems to exist, for example, between the length of the life cycle and the DNA content of herbaceous plants[12], *i.e.* annuals usually have less DNA than perennials. In many groups of plants a strong correlation exists between the DNA content (or cell volume) and ecological

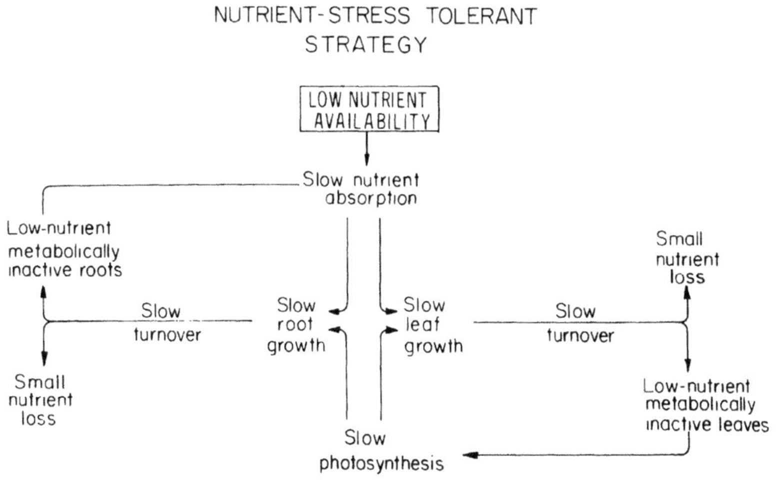

Fig. 2. Interacting characteristics of nutrient-stress tolerant strategy (reproduced, with permission, from ref.[13]).

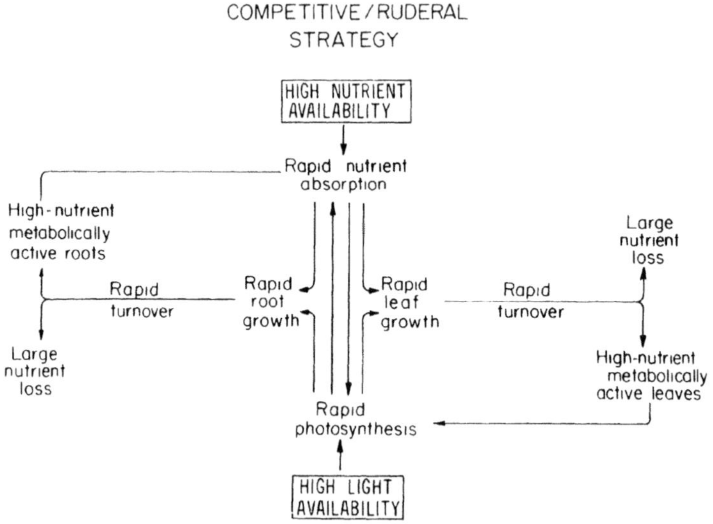

Fig. 3. Interacting characteristics of competitive/ruderal strategy (reproduced, with permission, from ref.[13])

adaptability[110]. In temperate regions, for example, plants that grow in the summer have smaller nuclei than those that grow in the winter. An interesting question is whether stress-tolerant species, which are characterized by slow growth and slow metabolic rates, have higher nuclear DNA contents than ruderals and competitors, which grow and metabolize faster, and whether the evolution of cultivated plants, which have a high growth rate, from slow-growing wild species is accompanied by a decrease in DNA content.

Arid and semiarid regions, which are usually characterized by saline soils and water, drought, and extreme temperatures, constitute about 40% of the world's land surface[32,69]. Some acquaintance with the ecology of their native plants is important not only because these regions represent a large, currently unexploited agricultural land reserve, but also because plants growing in these areas are potential sources of genes for improving the tolerance of crop plants to salt and other stresses. According to Solbrig[107], since such regions are characterized by large oscillations in environmental factors, 'successful plants need not be optimal in an absolute sense, but must be able to grow and reproduce over a variety of circumstances.' This strategy is the essence of the satisfaction principle[107], by which arid-zone plant species evolved various combinations of characteristics, each directed by natural selection towards the best compromise for survival.

Finley and Wilkinson[29] concluded that a breeding program designed to produce an ideal variety having maximum yield in the most favorable environment together with maximum phenotypic stability will almost certainly fall short of this goal. They found that barley varieties with high phenotypic stability under various conditions were unable to produce a high yield in favorable environments. On the other hand, varieties that gave high yields in favorable environments had low phenotypic stability, *i.e.* their yield was much lower under unfavorable conditions.

The kind of compromise between yield potential and stability in determining a high average performance of a generally adapted variety cannot be predicted, since it depends on the particular genotypes and range of environments under consideration[92]. In subsistance agriculture, which characterizes many Third World countries, consistent performance under varying conditions is more important than high yield in favorable environments. However, where advanced techniques can be used to minimize variation in the environment, high yield is more important than its stability.

In an article on the genetic improvement of drought resistance in crop plants, Blum[6] discusses the two major approaches to breeding for stress tolerance. The first states that a variety with superior yield under optimum growth conditions will also yield relatively well under suboptimum conditions, on the assumption that the same genes control high and stable yield. The advantage of this approach, if valid, is that selection for yield is more efficient under optimum conditions. However, the more commonly accepted approach in breeding for stress tolerance is that the maximum potential yield and stability of yield are largely independent of each other. Stress-tolerant cultivars should thus be selected under the stress to which tolerance is desired[9]. This assumes that yield stability in various environments, which may be an attribute desired by the breeder, can be transferred from wild stress-tolerant plants independent of their low rates of growth and metabolism.

As mentioned above, tolerances to salt, water, or other stresses are still considered 'complex' characteristics[87,135]. Since much of this 'complexity' stems from a lack of knowledge, it can be resolved by coordinated physiological-genetic research. However, since the genetic study is now lagging behind the physiological research, more emphasis must be placed on the genetic aspects. One of the few examples of a genetic analysis of physiological processes related to stress tolerance is the work of Quarrie[84,85]. In these studies attempts were made to identify the major genes that control the level of abscisic acid, which has a central role in tolerance to drought and probably to other stresses[67, 111]. Quarrie[84, 85] analyzed the frequency distributions of segregating generations obtained from the cross of sensitive and resistant lines, low and high in abscisic acid content. The results indicate that a small number of major genes control abscisic acid level.

At present, most scientists interested in stress in plants specialize in only one stress, which may be undesirable since: (1) Crop plants in commercial and especially in subsistence agriculture are usually exposed to more than one stress at once. (2) The wild plants to be used as gene sources may be tolerant to more than one stress. Wild plants that evolved in arid areas, for example, may be tolerant to water, salt, mineral, and temperature stresses. According to Ramage[87], salt-tolerant varieties are usually more drought and winter tolerant than salt-sensitive varieties. (3) A generally accepted idea is that tolerance to various stresses may be at least partly controlled by common physiological and molecular mechanisms[57, 115]. This may result from: (a) an effect that is common to different stresses, for example, dehydration, which occurs during water, salt, heat, and freezing stresses; (b) the existence of molecular sites or processes that may be common targets of various stresses, for example, the membrane, and abscisic acid activity. An integration between the study of the various aspects of stress phenomena in plants will increase the efficiency of the genetic, biochemical, and physiological research and of breeding for stress tolerance.

According to Woolhouse[135], only physiological studies of photosynthesis, leaf duration, and the metabolism of nitrogen and of reserve proteins have a good chance for immediate profit in plant improvement. He wrote that 'there seems little prospect for subjecting them [drought resistance characteristics, for example, which he feels are too complicated] to genetic analysis at the present time, and it would seem realistic to sustain the effort on water relations as long term strategic work at an academic level rather than to develop this as front line crop physiology with the expectation of immediate pay-offs.'

There is little doubt that an appreciable portion of the 'complexity' of tolerance will be more understandable and become more easily manipulated through coordinated physiological-genetic research. However, even with the present state of affairs, there is room for optimism about the chances for pay-off from physiological research on stress tolerance. Some contributions of physiological research on drought tolerance were discussed by Boyer[9]. Genetic improvement of soybean yield in Illinois was correlated with an increase in the average leaf water potential. He claims that if breeders had known of this change in water potential beforehand and had he intentionally used it in selection, improvement would have occurred much faster. Another example is the increase in wheat yield attributed to selecting for better osmotic adjustment to drought[9].

Salt tolerance

Excessive salinity has been considered to be the most important edaphic factor limiting the distribution of plants in natural habitats, and is an increasingly severe agricultural problem in many parts of the world[4,9,21,83,87]. In spite of its importance, the physiology of salt tolerance has been investigated less than that of tolerance to water and temperature stresses, and almost nothing is known about its inheritance. It is thus impossible to predict how much of an increase in salt tolerance is obtainable by genetic manipulation[69].

According to Epstein[24], one prerequisite for improving salt tolerance in crop plants is that vigorous plant growth must be possible on saline substrates. This is demonstrated by the existence of halophytic species, which were defined[48] as the flora native to saline habitats such as seashores, swamps, inland saline seeps and sinks, and saline deserts. These plants include approximately 550 species belonging to 220 genera in 75 families[69]. It is, however, difficult to draw a clear-cut distinction between halophytes and nonhalophytes[70,130]. Two systems of classification are based on (1) internal salt relations or salt tolerance mechanisms and (2) the salt content of the habitat. The main groups under the first classification comprise euhalophytes, which include salt-requiring and salt-tolerating halophytes, and pseudohalophytes, which include salt-avoiding plants. According to classification (2) plants are divided into euhalophytes, which grow in highly saline habitats and are characterized by low morphological and taxonomical diversity[69], and miohalophytes, which are found in less saline habitats along the border between highly saline and nonsaline upland soils. The latter are characterized by broad physiological diversity, which enables them to cope with both nonsaline and saline conditions. Because of their taxonomic isolation and limited representation, the potential of the euhalophytes for breeding is very limited[69]. In contrast, the miohalophytes are taxonomically diverse and include important agricultural species. According to Mudie[69], the potential existing in wild halophytes should be used for the improvement of salt tolerance in crop species, by even interspecific and intergeneric crosses.

Another prerequisite[24] for improving the salt tolerance of crop species is suitable genetic variability in the cultivated species or their wild relatives. Genotypic variation for salt tolerance in crop species has been reviewed[21,22,26,61,78,83,103,115]. Such variation seems relatively limited, probably because most cultivated species were selected in nonsaline environments where salt-tolerance traits would have been gradually lost from their gene pools[61,69]. As mentioned in the General introduction, salt-tolerant mutants selected in cell culture[109,115] or genes from wild salt-tolerant plants can be used to enrich these gene pools. In spite of the difficulties associated with the former approach, it appears promising[86]. Information on the genetic variability for salt tolerance in wild species that can be hybridized to crop plants is limited. A survey of the salt tolerance potential of wild plants and the maintenance of gene banks of promising wild species are therefore important.

A third prerequisite[24] is a method for screening large numbers of genotypes for salt tolerance. The advantages and disadvantages of various alternative methods for large-scale screening and of the study of salt tolerance under controlled and under field conditions have been discussed elsewhere[24,26,103].

The final requirement[24] is diagnostic criteria by which salt-tolerant genotypes can be easily identified in large segregating populations. This is a major limiting factor in breeding for salt tolerance[24,42,77,83,101,102,103]. Major potential selection criteria are summarized in Table 1.

Table 1. Potential selection criteria for salt resistance[77,101,103,105]

Growth	
Germination percentage Emergence percentage Seedling survival	Useful only for crops that are equally resistant during germination and later stages
Rooting	
Ions	
Accumulation	Indicates tolerance in salt includers
Exclusion	Indicates tolerance in salt excluders
Membranes	
Leakage of electrolytes Abnormal plasmolysis	Indicates sensitivity
Vital staining	May indicate disturbed or undisturbed metabolism
Chlorophyll fluorescence	Changes in its parameters may be used for screening for salt tolerance

Past and present efforts in breeding for salt tolerance have been discussed recently[25,26,78,83,87,102,103]. Their slow progress is a result of insufficient economic input, inadequate methods for measuring salinity, poor information on the effects of salt alone and in interaction with other environmental factors, incomplete understanding of plant responses to salinity, and a lack of reliable selection criteria[24,26,87,103].

The response of plants to salt appears complex, since it can vary with the salt concentration, the type of ions, other environmental factors, and the stage of plant development. Inhibition of growth is the typical effect of salt at concentrations that do not cause specific toxic symptoms[61,75,81,103]. Practical interest in the various aspects of plant growth under salt stress results from the fact that yield is usually positively correlated with at least some of the stages of vegetative growth. Salt tolerance is usually assayed in terms of absolute or relative growth or yield[77,103]. Since relative growth or yield can be used to express the stability of performance, both absolute and relative criteria may have important economic significance[29,92].

Key processes and targets that are or might be essential determinants of plant growth and yield under salt stress have been extensively reviewed[21,37,42,61,70,81] and will be mentioned only briefly in the following paragraphs.

1. *The growing region* Studies of the responses of the growing regions to salt stress have been limited. The results of Munns *et al.*[71] with barley and Taliesnik-Gertel *et al.*[125] with tomato suggest that the primary cause of reduced shoot growth under NaCl salinity is located in the growing tissues and not in the mature photosynthetic tissues. The inhibition by salt of cell division or enlargement (or both) in the growing region may be indirect or direct[53,61,66,70,75,100]. Salt may affect growth indirectly by decreasing the amount of photosynthates, water, or growth factors reaching the growing region[61,71]. The amount of photosynthates reaching the growing region may decrease because of inhibition of photosynthesis due to stomatal closure[104] or by direct effects of salt on the photosynthetic apparatus. In addition, transport of photosynthates in the phloem may be inhibited[61]. Water deficit in the growing region may occur by insufficient osmotic adjustment or increased resistance to water flow[70,80]. Results with barley[71] suggest that inhibition of growth involves a water deficit rather than a direct adverse effect of ions on metabolism. The greater inhibition by salt of the growth of shoot apices detached from salt-sensitive as compared with salt-tolerant tomatoes could result from an adverse water balance or from direct toxic effects of the ions[125]. Maas and Nieman[61] suggested that salt ions can damage growing cells indirectly, by depriving them of essential substances or conditions. Conclusive indications of a direct adverse effect of Na^+ or Cl^- on metabolism are limited. Leopold and Willing[56] list various symptoms of salt damage which may result from membrane damage by salt ions. Salt also affects cellular and nuclear volume[53,112]; induces endopolyploidy[53,112], which might be a normal aspect of differentiation in multicellular organisms[73]; and inhibits or

stimulates nucleic acid and protein synthesis[10, 53, 100, 112, 129]. It is important to determine whether the cytological, ultrastructural[106], and metabolic changes observed are general effects of salt, symptoms of damage, or adaptive responses. More studies should be done on the effects of salt on: (a) Protein synthesis and turnover in the meristem and its neighboring tissues, which is an essential component of metabolic control[46]. Wyn Jones *et al.*[137] suggested that several steps of protein synthesis are very sensitive to changes in the ionic environment. The results of Helal and Mengel[45] support the view that impairment of protein metabolism is an important aspect of salt stress. (b) Enzymes at branch points of metabolism and K^+-requiring enzymes[70]. (c) Gene expression. Additional study with both mature and growing tissues is required to determine whether specific genes are turned on or off under salt stress. Changes in the relative proportions of malate dehydrogenase isoenzymes in salt-treated plants[132], the appearance of new proteins in plant tissues after an increase in the external osmotic pressure[18, 35], and the synthesis of new proteins in tomato roots during recovery of plants from salinity-induced wilting[96] are possible examples of changes in gene expression. Munns *et al.*[70] claim that most existing data on the metabolism of plants under stress conditions describe the consequences rather than the causes of reduced growth and warn against the danger of falling in this trap in the future.

2. *'Book-keeping' of energy balance and cost* Information on the energy costs for the synthesis and maintenance of the mechanisms involved in salt tolerance is needed. One of the most accepted explanations for growth inhibition by salt is the diversion of energy from growth to maintenance[76]. The latter may include the regulation of ion concentrations in various organs and within the cell, the synthesis of organic solutes for osmoregulation or for protection of macromolecules, and the maintenance of membrane integrity. Accurate quantitative information is, however, still lacking. Yeo[140, 141] suggested that the halophytic method of osmoregulation, *i.e.* the accumulation and compartmentation of inorganic ions in the vacuole and of organic compounds in the cytoplasm, is energetically much more efficient than overall osmoregulation by organic solutes. A low permeability of the tonoplast, rather than an active ion pump, was suggested as an energetically inexpensive mechanism for maintaining a steep ion concentration gradient between the vacuole and the cytoplasm in halophytes. Additional reasoning as to the advantages of the halophytic over the large-scale organic method of osmoregulation was presented by Epstein[23]: (a) salt ions are readily available and their absorption does not require highly specialized mechanisms, (b) ion transport to the shoot in the

transpiration stream is metabolically inexpensive, (c) the utilization of ions as osmotica saves on photosynthates, which can instead be used for growth, and saves on their transport via the phloem from the shoot for osmoregulation in the root, a demanding process in terms of metabolic energy. However, conclusive proof for the advantage of this strategy for maintaining a high growth rate under salinity is still lacking[103].

3. *Membrane* Since membranes are the most important structures in the plant for regulating ion content in the cells[61], the analysis of membrane structure and function is an important part of the study of the mechanism of salt tolerance. The significance of the amount and composition of membranal lipids in the control of active and passive ion transport in salt-tolerant plants has been greatly appreciated in recent years[52,54,56].

4. *Regulation of ion content* The regulation of the transport and distribution of ions in the various organs of the plant and within the cells is an essential component of the mechanism of salt tolerance. The extensive knowledge on the regulations of ions in the whole plant has been reviewed[47,49,60,61,79]. According to Osmond *et al.*[79], various aspects of salt tolerance remain uncertain because of the lack of information on the concentration of salt ions in the various compartments of the cell.

5. *Osmoregulation and protection of macromolecules* Osmoregulation ensures the maintenance of adequate turgor in the cell. Organic compounds that accumulate in the cytoplasm may function as osmotica and in protecting the conformation of macromolecules in a changing ionic environment[8,138,139].

6. *Balance of essential nutrients* The energy crisis has been the catalyst for the recent interests in the genetic improvement of salt tolerance and in the uptake and utilization of essential nutrients[14,40,98]. Common interest in both has resulted because nutrient imbalance is a possible effect of high salt concentration[24,77,98]. In addition, the two subjects have many genetic and physiological features in common, and fertilization may affect salt tolerance[24,40].

7. *Hormones* The involvement of hormones in the mechanism of salt tolerance, especially of abscisic acid, whose participation in both salt and water balance in the plant is extensively supported[67,111], is not yet understood well and deserves much more attention.

As expressed by Shannon[103], 'we still do not have enough information to choose which characters to incorporate into a new salt-tolerant variety and which to discard.' The value of the physiological processes and targets as components of the mechanism of salt tolerance requires experimental verification in the field. The review by Fischer[31] describes the 'black box' and 'ideotype' approaches for evaluation of physiological traits in breeding for drought resistance in wheat and barley, which may be applied to breeding for salt tolerance as well.

For ecological, physiological, and molecular reasons, plants may be cotolerant to more than one stress at the same time. Cotolerance has been described, for example, in a clone of *Agrostis stolonifera*[44], a wild NaCl-tolerant grass native to salt marshes. The clone was cotolerant to high concentrations of alkaline metal chlorides as well as to osmotic stress. Wild plants that evolved in deserts may be cotolerant to drought, salt, and temperature stresses. This possibility was examined in three wild salt-tolerant species of tomato that evolved in arid areas of South America[120]. Their tolerance to dehydration and heat stresses was assessed by the leakage of electrolytes from leaf discs exposed to these stresses and by the integrity of naked leaf protoplasts exposed to heat stress[120]. Contrary to expectation, the membranes of the wild species seemed less stable than those of the cultivated species. In all species studied, discs taken from salt-treated plants leaked more under heat stress than those taken from control plants. However, the effect of salt was less in the wild species.

As mentioned before, genetic research on salt tolerance lags behind physiological research. Not only is nothing known on the genetics of the physiological mechanism of salt tolerance, but almost nothing has been done even with the crude traditional approach, which includes analysis of the genetic and environmental effects on yield or growth, to understand the inheritance of salt tolerance. Once the physiological components are defined, it will be possible to begin identifying the major genes that control them, as discussed in the second section. The identification of major genes controlling the basic components of salt tolerance may help the physiologist and the plant breeder, and is necessary to the genetic engineer. Mutations in major genes can provide a powerful tool for investigating control mechanisms and elucidating the relations among various processes and functions[40,116]. The value of this approach has been demonstrated with *scabrous diminutive* (*sd*), a wilty mutant of pepper, (*Capsicum annuum*), which is one of the few mutations known in higher plants in genes controlling the uptake or transport of Na^+ and K^+ [68,116]. A scheme summarizing the possible epigenetic sequence of gene action in this mutant is presented in Fig. 4. This study may strengthen the evidence for the relationship suggested between membrane-bound ATPase and active ion uptake in plants, a relation which may have a central role in salt tolerance.

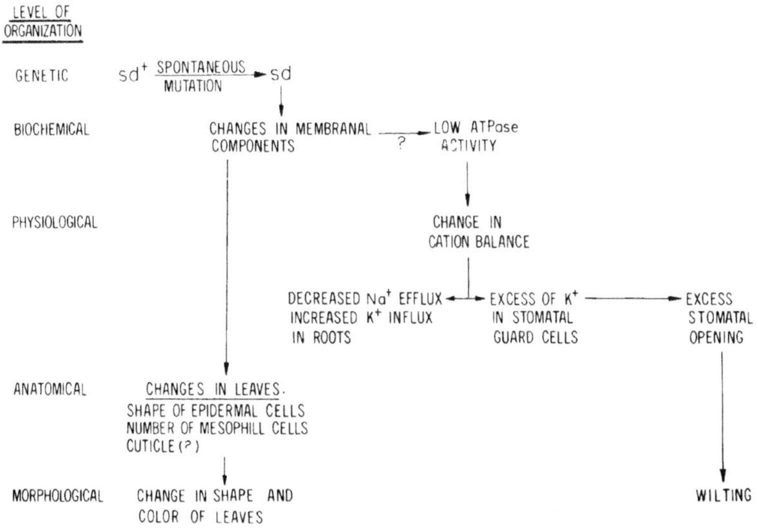

Fig. 4. Possible epigenetic sequence of gene action in *scabrous diminutive*, a wilty mutant of pepper (from ref.[68]).

According to Fischer[30], 'part of the growth in understanding in crop physiology has been appreciation of the need to fit our single-process-oriented research of the past into the whole picture by building integrative models of crops.' Interdisciplinary research concentrating on a number of representative model groups of species may make a great contribution towards (a) integration of the broad, but fragmented, knowledge on the mechanisms and sites related to salt tolerance; (b) a coordinated physiological-genetic analysis of the components of salt tolerance; and consequently, (c) an increase in the efficiency of plant breeding. Some of the prerequisites for such model groups are: representation of the major strategies of salt tolerance, existence of background physiological knowledge, a simple and well mapped genome, and economic importance. Tomato[116] and barley[22] seem to meet these criteria for use as experimental models.

Species in the genus Lycopersicon differ markedly in their degree of salt tolerance[26,116]. The wild relatives of tomato, *L. peruvianum* and *L. pennellii*, which are characterized by slower growth rates than the cultivated species *L. esculentum*, are less affected by salt with respect to their growth rates and relative water contents. These species are also more succulent and accumulate more Na^+ and Cl^- and less K^+ under saline conditions[17,91,114,118,119,121,122]. Similar results were reported for *L. cheesmanii* grown with seawater[93,94] or salinated

Hoagland solution[119]. In all three wild species the Na^+/K^+ and Cl^-/Na^+ ratios were much lower than in the cultivated one. Other responses of *L. peruvianum* and *L. pennellii* plants to salt, relative to the cultivated species, include: (a) better growth of excised shoot apices in saline medium[125] (in contrast, no difference was found with excised fully differentiated tissues); (b) greater stomatal closure and reduction of transpiration[118,123]; and (c) smaller increase in accumulation of free proline[122]. Both *L. peruvianum* and *L. pennellii* seem to have greater tonoplast resistance to K^+ transport[124]. Among the wild plants, *L. pennellii* has the highest succulence in its major vegetative parts and the largest amount of Na^+ and smallest amount of K^+ in the shoot[119]. The responses of the wild tomato species to high NaCl salinity, relative to those of the cultivated species, are summarized in Table 2.

According to Epstein[26], concentration in the past on systematically distinct groups of plants from the extremes of the spectrum of salt tolerance caused comparative physiological-genetic study not to receive the attention it merits. The wild salt-tolerant tomato species, especially *L. pennellii* and *L. cheesmanii*, can be easily hybridized with the cultivated species[89] and are therefore ideally suited for comparative physiological-genetic study. Attempts are now being made to estimate the number and identity of the major genes controlling components of the mechanism of salt tolerance in Lycopersicon by linkage studies and by analysis of the frequency distributions of progeny derived from crosses of the relatively sensitive cultivated species *L. esculentum* and the tolerant wild species *L. pennellii*. These two species differ in well-defined alleles at enzymatic loci that can be used for linkage tests[126]. Results of an analysis of frequency distributions indicate that a low leaf K^+ level, which characterizes the wild species, is dominant over a high level, and that the number of genes controlling the level of leaf K^+ in these species is low[117,119].

The tomato belongs to the Solanaceae family, which includes some of the most economically important plants and in which great progress has been made in developing tissue and cell culture techniques[27]. The application of these high-potential techniques for physiological study and genetic manipulation of stress tolerance is only starting to be realized[115]. Some of their potential has been demonstrated with calli originating from wild and cultivated tomato plants, which responded to high salt similar to the whole plant[121]. It was concluded, therefore, that better osmotic adjustment at the cellular level characterizes the wild plants under salt stress. Protoplasts isolated from the leaves of *L. peruvianum* plants divided and grew better in a saline medium than those of the cultivated species[93]. Rosen and Tal[93] proposed that this

Table 2. Responses of the wild salt-tolerant tomato species to high NaCl levels, relative to those of the cultivated tomato

Characteristic	Response	Reference
Growth		
Whole plant (including germination)	Less disturbed	17, 114, 119
Callus tissue	Less disturbed	121
Naked protoplasts	Less disturbed	91
Excised shoot apex	Less disturbed	125
Excised leaf	No difference	125
Ion balance		
Na^+, Cl^-	Higher in leaves and shoot apices	119
Cl^-/Na^+	Lower	119
K^+	Lower in leaves, higher in roots	124
K^+/Na^+	Lower	124
Regulation	Higher succulence, less K^+ uptake by the root, lower tonoplast permeability to K^+	18, 119 114, 124
Water balance		
Water potential, Relative water content	Smaller decrease	17, 114, 123
Stomatal conductance, Transpiration	Greater decrease	118, 123
Root resistance (no salt)	Higher	118
Membranes		
No salt	Greater sensitivity to dehydration and heat stresses	120
Salt treated	Less sensitivity to heat stress	120
Osmotic adjustment		
Proline	Smaller increase	122
Carbohydrates (myoinositol)	Greater increase?	95
Photosynthesis and respiration	No difference	123
Hormone balance	Lower abscisic acid concentration	118

difference be used to establish a selection system for somatic hybridization by which genes for salt tolerance will be transferred to *L. esculentum.*

With respect to the genetic makup, the tomato genome is relatively simple and well mapped[2].

The knowledge gathered on the physiology and genetics of salt tolerance in tomato is being applied to potato[33], which is more important economically but less amenable to experimentation.

It is necessary to decide which of the characteristics of the wild species to include and which to discard to efficiently breed salt-tolerant commercial varieties of tomato (Table 2). Tolerance is defined by Rosielle and Hamblin[92] 'by a small difference in productivity between stress and non-stress environments.' The slow-growing wild species of tomato are defined as salt tolerant since they are characterized by a smaller inhibition of growth under saline conditions than occurs in the cultivated species. According to Rosielle and Hamblin[92], selection for tolerance is worthwhile only when an increase in yield in stress environments is the objective, since such selection will decrease both mean productivity and yield in non-stress environments. This limitation should be taken into consideration if and when transfer of salt tolerance from the wild to the cultivated species is being contemplated. The characteristics that may play an important role in the mechanism of salt tolerance in the wild relatives of tomato and are, therefore, worth special attention during breeding are:

1. *Osmotic adjustment* As mentioned previously, better osmotic adjustment could be the mechanism responsible for the superior performance of some tomato species under salinity. The response of *L. cheesmanii* to salt, which is similar to that of *L. peruvianum* and *L. pennellii*, was interpreted by Rush and Epstein[94] as a typical halophytic response. They suggested that the positive correlation between Na^+ content and salt tolerance can be used as an index of salt tolerance in tomatoes. Another view is expressed by Sacher *et al.*[97], who worked with tomato hybrids. They compared the response to salinity of stable breeding lines derived from self-crossed progeny (F_9 generation) of the backcross of the F_1 (*L. esculentum* × *L. pennellii*) to the cultivated parent. No selection for salt tolerance was performed during preparation of the lines. A low ratio of foliar ion content in salt-grown plants to that in control plants (defined as the ion regulation index) was interpreted as indicating good regulation of the accumulation of a particular ion. A positive correlation was found in these plants between salt tolerance (expressed as the dry weight of salt-treated plants relative to that of control plants) and a low ion regulation index for Na^+. Unlike halophytic species, none of the lines took up salt in proportion to the amount applied. They suggested that the positive correlation in the hybrids resulted from an interaction between an ionic homeostasis inherited from *L. esculuntum* and a tolerance system contributed by *L. pennellii*, without the need for halophytic-type salt accumulation. Further research is needed to determine the significance of the 'halophytic' mode of osmoregulation in the wild tomatoes and the 'nonhalophytic' mechanism in the salt-tolerant progeny of the *L. esculentum* × *L. pennellii* cross.

2. *Succulence* Plants of the wild *L. pennellii* species are characterized by a large increase in the succulence of all their parts under salinity. This trait as well as the ability to retranslocate excess ions from shoot to root via the phloem were proposed as important components of the mechanism that regulates ion concentration in the shoots of halotypes that lack salt glands and bladders[47,48,70].

3. *Low Cl^-/Na^+ ratio* The lower Cl^-/Na^+ ratio in the wild tomato plants under high NaCl salinity may indicate a greater efficiency of NO_3^- uptake under saline conditions. Nitrate was found to compete successfully with Cl^- in the tomato[51].

4. *Low K^+/Na^+ ratio* The lower K^+/Na^+ ratio in the wild tomato plants under salinity may result from replacement of K^+ by Na^+ or from a greater affinity for K^+ in K^+-requiring enzymes[36].

5. *Water use efficiency* The smaller decrease in growth and the larger decrease in stomatal conductance and transpiration rate in the wild relatives of tomato under salinity may indicate an increase in the efficiency of water use under salt stress. The results of our (Tal, Itai, Bejerano – unpublished) recent study support this suggestion, which is not in agreement with the commonly accepted view[31] that water use efficiency is a constant and invariable attribute of herbaceous C3 plants. An additional example of an increase in the efficiency of water use under high salinity was reported in *Spartina alternifora*[59].

The value of these attributes for breeding salt-tolerant crops will no doubt be clarified further after the genomes of the wild and cultivated species meet and interact during breeding. Obviously, their applicability for breeding will also depend on how easy it is to measure them.

Concluding remarks

Although experience to date in the breeding and selection of salt-tolerant plants has been relatively limited, there is reason to be optimistic that, as with drought tolerance, genetic progress can be made by relying only upon empirical testing of yield under high salinity, without an understanding of the physiological, biochemical and genetic bases of salt tolerance. However, the efficiency of breeding programs can be increased only by adoption of new selection criteria based on knowledge of the physiological processes limiting crop production under conditions of stress and of their genetic control.

Overall prerequisites for efficient breeding programs in general and for increasing salt tolerance in particular are the availability of sufficient genetic variability and of reliable selection criteria. At present, neither prerequisite is met. Information on potential sources of genetic variability is incomplete and must therefore be supplemented with more descriptive research. Except in a few cases, the value of the various attributes proposed as selection criteria still requires experimental testing.

Progress from the stage of making general statements of this nature to their application depends, in my opinion, on adopting an interdisciplinary approach concentrating on representative model species that meet a number of defined criteria. The genus Lycopersicon was proposed as a model group of species for dicotyledons. Hordeum may be a useful model for monocotyledons.

Other essential requirements that have not generally been emphasized enough are: (a) greater appreciation and use of the physiological knowledge and breeding experience acquired for stresses other than salinity, and (b) greater appreciation of the ecological strategies of stress-tolerant species in general. Appreciation of these principles should make the planning of programs to increase stress tolerance more effective. It is especially important when wild species are to be used as gene sources.

Acknowledgement I thank Dr. M C Shannon for his valuable comments.

References

1 Altieri M A, Letourneau D K and Davis J R 1983 Developing sustainable agroecosystems. BioScience 33, 45–49.
2 Anonymous 1977 Linkage summary. Rep. Tomato Genet. Coop. 27, 5.
3 Barton K A and Brill W J 1983 Prospects in plant genetic engineering. Science 219, 671–676.
4 Batie S S and Healy R G 1983 The future of American agriculture. Sci. Am. 248, 27–35.
5 Bishop J A, Keill O C and MacNair M R 1981 The number of genes on the second chromosome of *Drosophila melanogaster* and a comment on the genic structure of eukaryotes. Heredity 46, 151–159.
6 Blum A 1979 Genetic improvement of drought resistance in crop plants: a case for sorghum. *In* Stress Physiology in Crop Plants. Eds. H Mussell and R C Staples. pp 429–446. Wiley, New York.
7 Borlaug N E 1983 Contributions of conventional plant breeding to food production. Science 219, 689–693.
8 Borowitzak L J 1981 Solute accumulation and regulation of cell water activity. *In* The Physiology and Biochemistry of Drought Resistance in Plants. Eds. L G Paleg and D Aspinall. pp 97–130. Academic Press, Sidney.
9 Boyer J S 1982 Plant productivity and environment. Science 218, 443–448.
10 Catarino F M and Trewavas A J 1970 Metabolic changes in nucleic acids associated with the development of succulence. Phytochemistry 9, 1807–1809.
11 Caten C E 1977 Genetical and physiological analysis of continuous variation in fungi. Heredity 39, 428.
12 Cavalier-Smith T 1978 Nuclear volume control by nucleoskeletal DNA, selection for cell volume and cell growth rate, and the solution of the DNA C-value paradox. J. Cell Sci. 34, 247–278.
13 Chapin F S III 1980 The mineral nutrition of wild plants. Ann. Rev. Ecol. System. 11, 233–260.

14 Chapin FS III and Bieleski R L 1982 Mild phosphorus stress in barley and a related low-phosphorus-adapted barleygrass: phosphorus fractions and phosphate absorption in relation to growth. Physiol. Plant. 54, 309–317.
15 Dambroth M and El Bassam N 1983 Low input varieties: definition, ecological requirements and selection. Plant and Soil 72, 365–377.
16 Davidson E H and Britten R J 1979 Regulation of gene expression: possible role of repetitive sequences. Science 204, 1052–1059.
17 Dehan K and Tal M 1978 Salt tolerance in the wild relatives of the cultivated tomato: responses of *Solanum pennellii* to high salinity. Irrig. Sci. 1, 71–76.
18 Dhindsa R S and Cleland R E 1975 Water stress and protein synthesis. I. Differential inhibition of protein synthesis. Plant Physiol. 55, 778–781.
19 Duffield R D, Croy L I and Smith E L 1972 Inheritance of nitrate reductase activity, grain protein, and straw protein in a hard red winter wheat cross. Agron. J. 64, 249–251.
20 Elfring C 1983 Beyond the green revolution. BioScience 33, 617–618.
21 Epstein E 1972 Mineral Nutrition of Plants: Principles and Perspectives. Wiley, New York.
22 Epstein E 1976 Genetic potentials for solving problem of soil mineral stress: adaptation of crops to salinity. *In* Workshop on Plant Adaptation to Mineral Stress in Problem Soils. Ed. M J Wright. pp 73–82. Beltsville, Maryland.
23 Epstein E 1980 Responses of plants to saline environments. *In* Genetic Engineering of Osmoregulation: Impact on Plant Productivity for Food, Chemicals, and Energy. Eds. D W Rains, R L Valentine and A Hollaender. pp 7–21. Plenum Press, New York.
24 Epstein E 1983 Crops tolerant of salinity and other mineral stresses. *In* Better Crops for Food. Ciba Found. Symp. 97, pp 61–82. Pitman Books, London.
25 Epstein E and Norlyn J D 1977 Seawater-based crop production: a feasibility study. Science 197, 249–251.
26 Epstein E, Norlyn J D, Rush D W, Kingsbu R W, Kelley D B, Cunningham G A and Wrona A F 1980 Saline culture of crops: a genetic approach. Science 210, 399–404.
27 Evans D A, Sharp W R, Ammirato P V and Yamada Y (Eds.) 1983 Handbook of Plant Cell Culture Vol. I. Techniques for Propagation and Breeding. Macmillan, New York.
28 Fedlman M 1983 Gene transfer from wild species into cultivated plants. Genetika 15, 145–161.
29 Finlay K W and Wilkinson G N 1963 The analysis of adaptation in a plant breeding programme. Aust. J. Agric. Res. 14, 742–754.
30 Fischer R A 1977 Physiology of yield improvement-past and future. *In* 3rd Conference SABRAO Canberra, pp 1–13.
31 Fischer R A 1981 Optimizing the use of water and nitrogen through breeding of crops. Plant and Soil 58, 249–278.
32 Fischer R A and Turner N C 1978 Plant productivity in the arid and semiarid zones. Annu. Rev. Plant Physiol. 29, 277–317.
33 Fischer S and Tal M Salt tolerance in the wild relatives of the cultivated potato: growth and ion accumulation in *Solanum tuberosum*, *S. kurzianum* and *S. polytrichon* under high salinity. (*In preparation*).
34 Flavell R B 1982 Chromosomal DNA sequences and their organization. *In* Encyclopedia of Plant Physiology, New series, Vol. 14B, Nucleic Acids and Proteins in Plants II. Eds. B. Parthier and D Boulter. pp 46–71. Springer-Verlag, Berlin.
35 Fleck J, Durr A and Hirth L 1983 Gene expression in freshly isolated protoplasts from *Nicotiana sylvestris*. *In* 6th Intern. Protoplast Symp. Poster Proc. Eds. I Potrykus, C T Harms, A Hinnen, R Hutter, P J King and R D Shillito, pp 240–241. Basel.
36 Flowers T J and Lauchli A 1983 Sodium versus potassium: substitution and compartmentation. *In* Encyclopedia of Plant Physiology, New series, Vol. 15B, Inorganic Plant Nutrition. Eds. A Lauchli and R L Bieleski. pp 651–681. Springer-Verlag, Berlin.
37 Flowers T J, Troke P F and Yeo A R 1977 The mechanism of salt tolerance in halophytes. Annu. Rev. Plant Physiol. 28, 89–121.
38 Gale J and Zeroni M 1983 Cultivation of plants in brackish water in controlled environment

agriculture. *In* Salinity Tolerance in Plants: Strategies for Crop Improvement. Eds. R C Staples and G A Toenniessen. pp 363–380. Wiley, New York.

39 Gallagher L W, Soliman K M, Qualset C O, Huffaker R C and Rains D W 1980 Major gene control of nitrate reductase activity in common wheat. Crop Sci. 20, 717–721.

40 Gerloff G C and Gabelman W H 1983 Genetic basis of inorganic plant nutrition. *In* Encyclopedia of Plant Physiology, New Series, Vol. 15B, Inorganic Plant Nutrition. Eds. A Lauchli and R L Bieleski. pp 453–480. Springer-Verlag, Berlin.

41 Gifford R M and Evans L T 1981 Photosynthesis, carbon partitioning, and yield. Annu. Rev. Plant. Physiol. 32, 485–509.

42 Greenway H and Munns R 1980 Mechanisms of salt tolerance in nonhalophytes. Annu. Rev. Plant Physiol. 31, 149–190.

43 Grime J P 1979 Plant Strategies and Vegetation Processes. Wiley, Chichester.

44 Hadson M J, Smith M M, Wainwright S J and Opik H 1981 Cation cotolerance in a salt tolerant clone of *Agrostis stolonifera* L. New Phytol 90, 253–261.

45 Helal H M and Mengel K 1979 Nitrogen metabolism of young barley plants as affected by NaCl-salinity and potassium. Plant and Soil 51, 457–462.

46 Huffaker R C and Peterson L 1974 Protein turnover in plants and possible means of its regulation. Annu. Rev. Plant Physiol. 25, 363–392.

47 Jennings D H 1968 Halophytes, succulence and sodium in plants – a unified theory. New Phytol. 67, 899–911.

48 Jennings D H 1976 The effects of sodium chloride on higher plants. Biol. Rev. 51, 453–486.

49 Jeschke W D 1983 Cation fluxes in excised and intact roots in relation to specific and varietal differences. Plant and Soil 72, 197–212.

50 Jinks J L and Towey P 1976 Estimating the number of genes in a polygenic system by genotype assay. Heredity 37, 69–81.

51 Kafkafi U, Valoras N and Letey J 1982 Chloride interaction with nitrate and phosphate nutrition in tomato (*Lycopersicon esculentum* L.). J. Plant Nutr. 5, 1369–1385.

52 Kuiper P J C 1984 Functioning of plant cell membranes under saline conditions: membrane lipid composition and ATPases. *In* Salinity Tolerance in Plants: Strategies for Crop Improvement. Eds. R C Staples and G A Toenniessen. pp 77–91. Wiley, New York.

53 Kulieva F B, Shamina Z B and Strogonov B P 1975 Effect of high concentrations of sodium chloride on multiplication of cells of *Crepis capillaris in vitro.* Sov. Plant Physiol. 22, 107–110.

54 Kylin A and Quatrano R S 1975 Metabolic and biochemical aspects of salt tolerance. *In* Plants in Saline Environments. Eds. A Poljakoff-Mayber and J Gale. pp 147–167. Springer-Verlag, Berlin.

55 Lande R 1983 The response to selection of major and minor mutations affecting a metrical trait. Heredity 50, 47–65.

56 Leopold A C and Willing R P 1984 Evidence for toxicity effects of salt on membranes. *In* Salinity Tolerance in Plants: Strategies for Crop Improvement. Eds. R C Staples and G A Toenniessen. pp 67–76. Wiley, New York.

57 Levitt J 1980 Responses of Plants to Environmental Stresses, 2nd Edition Vol. II. Water, Salt and other Stresses. Academic Press, New York.

58 Lewontin R C 1977 The relevance of molecular biology to plant and animal breeding. *In* Proc. Intern. Conf. Quant. Genet. (Aug. 16–21, 1976). Eds. E Pollak, O Kempthorne and T B Bailey Jr. pp 55–62. The Iowa State University Press, Ames.

59 Longstreth D J and Strain B R 1977 Effects of salinity and illumination on photosynthesis and water balance of *Spartina alterniflora* Loisl. Oecologia 31, 191–199.

60 Luttge U and Higinbotham N 1979 Transport in Plants. Springer-Verlag, New York.

61 Maas E V and Nieman R H 1978 Physiology of plant tolerance to salinity. *In* Crop Tolerance to Suboptimal Land Conditions. Ed G A Jung. pp 277–299. Am. Soc. Agron. Publ. 32.

62 MacNair M R 1983 The genetic control of copper tolerance in the yellow monkey flower *Mimulus guttatus.* Heredity 50, 283–293.

63 Mahon J D 1983 Limitations to the use of physiological variability in plant breeding. Can. J. Plant Sci. 63, 11–21.
64 Mather K and Jink J L 1971 Biometrical Genetics. Chapman and Hall, London.
65 Mayr E 1966 Populations, Species and Evolution. Harvard University Press, Cambridge, Massachusetts.
66 Meiri A and Poljakoff-Mayber A 1967 The effect of chloride salinity on growth of bean leaves in thickness and in area. Isr. J. Bot. 16, 115–123.
67 Moorby J and Besford R T 1983 Mineral nutrition and growth. *In* Encyclopedia of Plant Physiology, New series, Vol. 15B, Inorganic Plant Nutrition. Eds. A Läuchli and R L Bieleski. pp 481–527. Springer-Verlag, Berlin.
68 Morton R J, Tal M and Benzioni A Ion imbalance in *Capsicum annuum scabrous diminutive*, a wilty mutant of pepper. III. Partial characterization of ATPase activity associated with root plasma membranes (*In preparation*).
69 Mudie P J 1974 The potential economic uses of halophytes. *In* Ecology of Halophytes. Eds. R J Reimold and W H Queen. pp 565–597. Academic Press, New York.
70 Munns R, Greenway H and Kirst G O 1983 Halotolerant eukaryotes. *In* Encyclopedia of Plant Physiology, New series, Vol. 12C. Physiological Plant Ecology III. Eds. O L Lange, P S Nobel, C B Osmond and H Ziegler. pp 59–135. Springer-Verlag, Berlin.
71 Munns R, Greenway H, Delane R and Gibbs R 1982 Ion concentration and carbohydrate status of the elongating leaf tissue of *Hordeum vulgare* growing at high external NaCl. II. Causes of the growth reduction. J. Exp. Bot. 33, 574–583.
72 Murai N, Sutton D W, Murray M G, Slightom J L, Merlo D J, Reichert N A, Sengupta-Gopalan C, Stock C A, Barker R F, Kemp J D and Hall T C 1983 Phaseolin gene from bean is expressed after transfer to sunflower via tumor inducing plasmid vectors. Science 222, 476–482.
73 Nagl W 1982 DNA endoreduplication and differential replication. *In* Encyclopedia of Plant Physiology, New series, Vol. 14B. Nucleic Acids and Proteins in Plants II. Eds. B Parthier and D Boulter. pp 111–124. Springer-Verlag, Berlin.
74 Nagl W 1983 Nuclear organization-physical aspects of gene regulation. *In* Kew Chromosome Conference II. Eds. P E Brandham and M D Bennett. pp 55–61. George Allen and Unwin, London.
75 Nieman R H 1965 Expansion of bean leaves and its suppression by salinity. Plant Physiol. 40, 156–161.
76 Nieman R H and Maas E V 1978 The energy charge of salt-stressed plants. Sixth Intern. Biophysics Cong. Abs. p. 121.
77 Nieman R H and Shannon M C 1976 Screening plants for salinity tolerance. *In* Workshop on Plant Adaptation to Mineral Stress in Problem Soils. Ed. M J Wright. pp 359–367. Beltsville, Maryland.
78 Norlyn J D 1980 Breeding salt-tolerant crop plants. *In* Genetic Engineering of Osmoregulations. Impact on Plant Productivity for Food, Chemicals, and Energy. Eds. D W Rains, R C Valentine and A Hollander. pp 293–309. Plenum, New York.
79 Osmond C B, Bjorkman O and Anderson D J 1980 Physiological Processes in Plant Ecology. Springer-Verlag, Berlin.
80 Ownbey R S and Mahall B E 1983 Salinity and root conductivity: differential responses of a coastal succulent halophyte, *Salicornia virginica*, and a weedly glycophyte, *Raphanus sativus*. Physiol Plant. 57, 189–195.
81 Poljakoff-Mayber A 1982 Biochemical and physiological responses of higher plants to salinity stress. *In* Biosaline Research. A Look to the Future. Ed. A San Pietro. pp 245–269. Plenum, New York.
82 Ponnamperuma F N 1982 Breeding crop plants to tolerate soil stresses. *In* Plant Improvement and Somatic Cell Genetics. Eds. I K Vasil, W R Scowcroft and K J Frey. pp 73–97. Academic Press, New York.
83 Ponnamperuma F N 1983 Role of cultivar tolerance in increasing rice production in saline lands. *In* Salinity Tolerance in Plants: Strategies for Crop Improvement. Eds. R C Staples and G A Toenniessen. pp 255–271. Wiley, New York.

84 Quarrie S A 1981 Genetic variability and heritability of drought-induced abscisic acid accumulation in spring wheat. Plant Cell Environ. 4, 147–151.
85 Quarrie S A 1983 Genetic differences in abscisic acid physiology and their potential uses in agriculture. *In* Abscisic Acid. Ed. F T Addicott, pp 365–419. Praeger, New York.
86 Rains D W 1982 Developing salt tolerance. Calif. Agric. 36, 30–31.
87 Ramage R T 1980 Genetic methods to breed salt tolerance in plants. *In* Genetic Engineering of Osmoregulation. Impact on Plant Productivity for Food, Chemicals and Energy. Eds. D W Rains, R C Valentine and A Hollaender. pp 311–318. Plenum, New York.
88 Rice T B and Carlson P S 1975 Genetic analysis and plant improvement. Annu. Rev. Plant Physiol. 26, 279–308.
89 Rick C M 1979 Tomato germplasm resources. *In* Proceeding 1st Intern. Symp. Tropical Tomato, Oct. 23–27. Ed. R Cowell. pp 214–224. Asian Vegetable Research and Development Center, Pub. 78–59.
90 Rick C M 1982 The potential of exotic germplasm for tomato improvement. *In* Plant Improvement and Somatic Cell Genetics. Eds. I K Vasil, W R Scowcroft and K J Frey. pp 1–28. Academic Press, New York.
91 Rosen A and Tal M 1981 Salt tolerance in the wild relatives of the cultivated tomato: responses of naked protoplasts isolated from leaves of *Lycopersicon esculentum* and *L. peruvianum* to NaCl and proline. Z. Pflanzenphysiol. 102, 91–94.
92 Rosielle A A and Hamblin J 1981 Theoretical aspects of selection for yield in stress and non-stress environments. Crop Sci. 21, 943–946.
93 Rush D W and Epstein E 1976 Genotypic responses to salinity. Differences between salt-sensitive and salt-tolerant genotypes of the tomato. Plant Physiol. 57, 162–166.
94 Rush D W and Epstein E 1981 Comparative studies on the sodium, potassium and chloride relations of a wild halophytic and a domestic salt-sensitive tomato species. Plant Physiol. 68, 1308–1313.
95 Sacher R F and Staples R C 1984 Inositol and sugars in adaptation of tomato and *Lycopersicon esculentum* × *L. pennellii* breeding lines to salt. Plant Physiol. (*In press*).
96 Sacher R F and Staples R C 1984 Salinity shock proteins in tomato roots. *In* Third Intern. Workshop on Biosaline Research. Poster Section. March 19–23, 1984, Beer-Sheva, Israel.
97 Sacher R F, Staples R C and Robinson R W 1983 Ion regulation and response of tomato to sodium chloride: a homeostatic system. J. Am. Soc. Hort. Sci. 108, 566–569.
98 Saric M R 1983 Theoretical and practical approaches to the genetic specificity of mineral nutrition of plants. Plant and Soil 72, 137–150.
99 Scowcroft W R and Larkin P J 1982 Somatoclonal variation: a new option for plant improvement. *In* Plant Improvement and Somatic Cell Genetics. Eds. I K Vasil, W R Scowcroft and K J Frey. pp 159–178. Academic Press, New York.
100 Setter T L, Greenway H and Kuo J 1982 Inhibition of cell division by high external NaCl concentration in synchronized cultures of *Chlorella emersonii*. Aust. J. Plant Physiol. 9, 179–196.
101 Shannon M C 1979 In quest of rapid screening techniques for plant salt tolerance. Hort. Sci. 14, 587–589.
102 Shannon M C 1980 Testing salt tolerance variability among tall wheat grass lines. Agron. J. 70, 719–722.
103 Shannon M C 1984 Breeding, selection and the genetics of salt tolerance. *In* Salinity Tolerance in Plants – Strategies for Crop Improvement. Eds. R C Staples and G A Toenniessen. pp 231–254. Wiley, New York.
104 Shoe M G T and Gale J 1983 Effect of sodium chloride stress and nitrogen source on respiration, growth and photosynthesis in lucerne (*Medicago sativa* L.). J. Exp. Bot. 34, 1117–1125.
105 Smillee R M and Nott R 1982 Salt tolerance in crop plants monitored by chlorophyll fluorescence *in vivo*. Plant Physiol. 70, 1049–1054.
106 Smith M M, Hodson M J, Opik H and Wainwright S J 1982 Salt-induced ultrastructural damages to mitochondria in root tips of a salt-sensitive ecotype of *Agrostis stolenifera*. J. Exp. Bot. 33, 886–895.

107 Solbrig O T 1979 Life forms and vegetation patterns in desert regions. *In* Arid Land Plant Resources. Eds. J R Goodin and D K Northington. pp 82–95. Texas Tech. Univ., Lubbock, Texas.
108 Sparague G F, Alexander D E and Dudley J W 1980 Plant breeding and genetic engineering: a perspective. BioScience 30, 17–21.
109 Stavarek S J and Rains D W 1984 Cell culture techniques, selection and physiological studies of salt tolerance. *In* Salinity Tolerance in Plants: Strategies for Crop Improvement. Eds. R C Staples and G A Toenniessen. pp 321–334. Wiley, New York.
110 Stebbins G L 1966 Chromosomal variation and evolution. Science 152, 1463–1469.
111 Steveninck R F M 1976 Effect of hormones and related substances in ion transport. *In* Encyclopedia of Plant Physiology, New series, vol. 2, Transport in Plants. Eds. U Luttge and M G Pitman. pp 307–342. Springer-Verlag, Berlin.
112 Strogonov B P, Kabanov V V, Shevjakova N I, Lapina L P, Komizerko E I, Popov B A, Dostanova R K and Prykhodko L S 1970 Structure and Function of Plant Cells in Saline Habitats. New Trends in the Study of Salt Tolerance. Israel Program for Scientific Translations, Jerusalem. Wiley, New York.
113 Summerfield R J 1980 The contribution of physiology to breeding for increased yields in grain legume crops. *In* Opportunities for Increasing Crop Yields. Eds. R G Hurd, P V Biscoe and D Dennis. pp 51–69. Pitman Adv. Pub. Prog., Boston.
114 Tal M 1971 Salt tolerance in the wild relatives of the cultivated tomato: responses of *Lycopersicon esculentum*, *L. peruvianum* and *L. esculentum* minor to sodium chloride solution. Aust. J. Agric. Res. 22, 631–638.
115 Tal M 1983 Selection for stress tolerance. *In* Handbook of Plant Cell Culture, Vol. I. Techniques for Propagation and Breeding. Eds. D A Evans, W R Sharp, P V Ammirato and Y Yamada. pp 461–488. Macmillan, New York.
116 Tal M 1984 Physiological genetics of salt resistance in higher plants: studies on the level of the whole plant and isolated organs, tissues and cells. *In* Salinity Tolerance in Plants: Strategies for Crop Improvement. Eds. R C Staples and G A Toenniessen. pp 301–320. Wiley, New York.
117 Tal M Salt tolerance in the wild relatives of the cultivated tomato: responses of plants of F_2 and F_3 generations to high NaCl salinity (*In preparation*).
118 Tal M and Gavish U 1973 Salt tolerance in the wild relatives of the cultivated tomato: water balance and abscisic acid in *Lycopersicon esculentum* and *L. peruvianum* under low and high salinity. Aust. J. Agr. Res. 24, 353–361.
119 Tal M and Shannon M C 1983 Salt tolerance in the wild relatives of the cultivated tomato: responses of *Lycopersicon esculentum*, *L. cheesmanii*, *L. peruvianum*, *Solanum pennellii*, and F_1 hybrids to high salinity. Aust. J. Plant Physiol. 10, 109–117.
120 Tal M and Shannon MC 1983 Effects of dehydration and high temperature on the stability of leaf membranes of *Lycopersicon esculentum*, *L. cheesmanii*, *L. peruvianum*, and *Solanum pennellii*. Z. Pflanzenphysiol. 112, 411–416.
121 Tal M, Heikin H and Dehan K 1978 Salt tolerance in the wild relatives of the cultivated tomato: responses of callus tissues of *Lycopersicon esculentum*, *L. peruvianum* and *Solanum pennellii* to high salinity. Z. Pflanzenphysiol. 86, 231–240.
122 Tal M, Katz A, Heikin H and Dehan K 1979 Salt tolerance in the wild relatives of the cultivated tomato: proline accumulation in *Lycopersicon esculentum* Mill., *L. peruvianum* Mill. and *Solanum pennellii* Cor. treated with NaCl and polyethylene glycol. New Phytol. 82, 349–355.
123 Taleisnik-Gertel E and Tal M Salt tolerance in the wild relatives of the cultivated tomato: growth, water balance and photosynthesis in *Lycopersicon esculentum*, *L. peruvianum*, and *L. pennellii* under NaCl salinity. (*In preparation*).
124 Taleisnik-Gertel E and Tal M Stress tolerance in the wild relatives of the cultivated tomato: potassium balance in *Lycopersicon esculentum*, *L. peruvianum*, and *L. pennellii* (*In preparation*).
125 Taleisnik-Gertel E, Tal M and Shannon M C 1983 The response to NaCl of excised fully differentiated and differentiating tissues of the cultivated tomato *Lycopersicon esculentum*,

and its wild relatives *L. peruvianum* and *Solanum pennellii.* Physiol Plant. 59, 659–663.

126 Tanksley S D, Medina-Filho H and Rick C M 1982 Use of naturally-occurring enzyme variation to detect and map genes controlling quantitative traits in an interspecific backcross of tomato. Heredity 49, 11–25.

127 Thoday J M 1961 Location of polygenes. Nature 191, 368–370.

128 Thompson J N 1977 Analysis of gene number. Stadler Genet. Symp. 9, 63–82.

129 Tsenov E I, Strogonov B P and Kabanov V V 1973 Effect of NaCl on content and synthesis of nucleic acids in tomato tissues. Sov. Plant Physiol. 20, 40–46.

130 Waisel Y 1972 Biology of Halophytes. Academic Press, New York.

131 Walbot V 1980 Molecular biology of higher plants. *In* Biology of Crop Productivity. Ed. P S Carlson. pp 343–382. Academic Press, New York.

132 Weinberg R 1967 Effect of sodium chloride on the activity of a soluble malate dehydrogenase from pea seeds. J. Biol. Chem. 242, 3000–3006.

133 Wittwer S H 1978 The next generation of agricultural research. Science 199, 375.

134 Wittwer S H 1980 The shape of things to come. *In* The Biology of Crop Productivity. Ed. P S Carlson. pp 413–459. Academic Press, New York.

135 Woolhouse H W 1981 Crop physiology in relation to agricultural production: the genetic link. *In* Physiological Processes Limiting Plant Productivity. Ed. C B Johnson. pp 1–21. Butterworths, London.

136 Woolhouse H W 1983 Toxicity and tolerance in the response of plants to metals. *In* Encyclopedia of Plant Physiology. New series. Vol. 12C. Physiological Plant Ecology III. Eds. O L Lange, P S Nobel, C B Osmond and H Ziegler. pp 245–300. Springer-Verlag, Berlin.

137 Wyn Jones R G, Brady C J and Speirs J 1979 Ionic and osmotic relations in plant cells. *In* Recent Advances in the Biochemistry of Cereals. Eds. D L Laidman and R G Wyn Jones. pp 63–103. Academic Press, London.

138 Wyn Jones R G and Gorham J 1980 Osmoregulation. *In* Encyclopedia of Plant Physiology, New series, Vol. 12C. Physiological Plant Ecology III. Eds. O L Lange, P S Nobel and C B Osmond. pp 35–58. Springer-Verlag, Berlin.

139 Wyn Jones R G and Pollard A 1983 Proteins, enzymes and inorganic ions. *In* Encyclopedia of Plant Physiology, New series, Vol. 15B. Inorganic Plant Nutrition. Eds. A Lauchli and R L Bieleski. pp 528–562. Springer-Verlag, Berlin.

140 Yeo A R 1981 Salt tolerance in the halophyte *Suaeda maritima* L. Dum.: intracellular compartmentation of ions. J. Exp. Bot. 32, 487–497.

141 Yeo A R 1983 Salinity resistance: physiologies and prices. Physiol. Plant. 58, 214–222.

Plant and Soil 89, 227–241 (1985). Ms. BR 15

Principles and strategies in breeding for higher salt tolerance

M. C. SHANNON
USDA, ARS, U.S. Salinity Laboratory, Riverside, CA 92501, USA

Key words Salt tolerance Selection Stress

Summary Salinity is an environmental component that usually reduces yield. Recent advances in the understanding of salt effects on plants have not revealed a reliable physiological or biochemical marker that can be used to rapidly screen for salt tolerance. The necessity of measuring salt tolerance based upon growth in saline relative to non-saline environments makes salt tolerance measurements and selection for tolerance difficult. Additionally, high variability in soil salinity and environmental interactions makes it questionable whether breeding should be conducted for tolerance or for high yield. Genetic techniques can be used to identify the components of variation attributable to genotype and environment, and the extent of genetic variation in saline and nonsaline environments can be used to estimate the potential for improving salt tolerance. Absolute salt tolerance can be improved best by increasing both absolute yield and relative salt tolerance.

Introduction

The need to develop crops with higher salt tolerance has increased tremendously within the last decade; however, methods and approaches for accomplishing this goal are either not available or not well-defined because of a lack of information required for conducting successful research. The objective of this paper is to describe how some progress may be made toward obtaining plants with enhanced salt tolerance by meshing the theoretical aspects of breeding plants for salt tolerance with the present knowledge concerning the effects and mechanisms of salt stress. For brevity, it will be assumed that the readers are familiar with background information on the physiology and biochemistry of salt effects on plants. If not, there are several recent and thorough reviews on these subjects[6, 8, 12, 15]. Generally, successful breeding requires the existence of variability and a means of stable transfer of a character from one individual to another. A method of recognition or identification of the character is required so that a selection protocol may be devised. The choice of an easily identifiable character, or marker, is highly desirable to simplify the selection procedure. From this point, all approaches basically include selection and breeding, but there is much diversity of selection criteria and philosophies. The actual method of breeding and inducing recombination is dependent upon the breeding system of the crop (*i.e.*, inbreeding, outbreeding, *etc.*).

Variability in plant responses to salinity

Extremely high salt concentrations kill plants through the combined effects of ion toxicity and decreased water potential. More moderate, but still high, concentrations of salt may cause leaf burn and severe growth inhibition. At more moderate to low salinities, differences in growth rate are often the only visual symptom of salt effect, though in some species there is also a slight darkening and/or thickening of the leaf. Although this general description of salt effects applies to all higher plants, there is a great deal of variation among species and even cultivars within a species in the degree of response to any level of salinity.

The subject of genetic variability among plants in responses to salt has been reviewed on several occasions[14,24,25]. Variability in salt response, as with almost any character, increases through the progression from variety to species to genus to family. Although salt-tolerant species are found among a wide range of plant families, some families have greater numbers of tolerant species. For example, the Chenopodiaceae include many salt-tolerant species, ranging from wild species of Salicornia and Atriplex to the cultivated sugarbeet (*Beta vulgaris*). Representative salt-tolerance species in the Gramineae are wild Spartina species, tall wheatgrass (*Elytrigia pontica*), bermudagrass (*Cynodon dactylon*), barley (*Hordeum vulgare*) and sugarcane (*Saccharum officinarum*). Variability in salt tolerance within species has been reported with increasing frequency in recent years; however, the choice of criteria by which tolerance is measured has not been consistent among investigators[14,15,19,23,26].

Tolerant phenotypes

Despite the multitude of comparisons that have been made between salt-tolerant and salt-sensitive species, few distinctive phenotypic markers reliably distinguish sensitive from tolerant individuals. A useful phenotype for plant breeding is an observable or otherwise measurable character. The phenotype (P) of a plant is the expression of the plant's genetic constitution (G), or genotype, modified by the environment (E). This relationship is commonly stated as $P = G + E$. Some traits that investigators have tried to use as markers include the Na^+ exclusion mechanisms in soybean (*Glycine max*)[1], tomato (Lycopersicon species)[20], and wheatgrass (*Elytrigia pontica*)[23], Cl^- exclusion in Citrus[16], and pollen sterility in rice (*Oryza sativa*)[2]. Even these were applicable only within a limited range of salt concentrations and are species- or variety-specific. For example, Rush and Epstein[19,20] found ion accumulation in the wild tomato species *Lycopersicon cheesmanii* to be a reflection of the halophytic nature of that ecotype, whereas Sacher *et al.*[21,22] have found no relationship between ion accumulation and salt tolerance in crosses between *L. peruvianum* and its cultivated descendent *L. esculentum*.

Phenotypic markers have not been used successfully in any concerted breeding effort. One problem is that it is unlikely that salt tolerance is determined by a single gene. Salt tolerance is probably the expression of a number of genes, and the importance of the expression of each is dependent upon its interaction with other salt tolerance genes and the external salt concentration.

The study of the inheritance of any character centers around its variation. Components causing variation (V) must be separated from one another. For example, in a heterogeneous population Vp = Vg + Ve, whereas in a genetically homogeneous population Vg = 0 and Vp = Ve. If both homogeneous and heterogeneous populations can be tested in the same environment, then Ve can be measured with the homogeneous population, and the Vg of the heterogeneous population can thus be determined. Tomato is self-pollinating, and parental and F1 populations can be considered to be highly homogeneous. Leaf Na accumulation in parental and F1 populations of tomato had coefficients of variation (CV) of about 22% (Table 1). The F2 populations, which are heterogeneous, had CV values of 55%, an increase in genetic variance of about 33% for this character as a result of intraspecific crosses.

Measurement of salt tolerance

Maas and Hoffman[11] have proposed that salt tolerance be defined by a simple linear equation, described in general terms as y = a + b (x), where y = relative yield and x = salinity of the saturated soil extract of the root zone measured in units of electrical conductivity (dS/m). Actual salt response data are often nonlinear and follow a reverse sigmoid curve with a small but significant increase in yield at low salinities. However, a simple linear equation is applicable to most of the data obtained from salt tolerance studies[11]. Significant parameters of the equation include the slope (b), which is a representation of decline in yield per increment salinity increase, and the threshold, which is the lowest salinity at which a significant yield reduction occurs. Both yield decline and threshold are measured relative to nonsalinized plants. Such measurements are useful for comparing salt tolerance between species with large differences in yield potential, but carefully controlled conditions are necessary to obtain meaningful threshold and slope values.

Table 1. Comparative leaf Na ion contents in small populations of tomato (Lycopersicon) genotypes. Values determined on 6-mm-diameter leaf punches from 6-week-old plants grown in nutrient solutions containing 75 mol m^{-3} salt (5:1 NaCl: $CaCl_2$) (n = population size)

n	Genotype	Na[a]	CV[b]	
	Parental			
40	*L. esculentum* (Le)	288	16.9	
42	*L. cheesmanii* (Lc)	798	22.9	
39	*L. peruvianum* (Lp)	876	21.8	
33	*Solanum pennelli* (Sp)	1822	17.0	
				19.7
	F1			
35	Le × Lc	372	20.7	
36	Le × Lp	286	23.2	
35	Le × Sp	588	28.2	
				24.0
	F2			
84	Le × Lc	268	48.1	
96	Le × Lp	149	62.1	
				55.1

[a] Values given as mmol kg^{-1} dry weight as determined by dry weight measurements made on separate samples.
[b] CV Coefficient of variation.

Threshold values are especially sensitive to interaction with other environmental factors. Additionally, information is lost in comparing high- and low-yielding varieties of the same species on a relative basis. Such loss of information may be avoided if regressions are based on absolute yields in special circumstances in which plant breeding information is required.

Salinity variation within a field and interactions between environmental factors contribute to large standard errors and prohibit comparisons between individuals of segregating generations and between populations with high natural variability or environmental instability. Therefore, other criteria for the determination of salt tolerance are necessary.

Salinity interactions

Salt tolerance is usually expressed in terms of yield, a quantitative character that is also influenced by a wide array of factors other than salinity. For the breeder, yield is a simple phenotypic trait, and salinity is regarded as one component of the environment. There are two types of interaction, however, that may complicate this rather simple relationship. First, other environmental factors can positively or negatively modify the effect of salinity, as has been discussed elsewhere[11,12,24]. Second, genotype-environment interaction means that the 'best genotype' at one salt concentration may not be the 'best' at another. For instance, at a moderate salinity the criterion for selection may be significant yield reduction[26], while at high salinity survival might be the criterion[3,14]. Unfortunately, there is a fundamental gap in our understanding concerning the relationship between tolerance at low and high salinities. This is an important problem in choosing the salt concentration that will be used for screening. While it may be suspected that the physiological mechanisms that prevent damage at low salt concentrations are not the same as those that contribute to tolerance at extremely high concentrations, there is very little experimental evidence on which to base a conclusion. The principles of quantitative genetics could be used to study this problem. It has been demonstrated that a character measured in two different environments may be regarded as two characters rather than one[4]. By regarding tolerance at high and low salinities as different characters, variances of response at low and high salinity can be used to determine the genetic correlation between the characters. A low correlation would mean that different genes are involved and a high correlation would indicate the involvement of the same gene or genes.

Ontogenetic drift is another factor that can modify the relationship between phenotype and environment. Ontogenetic drift is actually a change in genotypic expression with plant development. During plant growth, the form and function of the various organs change. A plant's response and, consequently, its effective salt tolerance are influenced by its ontogenic stage. Thus, salinity effects may vary depending upon the growth stage at the time of stress. Indeed, salt tolerance at germination is not consistently related to tolerance during emergence, vegetative growth, flowering or fruiting[10]. The plant's ability to respond to salt stress depends upon the genes that are functioning at the stage of development during which the stress occurs. Salt-tolerance measurements over a short period of time during the early vegetative growth of muskmelon (*Cucumis melo*) indicated that salt tolerance gradually increases as the seedling matures (Fig. 1) and confirmed data obtained by other methods[11,12].

The effects of and responses to salt stress may also be modified by changes that have occurred due to previous stresses. For example, during rapid vegetative growth, the plant strives to maintain as large a photosynthetic area as possible and to maintain a root system that will support the plant and provide the required water and nutrients. Salinity affects this balance, typically reducing vegetative top growth more

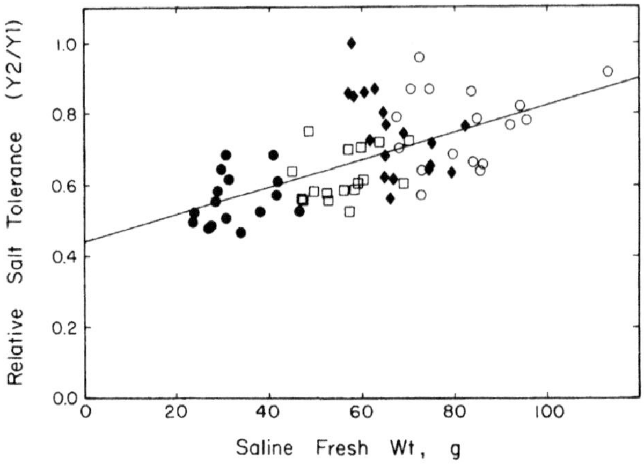

Fig. 1. Salt tolerance of 68 muskmelon (*Cucumis melo*) cultivars in four sand culture experiments in the greenhouse. In each experiment (shown by different symbols) cultivars were irrigated with control (1 dS/m) or saline (≃ 8 dS/m) nutrient solutions. Time between salination and harvest varied between 2 and 4 weeks with experiment and was directly related to the differences in average fresh weight. Y1 is fresh weight of controls and Y2 is fresh weight of salinized plants.

than root growth[12]. Such changes may affect the severity of and response to subsequent stresses.

Selection criteria

In plant breeding the most widely utilized criteria for selection are (1) mean yield, (2) regression response on site mean yield, and (3) uniformity of regression. Selection for mean yield is fundamentally a selection for the most high-yielding variety. Selection for the latter two criteria are based on stable performances across a range of environments. In the commercial agriculture characteristic of highly developed countries, high yield is the most important objective, whereas stable performance is more important in subsistence agriculture.

Regression response on site mean yield has been employed by breeders to measure yield stability across environments[5] and is similar to the regression method used to measure salt tolerance. A linear regression calculation is made of variety yield against the mean yield of all varieties at each location or environment. The use of mean yield allows a numeric grading of environments across sites and seasons. Because simple variety yields are plotted against the mean of all varieties, the population mean has a slope of 1.0. Slopes greater than unity indicate less stability than slopes smaller than 1.0. A small slope is also an indication of phenotypic stability in the salt tolerance regression. In a range of saline environments there are fundamentally two ways to change this slope. First, increasing the yield at high salinities would effectively reduce the slope, increase environmental stability, and improve salt tolerance, provided that the threshold remains stable. Alternatively, an independent increase in the threshold would make the slope greater and decrease environmental stability. The concept of selection for salt tolerance is actually to select for high yields across a range of saline environments, because salinity is a continuous rather than a discrete variable and is highly variable, even in the same field. These factors make it imperative that varieties are selected that have stable performance at more than a single salt concentration. Recent efforts to study salt tolerance of wheat and barley varieties at field locations have led Richards[17] to conclude that salt tolerance may best be improved by breeding for high yield in nonsaline environments. The main reason for this conclusion was high variability in soil salinity throughout the field. The benefits of high yield versus those of relative salt tolerance have been discussed previously[24]. At low salinities high-yielding varieties may produce better yields than lower-yielding varieties that may have a better degree of salt tolerance. For example, in a 1976 field study the muskmelon cultivar 'Top Mark' was higher-yielding than the cultivars 'Hale's Best' and 'PMR 45' up to saturated soil extract electrical conductivities (EC_e) of about 5 dS/m (Fig. 2). At higher salinities, above an EC_e of 8 dS/m, the lower-yielding varieties had higher yields than 'Top Mark.' Thus, it appears that 'Hale's Best' and 'PMR 45' have more environmental stability, lower yield reductions for corresponding increases in salinity, and higher relative salt tolerance. Although the yields of muskmelon beyond an EC_e of 8 dS/m cannot be considered to be economically important, the genetic characteristics that make melons tolerant to salinity may be a serious consideration to the plant breeder.

The point made by Richards[17] is that high variability in salinity within a field allows anywhere from maximal to zero yields within different portions of the same field. He estimated that fields with high, medium and low salinity problems consist of about 25, 50, and 75 percent, respectively, of nonsaline areas in which 100% yields can be obtained. Thus, high-yielding varieties can contribute substantially to the total yields in these areas. For example, in Table 2 fields similar to those previously reported[17] are classified within salinity ranges and an overall weighted mean salinity is given. The yields of the three muskmelon cultivars from Figure 2 were predicted based on their salt-tolerance thresholds and slope values for each area within these fields

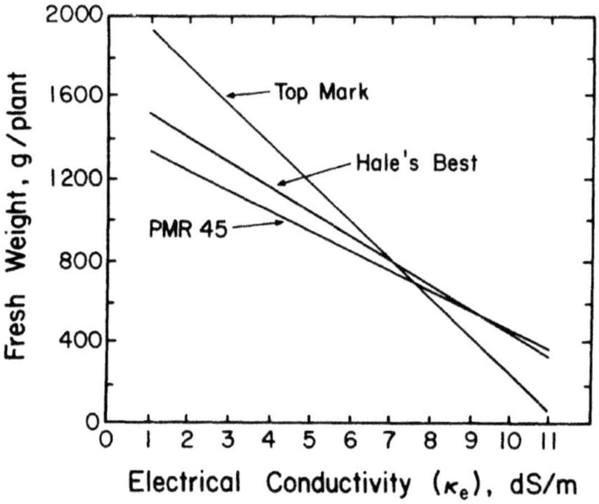

Fig. 2. Variation in absolute salt tolerance of 3 muskmelon (*Cucumis melo*) cultivars. ($\kappa_e = EC_e$).

Table 2. Approximate distributions of saline areas in four model fields as percent of the total field area

	Soil salinity (EC_e) (dS/m)				
Field salinity	0–4	4–8	8–12	> 12	Wt mean EC_e
Nonsaline	100	0	0	0	2.0
Low	75	15	5	5	3.6
Medium	50	20	10	20	6.0
High	25	25	15	35	8.4

(Table 3). It is apparent that 'Top Mark,' because of its high-yielding characteristics, has greater yields up to about 5 dS/m. Based on calculations made according to the Maas and Hoffman convention, 'Top Mark' has the highest threshold value of the three cultivars and the lowest slope value (− 8.8).

Tolerance to stress may be defined in a variety of ways, for example as the difference between yield in a nonsaline (Y1) and in a saline environment (Y2), or Y1 − Y2. This is a measure of tolerance on an absolute scale, since no account is made of differences in normal growth rates. Relative tolerance may be calculated by dividing by Y1. Reducing this equation, we obtain 1 − (Y2/Y1). In both definitions the highest tolerance is indicated by the smallest number. Another definition of tolerance is the mean productivity, which may be mathematically expressed as Y1 + Y2/2.

An example of the results obtained using these tolerance measurements is given for data from greenhouse experiments in which 48 lettuce (*Lactuca sativa*) cultivars were grown under nonsaline (control,

Table 3. Marketable fruit yields of the muskmelon cultivars 'Top Mark,' 'Hale's Best,' and 'PMR 45' in g/plant that would be harvested from fields of variable salinity as described in Table 2

Field salinity	Soil salinity (EC_e) (dS/m) 0–4	4–8	8–12	> 12	Total yield
	'Top Mark'[a]				
Nonsaline	1969	0	0	0	1969
Low	1477	180	22	0	1679
Medium	985	240	44	0	1269
High	492	301	65	0	858
	'Hale's Best'[b]				
Nonsaline	1482	0	0	0	1482
Low	1112	150	26	1	1289
Medium	741	199	51	5	996
High	371	249	77	9	706
	'PMR 45'[c]				
Nonsaline	1317	0	0	0	1317
Low	987	140	27	8	1162
Medium	658	186	55	32	931
High	329	233	82	56	700

[a] Maximum yield = 2178, threshold = 0.91, and slope = − 8.8
[b] Maximum yield = 1662, threshold = 0.52, and slope = − 7.3
[c] Maximum yield = 1439, threshold = 0.73, and slope = − 6.7

1.0 dS/m) and saline (9.3 dS/m) conditions in sand cultures[26]. Fresh leaf weights were compared on the bases of absolute and relative salt tolerance and mean yields. There was high variability in fresh weight in both control and saline environments, but generally variability in nonsaline environments was higher than in saline ones (Fig. 3a). This is a common finding for any environmental stress. It has been found that decreased stress permits a greater range of phenotypic expression and increases variances among lines in 17 out of 20 cases in oats (*Avena* species)[9]. Variation in tolerance as measured by Y1 − Y2 shows less variability in fresh weight than is shown by Y1 alone (Fig. 3b). The correlation between tolerance measured in this manner and control fresh weight shows the direct effect of high-yield potential. Relative salt tolerance is highly variable among low-yielding lines and less variable among high-yielding lines (Fig. 3c). There is a general decrease in relative salt tolerance as the yield potential increases. Conversely, relative salt tolerance is generally higher among lines that have high yields in saline conditions (Fig. 3d). These data indicate that selection for high yield under nonsaline conditions would slightly favor varieties with low relative salt tolerance, and selection for high yield in saline conditions would favor varieties with higher relative salt tolerance.

The question of whether selection for tolerance to stress is even

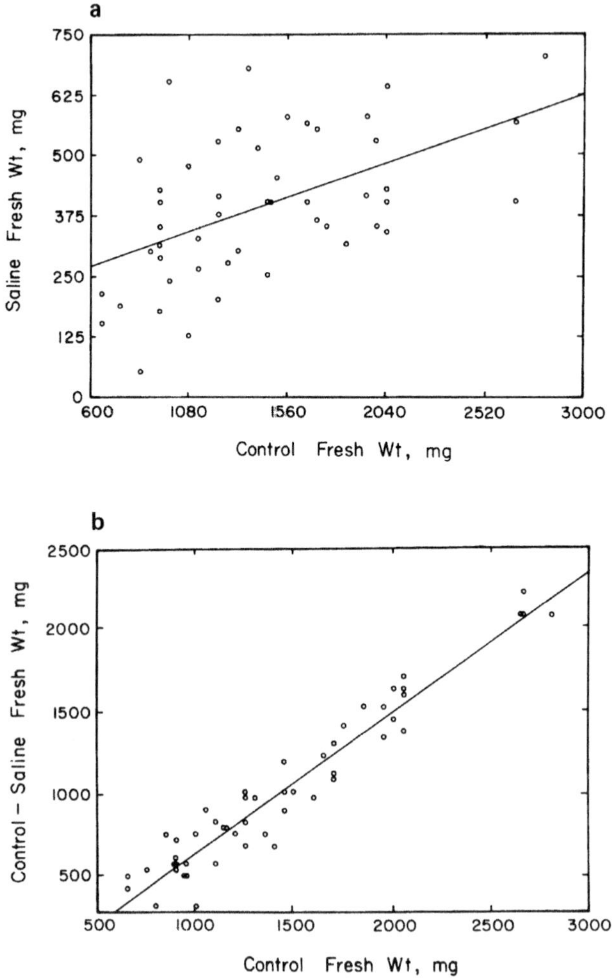
a
Saline Fresh Wt, mg
750
625
500
375
250
125
0
600
1080
1560
2040
2520
3000
Control Fresh Wt, mg
b
Control - Saline Fresh Wt, mg
2500
2000
1500
1000
500
500
1000
1500
2000
2500
3000
Control Fresh Wt, mg

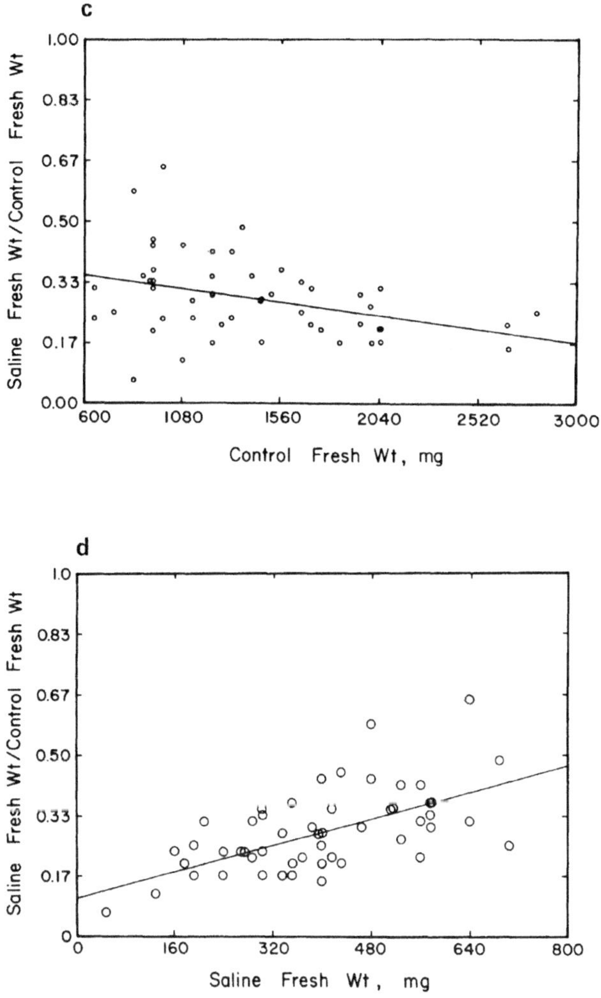

Fig. 3. (a) Correlations between fresh weight yields for 48 lettuce (*Lactuca sativa*) cultivars grown in greenhouse sand cultures and irrigated with either control (1 dS/m) or saline (10 dS/m) nutrient solutions.

(b) Correlation between control fresh weights and salt tolerance as measured by the difference between average fresh weight yield under control and saline conditions.

(c) Correlation between control fresh weights and relative salt tolerance (fresh weight of salinized plants/control).

(d) Correlation between saline fresh weights and relative salt tolerance.

worthwhile has recently been proposed[18]. It was shown that selection for mean productivity will normally increase yields in both stress and nonstress environments, provided that genetic variance in the stress environment is less than in the nonstress environment and that the genetic correlation of yields in both environments is highly negative[18]. Alternatively, selection for tolerance under these same conditions would decrease the mean productivity. This implies a negative correlation between mean yield and tolerance. Indeed, the data in Fig. 3c support this contention. For a positive relationship to exist between tolerance and mean productivity the genetic variance in stress environments must be greater than in nonstress environments. This is not common, and it has been pointed out that varieties with high phenotypic stability usually have low mean yields[5,7,13].

Breeding method

Breeding approaches include backcrossing, progeny testing, and pure line selection. The main factor that determines the method of plant breeding, aside from the induction of useful variation, is the breeding system. Two-thirds of the world's food supply is provided by inbred species. Two of the major cereals, rice (*Oryza sativa*), and wheat (*Triticum* species), and most of the important oil and protein crops are inbred species. Outbreeding crops include most of the forages, sugarcane, and most of the tree fruits.

Inbreeding crops are developed by pedigree selection. In this process, one parent is chosen for genetic properties that are useful (*e.g.*, salt tolerance), while the other is chosen for its complementing and desirable agronomic characteristics. Self pollination leads rapidly to homozygous lines which can then be tested for overall improvements. This system is self perpetuating in that it generates some improved lines that can be used as parents in subsequent hybridizations. A disadvantage of this system is that early generations can be expected to be heterogeneous and quite unstable in their response to environmental interaction. This makes early selection difficult, because the individual is the unit of selection and the effects of variation in salinity across a field plot or greenhouse bench can result in some loss of segregates with genes for salt tolerance. Environmental effects during early generations may be reduced by selection under controlled conditions in the laboratory or through the use of saline irrigation water in field selections.

Other practices can be used to reduce environmental effects when screening is conducted in the field. Selection may be delayed until the F6–F8 generation. Bulk generations can be maintained to assure the maximum number of recombinants. The increased homozygousity of these generations will increase the correspondence between genotype and phenotype. However, increases in selection efficiency (in the field) cannot be expected, because environmental interactions may still be rather large. Another method that may be used to reduce environmental effects in inbreeding species is single-seed descent. Individual lines derived through a single unselected seed saved in each generation through F6–F8 may be used to test for a particular character. The advantages of this method are that many populations can be grown through several generations under conditions favoring reduced generation time, and a good correspondence can be expected between genotype and phenotype. Additionally, the F2 population can be used to predict a potential range of variation.

The above systems may be adapted to salt tolerance. Modifications that may improve selection efficiency include the use of saline irrigation water to reduce natural variation in soil salinity and the use of numerous salt concentrations so that regressions against mean yields can be used to determine salt resistance with more reliability.

The production of F1 hybrids takes a prominent place in the improvement of outcrossed species. Genetic advances among hybrids are usually the result of making crosses between selected inbred lines that have been chosen for their individual characters as well as their combining abilities. In species that do not have economical F1 seed production systems, the concept of population improvement is usually adopted. Most forms of population improvement include both mass selection and various types of progeny testing designed to accumulate alleles through recurrent selection over generations. The details of this plan have been the subject of much research and controversy, but the average gain by most of the methods indicates that they are all almost equally effective. The complication of salt tolerance would not seem to make one any more effective than another, but a particular selection procedure for salt tolerance may favor one method over another. For instance, recurrent selection within families includes replicative tests at several locations. This procedure could be developed to minimize the interactions between salinity and environment as well as those between genotype and phenotype. The reconstitution of advanced generations from parents with high scores for salt tolerance at several locations could eventually lead to a quantitative improvement over time.

Tissue culture techniques may eventually improve both of the above mentioned methods if haploid isolation procedures can be developed to allow the single-step creation of homozygous diploids from F1 hybrids. Another possibility is the use of somaclonal variation to induce the needed variability to select for salt tolerance.

Ultimately, the objective should be to increase tolerance by increasing both mean yield and yield stability. An understanding of the basic principles of both salt tolerance and plant breeding will be required to do this.

References

1 Abel G H 1963 Inheritance of the capacity for chloride inclusion and chloride exclusion in soybeans. Crop Sci. 9, 697–698.

2. Akbar M and Yabuno T 1977 Breeding for saline-resistant varieties of rice. IV. Inheritance of delayed-type panicle sterility induced by salinity. Jap J. Breed. 27, 237–240.

3 Epstein E, Norlyn J D, Rush D W, Kingsbury R W, Kelly D B, Cunningham G A and Wrona A W 1980 Saline culture of crops: a genetic approach. Science 210, 399–404.

4 Falconer D S 1952 The problem of environment and selection. Am. Nat. 86, 293–298.

5 Finlay K W and Wilkinson G N 1963 The analysis of adaptation in a plant breeding programme. Aust. J. Agric. Res. 14, 742–754.

6 Flowers T J, Troke P F and Yeo A R 1977 The mechanism of salt tolerance in halophytes. Annu. Rev. Plant Physiol. 28, 89–121.

7 Frey K J 1964 Adaptation reaction of oat strains selected under stress and non-stress environmental conditions. Crop Sci. 4, 55–58.

8 Greenway H and Munns R 1980 Mechanisms of salt tolerance in non-halophytes. Annu. Rev. Plant Physiol. 31, 149–190.

9 Johnson G R and Frey K J 1967 Heritabilities of the quantitative attributes of oats (Avena sp.) at varying levels of environmental stress. Crop Sci. 7, 43–46.

10 Maas E V 1985 Salt tolerance of plants. *In* Handbook of Plant Science in Agriculture. Ed. B R Christie, CRC Press Inc., Boca Raton, Florida (*In press*).

11 Maas E V and Hoffman G J 1977 Crop salt tolerance – current assessment. J. Irrig. Drainage Div., ASCE 103(IR2), 115–134.

12 Maas E V and Nieman R H 1978 Physiology of plant tolerance to salinity. *In* Crop Tolerance to Suboptimal Land Conditions. Ed. G A Jung. pp 277–299. Am. Soc. Agron. Spec. Publ. 32. Madison, WI.

13 Mederski H J and Jeffers D L 1973 Yield response of soybean varieties grown at two soil moisture stress levels. Agron. J. 65, 410–412.

14 Norlyn J D 1980 Breeding salt-tolerant crop plants. *In* Genetic Engineering of Osmoregulation – Impact on Plant Productivity for Food Chemicals and Energy. Basic Life Sciences. Vol 14. Eds. D W Rains, R C Valentine and A Hollaender. pp 293–309. Plenum Press, New York.

15 Pasternak D, Twersky M and De Malach Y 1979 Salt resistance in agricultural crops. *In* Stress Physiology in Crop Plants. Eds. H W Mussell and R C Staples. pp 127–142. Wiley Interscience, New York.

16 Ream C L and Furr J D 1976 Salt tolerance of some Citrus species, relatives, and hybrids tested as rootstocks. J. Am. Soc. Hort. Sci. 101, 265–267.

17 Richards R A 1983 Should selection for yield in saline regions be made on saline or non-saline soils? Euphytica 32, 431–438.

18 Rosielle A A and Hamblin J 1981 Theoretical aspects of selection for yield in stress and non-stress environments. Crop Sci. 21, 943–946.

19 Rush D W and Epstein E 1976 Genotypic responses to salinity. Differences between salt-sensitive and salt-tolerant genotypes of tomato. Plant Physiol. 57, 162–166.

20 Rush D W and Epstein E 1981 Comparative studies on the sodium, potassium, and chloride relations of a wild halophytic and a domestic salt-sensitive tomato species. Plant Physiol. 68, 1308–1313.

21 Sacher R F, Staples R C and Robinson R W 1982 Salt tolerance in hybrids of *Lycopersicon*

esculentum × *Solanum pennellii* and selected breeding lines. *In* Biosaline Research: A Look to the Future. Ed. A San Pietro, pp 325–336. Plenum Press, New York.

22 Sacher R F, Staples R C and Robinson R W 1983 Ion regulation and response of tomato to sodium chloride: a homeostatic system. J. Am. Soc. Hort. Sci. 108, 566–569.

23 Shannon M C 1978 Testing salt tolerance variability among tall wheatgrass lines. Agron. J. 70, 719–722.

24 Shannon M C 1982 Genetics of salt tolerance: new challenges. *In* Biosaline Research. A Look to the Future. Ed. A San Pietro. pp 271–282. Plenum Press, New York.

25 Shannon M C 1984 Breeding, selection and the genetics of salt tolerance. *In* Salinity Tolerance in Plants – Strategies for Crop Improvement. Eds. R C Staples and G H Toenniessen. pp 313–331. Wiley International, New York.

26 Shannon M C, McCreight and Draper J H 1983 Screening tests for salt tolerance in lettuce. J. Am. Soc. Hort. Sci. 108, 225–230.

Plant and Soil 89, 243–252 (1985). Ms. BR 16

Selection and breeding for salinity tolerance *in vitro**

P. SPIEGEL-ROY and GOZAL BEN-HAYYIM
Institute of Horticulture, Agricultural Research Organization, The Volcani Center, Bet-Dagan, Israel

Key words *Citrus sinensis* *Citrus aurantium* Cloning Ovular callus Sodium chloride Stable variants

Summary Selection for tolerance to NaCl in *Citrus sinensis* and *C. aurantium* has been carried out in agar and suspension cultures. Callus was subjected to culture media containing up to 0.17 *M* NaCl for ten passages. Selected cell lines were grown for three passages on media without salt before further tests on saline media. Four stable tolerant cell lines, differing in degree of tolerance, have been selected from *C. sinensis*. Four lines of similar tolerance have been selected from *C. aurantium*. The stability of most lines was very satisfactory. Most *C. sinensis* lines grew well in media containing up to 0.2 *M* NaCl, and *C. aurantium* lines in media of up to 0.15 *M* NaCl.

Embryos were regenerated in most selected cell lines from *C. sinensis* and, more sporadically, from *C. aurantium*. Addition of 0.5–0.6% NaCl to the media often enhanced embryogenesis. Embryos from a selected line of *C. sinensis* showed higher tolerance to NaCl in the medium than comparable embryos from an unselected line.

Single embryos derived from both selected and unselected cell lines of *C. sinensis* were successfully cloned. A limited comparison of plantlets from one tolerant line (R14) with plantlets from unselected control lines showed better adaptation of the former to salt (0.085 to 0.12 *M* NaCl in the medium), and a lesser degree of leaf burn symptoms.

Introduction

A major advantage of cell culture for genetic manipulation with higher plants is the opportunity to select for new phenotypes from large cell populations cultured under defined conditions. For a trait to prove agriculturally useful, the characteristics selected from testing *in vitro* should be stable in the absence of the stress agent, and the cell lines carrying that trait should be easy to propagate, *i.e.* show satisfactory initiation and development of embryos. Moreover, one has to be aware that selection for the novel phenotype is conducted at a level of differentiation different from that at which expression of the phenotype is desired[3].

Cell lines exhibiting resistance or, rather, relative tolerance to salt stress have been reported[5,6,8,19,23]. The relative degree of salt tolerance of callus cultures derived from two barley species (*Hordeum vulgare* and *H. jubatum*) seemed to correspond to that found in the whole plants[20]. Salt resistant plants from a cell line of tobacco have been regenerated. Progeny obtained by selfing of regenerated plants also

* Contribution No. 1045-E, 1984 series.

proved tolerant to sodium chloride. In recent experiments a selected NaCl-tolerant tobacco line (after 50 generations in cell culture) lacked stability[8].

Fruit trees, except for the date palm and to a lesser extent also pistachio, olive and pomegranate, are very sensitive to irrigation water with a high salt content[15]. Citrus species rank among the most sensitive to saline water[4]. Thus, breeding for rootstocks or, possibly, own-rooted cultivars with a high tolerance to salt could prove very important in many arid areas where irrigation is needed. Selection for genetic variants derived from cultured material might also serve as an important step for further conventional breeding. In the present paper, isolation of several stable salt-tolerant callus lines, from both 'Shamouti' orange (*Citrus sinensis*) and 'Sour orange' (*C. aurantium*), will be described, as will the performance of embryos and plantlets derived from two of these callus lines. The ovular (nucellar) callus system seems well adapted for mutant selection because of the following characteristics: growth and maintenance of callus without exogenous growth substances (auxins and cytokinins), embryogenic capacity, stable diploid state, and ease of cloning.

Materials and methods

Callus cultures were initiated from nucelli and ovules of 'Shamouti' (*Citrus sinensis*) and 'Sour orange' (*C. aurantium*) by methods developed in our laboratory[11,14]. Further details of the technique and of the selection procedure devised have been described recently[21]. For *C. sinensis*, growth and selection for tolerance to salt were done on a solid medium (agar), while for *C. aurantium* a suspension culture was used. Selection started with the inoculation of 1000 explants (40 mg fresh weight each) on an agar medium containing 0.085 *M* NaCl. After 5 weeks of growth the best-growing cultures were selected to initiate cell lines (ten replicates each). Recurrent selection was performed over ten passages of 5 weeks each, with 0.085 *M* NaCl in the medium. The five best explants were then selected on the basis of growth or embryogenic capacity. To test stability, the selected tolerant lines were subcultured for three passages in the absence of NaCl, then recultured in the presence of salt. The lines selected were grown on various media to determine the most suitable medium for embryo formation. Embryos were counted at 10× or 40× magnification. Small globular embryos were developed into cotyledonary embryos in suspension culture containing 2% sucrose, 1 g l^{-1} malt extract, and 1 mg l^{-1} gibberellic acid (GA_3). A procedure for cloning individual embryos was also developed, to provide uniform material for comparison of salt tolerance between plants derived from selected cell lines and plants of the wild type.

Ovular callus from 'Sour orange' was grown in suspension on a gyratory shaker (100 cycles/min), starting with 250-mg explants. For measurement of fresh weight, cells were collected by vacuum filtration through Whatman No. 1 filter paper after 3 weeks of growth. Selection was done in a manner similar to that described for 'Shamouti.' Because of the faster rate of growth of 'Sour orange,' transfers from the liquid medium were made every 21 days. Selection was done in the presence of 0.12 *M* NaCl. Because of partial loss of embryogenic capacity in the selected 'Sour orange' cell lines, no whole-plant comparisons have yet been made with this species between plantlets derived from selected cell lines and from a cell line of the wild type. Experiments to improve embryogenesis, as well as subsequent embryo growth and development in these cell lines, are still in progress.

Results

Salt tolerance and embryogenesis in selected callus

On the basis of the selection procedure outlined, ten 'Shamouti' orange callus lines with significantly better growth on NaCl than that of the sensitive control were isolated. Of these, four have been retained for further tests. Growth of these lines in the presence of 0.1 to 0.2 *M* NaCl, relative to the control, is shown in Figure 1. The four tolerant lines retained (R4, R7, R10 and R13) differ in their performance. Line R10 shows no decline in growth at 0.1 *M* NaCl relative to growth in a medium with no salt. This line retains over 35% relative growth at 0.2 *M* NaCl. The other three selected lines performed less satisfactorily, especially in the presence of salt concentrations exceeding 0.1 *M* NaCl. Still, line R4 showed 28% relative growth at 0.2 *M* NaCl. By comparison, the unselected control line achieved only 33% relative growth at 0.1 *M* NaCl and very little growth at 0.2 *M* NaCl. Stability tests for the four selected lines (results not shown) indicated that all lines had much better growth in the presence of 0.1 *M* NaCl than did the control when the test was performed after three passages on medium without salt. However, growth on 0.2 *M* NaCl after three passages without salt was decidedly better for some lines (R10) than for others (R4)[9].

Figure 2 presents the relative growth (percent of control) of the four selected cell lines of 'Sour orange' in the range of 0–0.2 *M* NaCl. All four lines were distinctly superior in growth to the unselected line in the presence of salt. The unselected line ceased to grow in the presence of 0.15 *M* NaCl. Differences in growth response of the selected lines were comparatively slight. One line (01-7-3) attained as much as 50% growth in the presence of 0.2 *M* NaCl, compared with the control. In stability tests (Fig. 2B) performed after three passages on a medium without salt, growth of line 01-7-3 was definitely superior to that of the unselected control, but inferior to that attained by the three selected lines 01-7-2, 01-7-4, and B-5-1. These three lines performed similarly, manifesting relatively good growth (50% of control) with 0.15 *M* NaCl, and very little growth with 0.2 *M* NaCl.

The selection of stable tolerant cell lines resulted, however, in impairment of the embryogenic capacity in certain lines, especially in the selected lines of 'Sour orange.' In 'Shamouti' orange, with galactose as a carbon source, only the unselected line L1 and the selected lines R4 and R13 maintained a fairly high embryogenic capacity. In further experiments the addition of 0.085 or even 0.17 *M* NaCl to the galactose medium significantly enhanced the capacity of lines R13 and R14 to form embryos. Line R14 at present forms embryos on both galactose

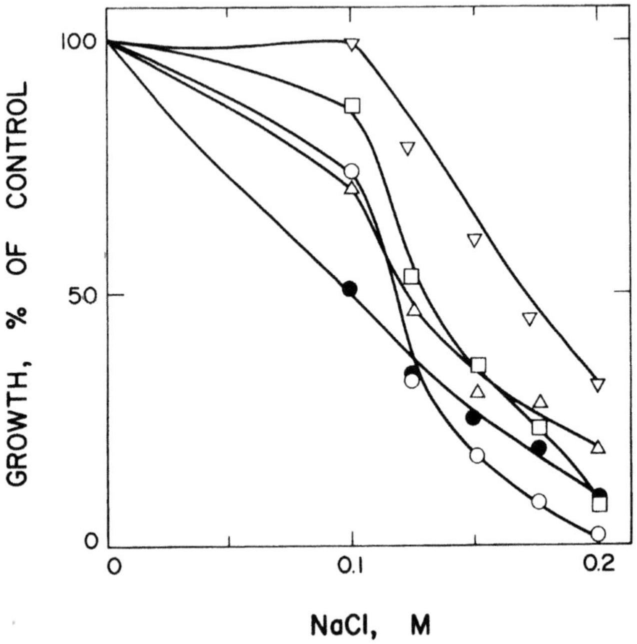

Fig. 1. Relative growth of the salt-sensitive and four salt-tolerant cell lines of 'Shamouti' orange (*Citrus sinensis*) after recurrent selection (ten passages) in the presence of 0.085 *M* NaCl. ● Control line; △ R4; ○ R7; ▽ R10; □ R13. Cells were collected after one month of growth. 100% growth for control line, R4, R7, R10 and R13: 430, 565, 650, 695 and 515 mg, respectively.

and sucrose, even without the addition of NaCl. Experiments performed recently showed that the latter line forms embryos quite readily also in media with sucrose as a carbon source. In contrast, line L1 (unselected control) forms no embryos on sucrose. Enhancement of the embryogenic capacity of line R14 was achieved also by including antigibberellins such as Alar (succinic acid, 2,2-methylhydrazide) or PP333 (1-(4-chlorophenyl)4,4-dimethyl-2(1,2,3-triazole-lyl)pentan-3-ol) in the galactose medium.

A more definite loss of embryogenic capacity and a greater difficulty in regaining this capacity was evident with selected salt-tolerant cell lines of 'Sour orange.' However, the control line also shows a decrease in embryogenic capacity. With 'Sour orange,' too, galactose proved the most suitable carbon source for the manifestation of embryogenic capacity, which was enhanced by the addition of 0.1 *M* NaCl to the medium. The possible role of antigibberellins in cell lines of 'Sour orange' is still unclear. None of these lines formed embryos in the presence of sucrose.

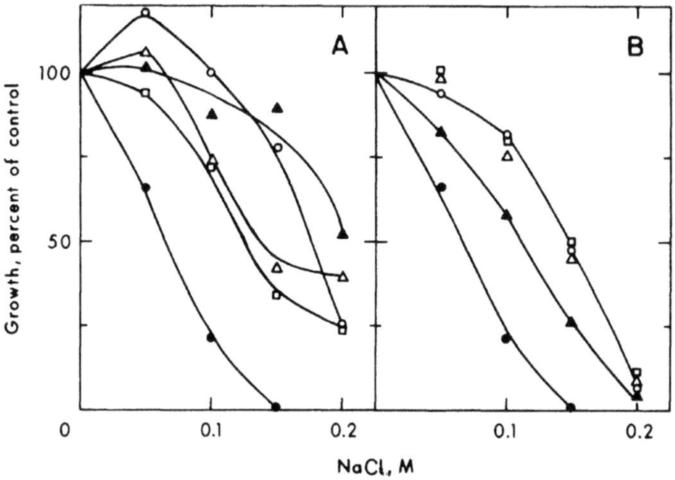

Fig. 2. Relative growth of salt-sensitive and four salt-tolerant cell lines of 'Sour orange' (*Citrus aurantium*) in the presence of NaCl. *A*: relative growth after recurrent selection (ten passages) on 0.17 m*M* NaCl. *B*: Relative growth after recurrent selection and three further passages in the absence of NaCl. ● Control line; ▲ 01-7-3; ○ 01-7-2; △ 01-7-4; □ B-5-1. 100% growth for control line, 01-7-3, 01-7-2, 01-7-4 and B-5-1: 309, 453, 530, 437 and 398 mg, respectively.

Salt tolerance of selected embryos

Small, globular embryos from a salt-tolerant 'Shamouti' orange line (R13) and from an unselected control line (L1) were placed in Murashige and Skoog's[17] basal medium (containing 5% sugar, without phytohormones), as modified by Murashige and Tucker[18], to which 0–0.25 *M* NaCl had been added. The relative weights of the embryos after 6 weeks of growth (compared with their weight on basal medium with no added salt: 312 mg) are shown in Table 1. Differences in final weight between

Table 1. Inhibition of growth of embryos derived from a salt-tolerant 'Shamouti' orange line and a non-selected susceptible line, on media with different NaCl concentrations. Results expressed as relative weight of embryos weighed after 6 weeks of growth (100% = 312 mg)

NaCl in medium (*M*)	R13 tolerant line (% of control)	L1 susceptible line (% of control)
0	100	100
0.085	51	40
0.130	39	40
0.170	20*	10**
0.210	16	8
0.255	11	6

* 60% browning
** 100% browning

embryos of the two lines were slight. However, while globular embryos from the salt-tolerant line developed into cotyledonary embryos (3–5 mm long) and remained green on media containing up to 0.13 *M* NaCl, embryos from the unselected cell line L1 turned brown in such media. On 0.17 *M* NaCl, 60% of embryos from the selected cell line turned brown, as compared with 100% of those derived from the unselected line.

Salt tolerance of selected rooted shoots and plantlets

A method of cloning embryos (P. Spiegel-Roy and S. Saad, unpublished) has been developed to improve the genetic uniformity and to obviate the effects of variations in embryo size and shape. Cloning was achieved by a series of steps involving a cytokinin for shoot proliferation and an auxin-like substance for rooting. With this procedure, uniformly rooted shoots could be obtained from both selected cell lines and the unselected control. These lines were compared with respect to their performance on media or in planting mixtures containing added NaCl.

Rooted shoots of two salt-tolerant lines, R7 and R14, were tested for their response to the presence of 0.12 *M* NaCl in the medium. The performance of these shoots was compared with that of shoots from an unselected line for over one month. Shoots of comparable size and vigor were selected. All shoots from plants derived from the unselected cell line showed severe stunting, leaf burn, and progressive wilting; however, 21 shoots of comparable size derived from plants obtained from the selected cell line R7 survived (out of a total of 55 rooted shoots, *i.e.* 40%). All of the latter showed no damage, while ten rooted shoots from R7 showed some leaf burn.

In another experiment, well-rooted plantlets about 3 cm tall were used. Plants from an unselected cell line, C32, served as a control; and comparable plants from a selected cell line, R14, were tested under the same conditions. The experiment was continued for 10 weeks and the results are summarized in Figure 3. Performance was rated on a scale of three: 1, no leaf burn; 2, partial leaf burn; and 3, total leaf burn. Altogether, 30 plants (15 from each line) were used. These were divided into three lots and grown in test tubes on basal medium, without NaCl, and with 0.085 and 0.12 *M* NaCl. No damage to either line was evident on basal medium without the addition of NaCl. In media containing 0.085 or 0.12 *M* NaCl, a larger proportion of plants from the unselected control showed leaf burn, and to a higher degree. The tolerance of plantlets derived from the selected cell line can be considered satisfactory, especially with 0.085 *M* NaCl. A large-scale comparison of plants in a potting medium of peat and perlite is now in progress in the greenhouse.

Discussion

There is increasing awareness of the possibilities and limitations in the use of techniques of callus and cell suspension culture for the development of new cell lines with valuable characters for agricultural use, particularly in relation to resistance to stress and to herbicides[2,3,16]. Cell lines from crop plants showing resistance to salinity have been reported[5,6,8,23], and in a few cases plants regenerated and were tested[19]. In one case a stepwise increase in NaCl in the medium resulted in a salt-tolerant Nicotiana line stable through 24 generations in the absence of added NaCl. Gene amplification was postulated as an explanation of the phenomenon[23]. Another possibility with our citrus material is that the enhanced ability to grow in the presence of comparatively high NaCl concentrations (0.15–0.2*M*) may have originated from pre-existing resistant cells in the ovular callus[9]. We isolated four tolerant, stable cell lines from *C. sinensis*, and four from *C. aurantium*. While the isolated lines of *C. sinensis* cv. 'Shamouti' differed in their degree of tolerance, those isolated from 'Sour orange' (*C. aurantium*) were quite similar to one another. The stability of most of the lines was satisfactory. No similar degree of stability was attained in experiments on the selection of a NaCl-tolerant tobacco cell line[8].

Maintenance of tolerance in cell lines throughout all subsequent

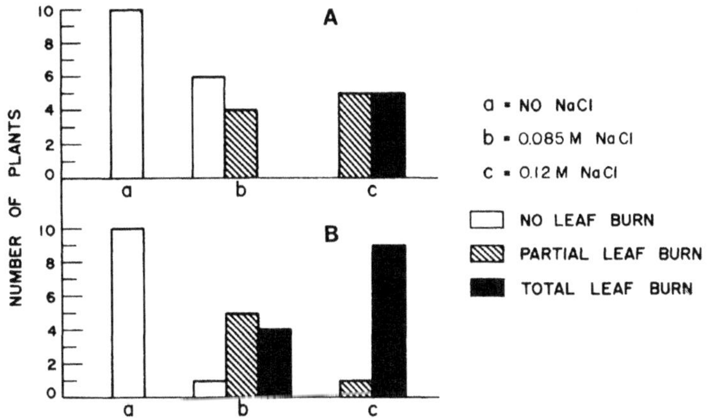

Fig. 3. Effect of sodium chloride in the medium on leaf burn in 'Shamouti' plantlets cloned from salt-tolerant (A) and salt-sensitive (B) ovular callus.

stages of plant development is considered to pose a problem. The theoretical basis involved has been discussed recently[3]. The morphogenetic characteristics of cells from NaCl-tolerant populations differ from those of cells never exposed to salt[7]. Cold-resistant carrot cell lines, although stable in resistance at the callus level, did not exhibit comparable cold resistance at the embryo stage[22]. However, in our case embryos from a selected cell line grew and developed more satisfactorily than wild-type embryos of comparable size and stage on media containing various concentrations of NaCl. Moreover, a definite enhancement in the presence of 0.085 or 0.1 *M* NaCl and, in some cases even 0.17*M* NaCl, of embryo formation was noted in both *C. sinensis* and *C. aurantium*. In the latter, adding NaCl to the medium has proved to be the only effective means of achieving even a moderate degree of embryogenesis[1,13]. It is not yet clear whether the effect of NaCl is specific, or if NaCl enhances embryo formation in a manner similar to that attained by use of antigibberellins[12] or sucrose starvation[10].

The lack of uniformity of plant material derived from callus has been pointed out repeatedly[16]. The presence of embryos of different developmental stages creates further difficulties in comparing tolerance of selected material derived from callus or cell suspensions with the unselected control. The method of cloning individual embryos developed in our laboratory has enabled us to initiate comparisons on the basis of fairly uniform rooted shoots.

Two of these comparisons, each with a different selected salt-tolerant line of *C. sinensis*, have revealed better adaptation of plantlets derived from selected tolerant cell lines to salt in the medium than of plantlets from the unselected control. No such comparison has yet been performed with *C. aurantium*. Lengthy selection procedures and growth in the presence of high concentrations of the stress agent resulted in both a reduction in the embryogenic capacity of cell lines and a decrease in the regularity of embryo development in at least some of the lines involved. Resuppression of embryogensis in citrus by certain medium components probably involved not only sugars[13] or glycerol[1] but also NaCl[9] and possibly antigibberellins in the medium. These measures, combined with perfecting the procedure for cloning individual embryos, will enable us to adequately test the transmission of the trait of salt tolerance from cell lines to whole plants.

Acknowledgements This study was supported by a research grant from the Gesellschaft fur Strahlenforschung, GSF, Munchen, BRD, and by the National Council for Research and Development, Jerusalem, Israel. Part of this work was carried out by the late J Kochba.

The authors wish to thank Miss S Saad and Mrs. H Neumann for their most valuable help in the experiments.

References

1 Ben-Hayyim G and Neumann H 1983 Stimulatory effect of glycerol on growth and somatic embryogenesis in *Citrus* callus cultures. Z. Pflanzenphysiol. 110, 331–337.

2 Chaleff R S 1981 Genetics of higher plants: application of cell culture. Cambridge Univ. Press, New York.

3 Chaleff R S 1982 Considerations of developmental biology for the plant cell geneticist. *In* Genetic Engineering of Plants. Basic Life Sciences, Vol. 26. Eds. T Kosuge, C P Meredith and Hollaender, pp 257–270. Plenum Press, NY.

4 Cooper W C, Goyron B C and Edwards C 1951 Salt tolerance of various citrus rootstocks. Proc. Rio Grande Val. Hort. Inst. 5, 46–52.

5 Crougham T P, Stavarek S J and Rains D W 1978 Selection of NaCl tolerant line of cultured alfalfa cells. Crop Sci. 18, 959–963.

6 Dix P J and Street H E 1975 Sodium chloride-resistant cultured cell lines from *Nicotiana sylvestris* and *Capsicum annuum*. Plant. Sci Lett. 5, 231–237.

7 Handa A K, Bressan R A, Handa S and Hasegawa P M 1982 Tolerance to water and salt stress in cultured cells. *In* Plant Tissue Culture. Proc. 5th Int. Con. Plant Tissue and Cell Culture (Lake Yamanaka, 1982) Ed. A Fujiwara, pp 471–474. Japan Assoc. for Plant Tissue Culture, Tokyo.

8 Hasegawa P M, Bressan R A and Handa A K 1980 Growth characteristics of NaCl selected and non-selected cells of *Nicotiana tabacum* L. Plant Cell Physiol. 21, 1347–1355.

9 Kochba J, Ben-Hayyim G, Spiegel-Roy P, Saad S and Neumann H 1982 Selection of stable salt-tolerant callus cell lines and embryos in *Citrus sinensis* and *C. aurantium*. Z. Pflanzenphysiol. 106, 111–118.

10 Kochba J, Button J, Spiegel-Roy P, Bornman C H and Kochba M 1974 Stimulation of habituated ovular callus from the 'Shamouti' orange (*Citrus sinensis*) as affected by tissue age and sucrose concentration. Z. Pflanzenphysiol. 73, 415–421.

11 Kochba J and Spiegel-Roy P 1973 Effect of culture media on embryoid formation from ovular callus of Shamouti orange (*Citrus sinensis*). Z. Pflanzenzucht. 69, 156–162.

12 Kochba J, Spiegel-Roy P, Neumann H and Saad S 1978 Stimulation of embryogenesis in citrus ovular callus by ABA, ethephon, CCC and alar and its suppression by GA_3. Z. Pflanzenphysiol. 89, 427–432.

13 Kochba J, Spiegel-Roy P, Saad S and Neumann H 1982 Carbohydrate control of somatic embryogenesis in subcultured nucellar callus of citrus cultivars. Z. Pflanzenphysiol. 105, 359–365.

14 Kochba J, Spiegel-Roy P and Safran H 1972 Adventive plants from ovules and nucelli in Citrus. Planta 106, 237–245.

15 Kovda VA, Berg C and Hagan R M Eds. 1967 Plants in relation to waterlogging and salinity. IX, *In* Irrigation and Drainage of Arid Lands in Relation to Salinity and Alkalinity. pp 343–391. FAO, Roma and UNESCO, Paris; Imprimerie Rolland, Paris.

16 Larkin P J and Scowcroft W R 1981 Somaclonal variation – a novel source of variability from cell cultures for plant improvement. Theor. Appl. Genet. 60, 197–214.

17 Murashige T and Skoog F 1962 A revised medium for growth and bioassays with tobacco tissue cultures. Physiol. Plant. 15, 473–497.

18 Murashige T and Tucker D P H 1969 Growth factor requirement of citrus culture. *In* Proc. First Int. Citrus Symp. 3, 1155–1161. University of California, Riverside, California.

19 Nabors N W, Danies A, Nadolny L and Brown C 1975 Sodium chloride tolerant lines of tobacco cells. Plant Sci. Lett. 4, 155–159.

20 Orton T J 1980 Comparison of salt tolerance between *Hordeum vulgare* and *H. jubatum* in whole plants and callus cultures. Z. Pflanzenphysiol. 98, 105–118.

21 Spiegel-Roy P and Vardi A 1984 Citrus tissue culture. *In* Handbook of Plant Cell Culture Vol. 3. Eds. P V Ammirato, D A Evans, W R Sharp and Y Yamada. Macmillan, New York. pp 355–372.

22 Templeton-Somors K M, Sharp W R and Fister R M 1981 Selection of cold resistant cell lines of carrots. Z. Pflanzenphysiol. 103, 139–148.

23 Watad A E A, Reinhold L and Lerner H R 1983 Comparison between a stable NaCl-selected Nicotiana cell line and the wild type. K^+, Na^+ and proline pools as a function of salinity. Plant Physiol. 73, 624–629.

Plant and Soil 89, 253–271 (1985). Ms. BR 17

Crop production and management under saline conditions*

A. MEIRI and Z. PLAUT
Inst. of Soils and Water, ARO, Bet-Dagan 50250, Israel

Key words Climate CO_2 concentration Drip Fertilization Irrigation interval Leaching Salt tolerance Salinity Sprinkler Stand

Summary This review evaluates management practices that may minimize yield reduction under saline conditions according to three strategies: (I) control of root-zone salinity; (II) reduced damage to the crop; (III) reduced damage to individual plants. Plant response to salinity is described by an unchanged yield up to a threshold soil salinity (a), then a linear reduction in relative yield (b), to a maximum soil salinity that corresponds to zero yield (Y_0). Strategies I and II do not take into consideration any change in the parameters of the response curve, while strategy III is aimed at modifying them.

Control of root zone salinity is obtained by irrigation and leaching. From the review of existing data it is concluded that the effective soil salinity parameter should be taken as the mean electrical conductivity of the saturated paste extract or of the soil solution over time and space. Several irrigation and leaching practices are discussed. It is shown that intermittent leaching is more advantageous than leaching at each irrigation. Specific cultivation and irrigation practices that result in soil salinity reduction adjacent to young seedlings and the use of water of low salinity at specifically sensitive growth stages may be highly beneficial. Recent data do not show that reduced irrigation intervals improve crop response more under saline than under nonsaline irrigation. Alternate use of water of different salt concentrations results in mixing in the soil and the crop responds to the mean water salinity.

Reduced damage at the field level when soil or irrigation water salinity is too high to maintain full yield of single plants requires a larger crop stand. For row crops reduced inter-row spacing is more effective than reduced intra-row spacing.

Reduced damage at the plant level while the salinity tolerance of the plants remains constant shows up in the response curve parameters as larger threshold and slope and constant salinity at zero yield. This is the effect of a reduced atmospheric water demand that results in reduced stress in the plant under given salinity. Management can also change the salt tolerance of the crop. This will show up as higher salinity at zero yield, as well as changes in threshold and slope. Such changes in the response curve were found at different growth stages, under different atmospheric CO_2, under different fertilization, and when sprinkler irrigation was compared with drip irrigation.

Introduction

Sound agricultural management consists of a number of practices aimed at optimizing crop production. These practices may need to be modified and new ones introduced when brackish water is employed. A comprehensive list which relates yield to total soluble salts in the root medium was compiled for many crops by Maas and Hoffman[29]. These data are, however, not suitable for specific salt damage resulting from canopy wetting with saline water under sprinkler irrigation, or

* Contribution from the Agricultural Research Organization, The Volcani Center, Bet Dagan, Israel. No. 1111-E 1984 series.

for specific toxic effects. Most of these data were obtained from experiments in which salinity was imposed after seedling establishment and do not necessarily apply to germination and the early seedling stages. The plots were rapidly salinized to obtain salinity that was essentially uniform with depth by irrigation with water of different salinity levels at a high leaching fraction (LF). Therefore, after the rapid initial salinization, temporal and spatial variations in salt content in the soil profile were small. Because of frequent irrigations, fluctuations in soil solution salinity (EC_{sw}) and soil matric potential (ψ_m) were also minimized. Optimum fertilization for nonsaline conditions was used, assuming no interaction between the effects of fertilization and salinity. To assemble data from experiments reporting different absolute yields and conducted in various soil types, both plant response and soil salinity were normalized, the former by using relative rather than absolute yields, and the latter by using the mean electrical conductivity of the saturated paste extract for the soil of the root zone ($\overline{EC_e}$). Thus, the possible effects of variations in salt content in the profile, solute chemistry, exchange reactions in the soil, and soil moisture management were not considered. The steady-state conditions and the normalization of the results enabled standardization of the reported data. However, this approach ignores the influence of different environmental conditions on the crop response. Knowledge and control of these are our management tools.

This review examines management practices that may minimize damage under saline conditions. These practices were classified according to three strategies: (I) control of root zone salinity below harmful levels by irrigation (method, amount, interval), leaching (amount and timing), and special cultivation practices (pre-planting leaching, seedbed shape, and placement of seeds); (II) accepting damage at the single-plant level and changing cultivation practice (inter- and intrarow spacing of plants) to reduce salinity damage at the field level; and (III) changing management factors to reduce salinity damage at the plant level (irrigation method, climatic conditions, atmospheric CO_2 concentration, fertilizer application).

Most of the work on management under saline conditions was devoted to the control of soil salinity. Information on the other aspects is scarce, was gained recently, and is much less understood. Selection and breeding for salt tolerance are not included in the present review.

Control of root-zone salinity below harmful levels

The control of root-zone salinity is mostly the outcome of suitable irrigation and leaching management. The salt content of a given soil volume at time T (S_T) is the sum of the initial salt content (S_0) and the salt introduced (C_iV_i) minus the salt removed (C_dV_d):

$$S_T = S_0 + C_iV_i - C_dV_d \tag{1}$$

where C = concentration, V = volume, i = irrigation, d = drainage water.

The change in concentrations and volumes between irrigation and drainage result mainly from evapotranspiration, which can, in practice, be considered a distillation process that leaves the salts behind. The relative volume of water that carries salts out of the root zone is defined as the leaching fraction (LF)[48]. The minimum LF that will keep the soil salinity below a required level is the leaching requirement (LR)[48]. The best available reference for this salinity level for various crops is the salt tolerance table presented by Maas and Hoffman[29]. This table assumes a response curve consisting of a plateau at maximum yield below a threshold salinity a and a linear descending line with slope b relating decreasing yield to increasing salinity, until zero yield is reached at a salinity of Y_0 (Fig. 1).

The soil salinity corresponding to any relative yield can be obtained from equation (2):

$$Y/Y_{max} = 1 - b(\overline{EC}_e - a) \tag{2}$$

where Y = yield, Y_{max} = yield of nonsaline control, a = threshold, b = slope.

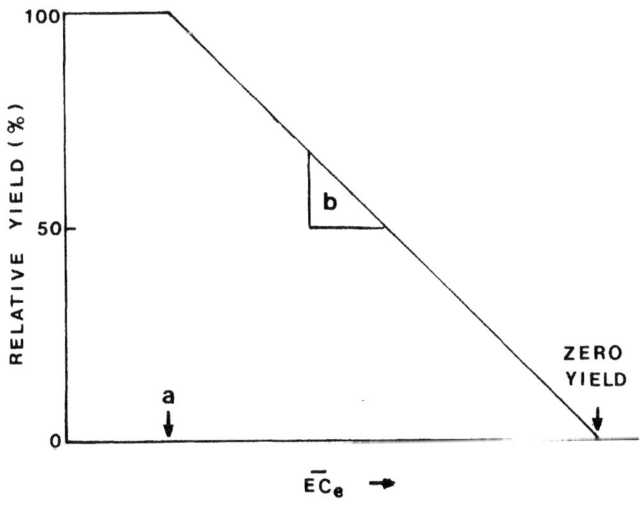

Fig. 1. Schematic response curve to salinity (after Maas and Hoffman[29]).

The mean salinity ($\overline{EC}_e$) of the saturated paste extract from the root zone was used as the effective salinity, *i.e.* that salinity parameter which is best correlated with the crop response. In 1973 Bernstein and Francois[3] challenged the use of $\overline{EC}_e$ and recommended using the weighted mean salinity of the absorbed water $\hat{EC}_{sw}$ as the effective salinity. They argued that the effective salinity is that against which water uptake actually occurs and that most of the water is absorbed from the soil volumes of lower salinity. This approach, which seems very logical, was based on observations when the salt distribution in the soil was at steady state. Under these conditions, the roots may not explore the more saline soil volumes. However, under most practical field conditions, root growth and spatial and temporal variations in salinity occur simultaneously. The variations in salinity are, therefore, in part, imposed on already-developed root systems.

The final cotton root distribution (Fig. 2) for two qualities of water supplied by drippers spaced 50 cm from the plant axis[32] is the outcome of non-steady-state conditions. It is evident that with nonsaline water most of the roots were evenly distributed between the plant stem and the dripper. When nonsaline soil was irrigated with saline water there was a high-salt front which moved, during the irrigation season, near the soil surface from the dripper towards the plant row. The larger horizontal component of root distribution below the plant

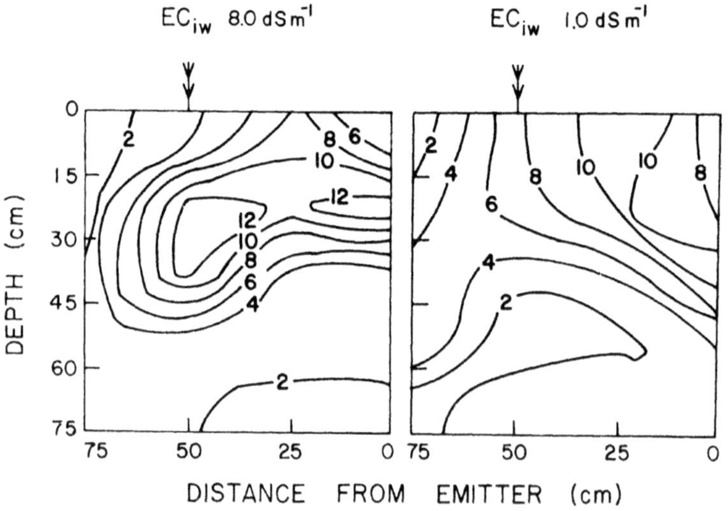

Fig. 2. Influence of water quality on root distribution of fully mature cotton plants irrigated through drippers spaced 50 cm from the crop row. The numbered contours indicate percent of roots in the soil by weight[32].

row and its larger fraction at 30 cm depth were the result of root growth that by-passed the region of the high-salt front. However, with the progress of the irrigation season the salt front moved further into the root zone and the major part of the root system was exposed to this salinity. Such variations in salinity, which are imposed on already-developed root systems, suggest that the mean root-zone salinity is the important factor.

Recent analysis of experiments in pots, lysimeters, and fields showed that the effective salinity is the temporal and spatial mean in the root zone[24,33]. With flood and sprinkler irrigation, the roots occupy the entire soil volume to the rooting depth. With furrow and drip irrigation, only the wetted part of the soil volume occupied by roots should be considered as the root zone.

The observations that the mean salinity is the effective one can be explained if the water status in the entire root zone determines the plant water status and yield. It is well documented that plants need to maintain full transpiration for full growth[10,11]. A higher rate of water uptake by the roots exposed to lower salinity will compensate for the reduced water uptake by the roots exposed to higher salinity. For constant root resistance the plant water potential will decrease, resulting in reduced growth. The weighted mean water salinity approach also does not account for the influence that water uptake distribution may have on the matric component of the soil water potential. This point will be extended in the discussion on the effects of the interval between irrigations.

Timing of leaching

To maintain a steady-state level of salinity, leaching is necessary at every irrigation. Accepting the temporal and spatial mean root-zone salinity as the effective salinity may lead to a modified leaching practice. The adoption of LR as excess water to be added with every irrigation is most undesirable when saline water is introduced into a field of lower salinity than the maximum allowed for the crop and salinity build-up occurs. Such management may aggravate the salinity effect as it will result in faster salinization. For a crop that is irrigated only a few times it may also result in a higher $\overline{EC}_e$ value[45] (Fig. 3).

Not only the mean but also the maximum permissible salinity must be considered. Bernstein and Pearson[6] studied the response of tomato and pepper plants to various salinity regimes. They showed that pepper was less tolerant of constant salinity, but could tolerate larger fluctuations in salinity than tomatoes. A build-up of salinity in the root zone results from water uptake by the plants in the entire root zone and

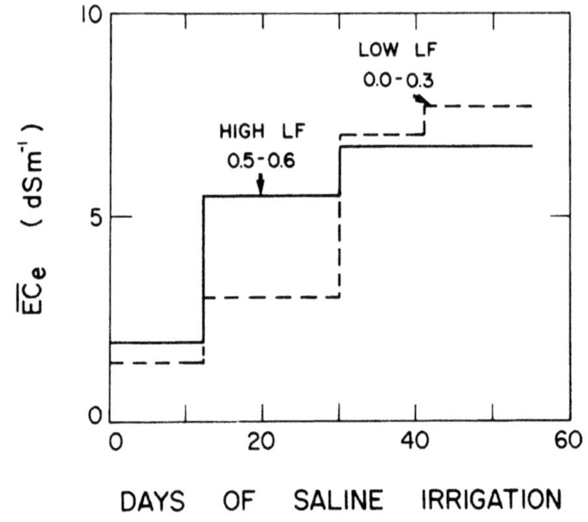

Fig. 3. Salt accumulation in the soil (mean, 0–90 cm) as a function of time, for two leaching fractions (LF)[45].

evaporation mostly near the surface. The maximum salinity at the bottom of the root-zone depends on the plants' ability to absorb water against the soil water osmotic potential. Thus, water uptake will stop when the osmotic potential of the soil solution decreases to the minimum water potential that can develop in the plant roots. A conclusion based on observations with alfalfa was that this salinity value is that corresponding to zero yield on the crop response curve. Reduction in LF cannot increase salinity above this value[3,49] (Fig. 4).

The evaporation from the soil surface may further concentrate the soil solution, resulting in a high salinity at the soil surface under furrow or drip irrigation. This high surface salinity may cause salinity shock when the surface salts are flushed through the root zone[2].

Under management for maintaining steady-state salinity, it was first recommended to allow – at the bottom of the root-zone, where maximum salinity exists – the salinity corresponding to 50% yield reduction on the crop response curve. Recently, however, it was shown that leaching can be decreased to allow for a maximum salinity corresponding to zero yield. This reduced LR is the minimum leaching requirement[49]. Under non-steady state, when leaching is delayed[5,16], this zero yield salinity can be stored in some portion of the deeper soil volume that was previously occupied by roots with no significant yield reduction. Thus, leaching may be intermittent. When rainfall occurs it may take care of part of the leaching. The amount of irrigation water

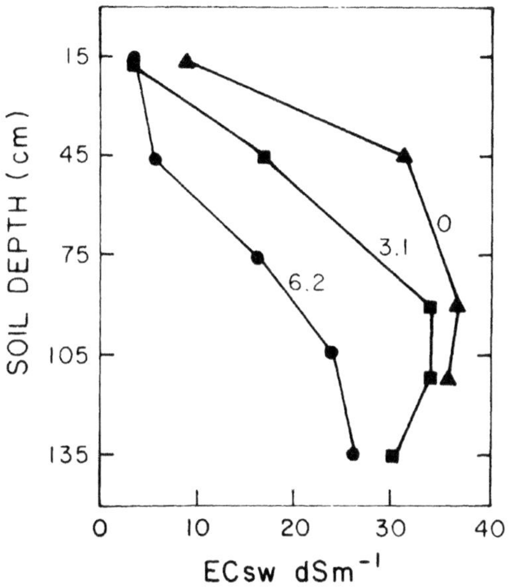

Fig. 4. Effect of reducing LF below salt balance limits for water of 2 dS m^{-1} on the build-up of salinity in the root zone. Numbers near lines indicate the LF[49].

available at specific seasons may also favor intermittent heavy leachings. Allowing higher salt accumulation in the root zone will result in higher salt concentrations in the drainage water and more effective leaching[7].

The irrigation dose under saline conditions should satisfy both the ET and the LR. Since LR is usually a small fraction of the irrigation, a small error in the estimate of ET may introduce a considerable error in the intended extent of leaching. The ET for a given crop under given conditions can be adjusted to evaporation demand by using a known pan evaporation factor or calculations based on meteorologic data. It is well documented that salinity reduces transpiration. Therefore, the estimation of ET according to these methods should be modified for saline conditions. A second way to estimate ET is to use the soil moisture deficit from field capacity (FC). This method may introduce an error if a constant FC is assumed for different irrigation doses. For frequent irrigations it was shown both experimentally[20] and by calculations[8] that the increase in irrigation depth intended to cause higher LF also results in an increase in the soil moisture content. This change in soil moisture content was shown theoretically[9] to be large enough to modify the yield response to salinity under non-steady-state conditions (Fig. 5).

Another error when using ET estimates based on soil moisture

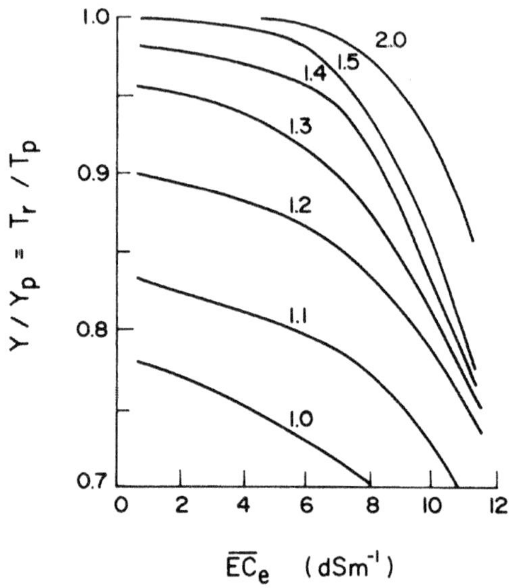

Fig. 5. Relative crop yield ($Y/Y_p = T/T_p$) as a function of electrical conductivity (ECe) and relative water application (numbers labeling the curve)[9].

deficit results from the fact that water is taken up during infiltration and redistribution. A comparison between the actual and estimated (from the soil moisture deficit) leaching volumes in a pot experiment[35] (Fig. 6) illustrates this error.

The data indicate an increased underestimation of the leaching with increased intended leaching. It is interesting to note that some drainage occurred in treatments that were aimed at having no leaching. The suppression of ET by salinity applies also to this level of irrigation. Therefore, the error will be larger at low salinity, when drainage is underestimated, and at high salinity, when drainage is overestimated.

Table 1 presents data from a field plot experiment by Meiri and Shalhevet[37] that show the greater efficiency of intermittent leaching in the control of soil salinity.

The outcome of these considerations is the approach already proposed by Ayers and Westcot[2]: 'under non-steady state, delay leaching and apply it when needed.' We would like to recommend the non-steady-state approach to leaching under most conditions. With such an approach one may reduce the amount of water needed for leaching and do the leaching at the best time, according to the salt tolerance of the crop at different growth stages and the availability of water. Smaller drainage volumes result in increased salt precipitation and reduced soil

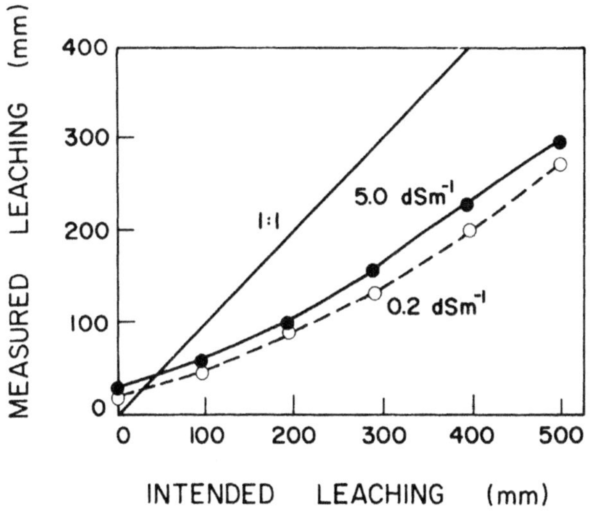

Fig. 6. Actual amount of leaching water compared with the intended leaching, assuming no water uptake during leaching, for two water qualities[35].

Table 1. Influence of the number of leaching irrigations as a fraction of the total number of irrigations (R) on Cl accumulation in the root zone (0–60 cm) of flood-irrigated pepper for two water qualities (EC_{iw})(Meiri and Shalhevet[37])

	EC_{iw} (dSm^{-1})			
	5.11		8.63	
R	LF^y	RCLC^z	LF	RCLC
0:1	0	1.00	0	1.00
1:1	0.31	0.76	0.26	0.82
1:7	0.15	0.47	0.24	0.42
1:13	0.21	0.36	0.23	0.68

y – Leaching fraction
z – Relative chloride content of soil

weathering. Thus, the salt load of the drainage water is lower, and secondary salinization of water resources is reduced (these aspects are not discussed in this review).

Special cultivation practices

High salinity in the soil volumes closest to the plants during seedling emergence and the initial growth stages may affect the entire production. Therefore, leaching of the soil volume adjacent to the seed, which constitutes the entire root-zone for the seedling, is most important. This leaching may be most critical with furrow and drip irrigation, where high salinity levels build up near the soil surface at the wetting front[9,50,52]. Adjustment of the soil surface contour and seeding or planting position (Fig. 7) according to the expected salt distribution is a rather simple solution with furrow irrigation[1,50]. With drip irrigation the same strategy is being used. Locating the emitters in a shallow slit in the soil alongside the seed rows will confine the water distribution and direct leaching of salts away from the plants (D. Pasternak and Y. De Malach, personal communication).

Irrigation interval

Reduced growth under extended irrigation intervals and nonsaline conditions results mainly from a low matric potential of the soil water. Under saline conditions, reduction in matric and osmotic potentials occur simultaneously. An experiment with guayule[51] showed a stronger salinity effect with longer irrigation intervals. Therefore, increasing the irrigation frequency was recommended under saline conditions[1]. Several experiments were conducted recently with alfalfa[3], eggplant[43], corn[44] and tall fescue[21] to study the effect of irrigation interval on response to salinity.

All experiments confirmed the conclusion that extended irrigation intervals may reduce absolute yields but did not support the second conclusion that it aggravates the relative salinity damage. With alfalfa the relative salinity damage was even smaller under a longer interval[3].

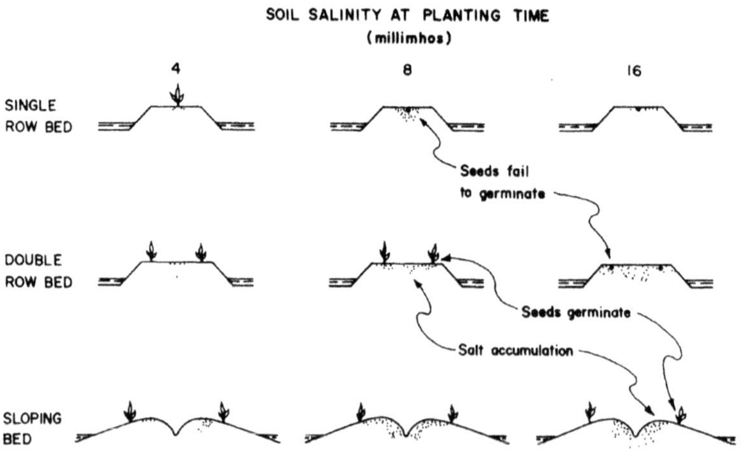

Fig. 7. Various bed shapes and resulting salt accumulation zones, illustrating the best positions for planting[1].

Salinity reduces ET. Thus, in experimental comparisons of the effect of irrigation interval with saline and nonsaline water, the salt-treated plots will retain more water than the controls between irrigations. This will moderate the rate of change in total soil moisture stress, resulting in reduced inhibitory effect of the interval on growth[47]. These effects are illustrated in Figs. 8 and 9. Extension of the irrigation interval results in a larger root zone and usually in reduced ET, and therefore a longer time and a larger amount of saline water will be required to salinize the root zone of the crop. In addition, the larger amounts of water per irrigation and the lower residual moisture in the upper soil layers will displace the more concentrated salt solution to a deeper soil layer, thus allowing larger root-zone volumes of lower salinity.

Multisource water supply operation

Water of various qualities from different sources can be applied separately or together. For separate applications, the types of water are allocated to different fields and the selection of crops for each field is determined by their salt tolerance. Another option is to use the different sources for partial supply of the crop water requirement. There are several operational practices, influenced by the expected crop response and practical operational limitations, which can be arranged in three groups: (a) To use the different sources at different

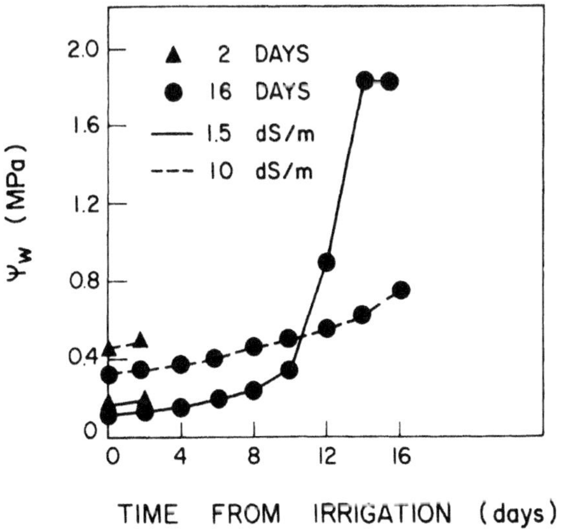

Fig. 8. Change in total soil water potential (ψ_m) with time from irrigation for two irrigation intervals (2 and 16 days) and two irrigation water salinities (1.5 and 10.0 dS m^{-1})[43].

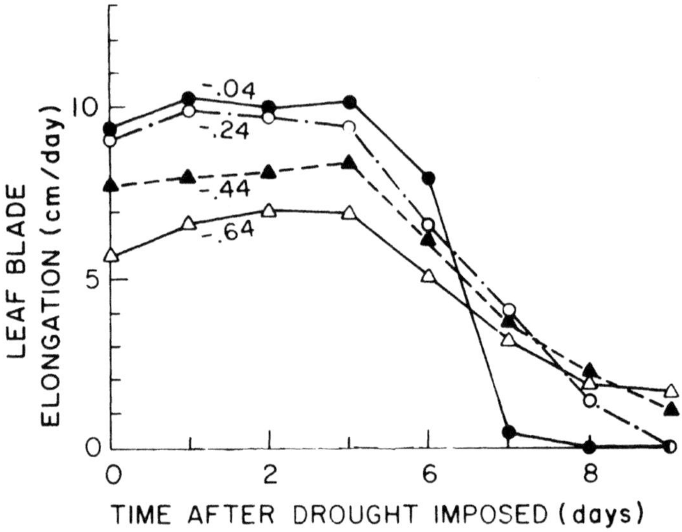

Fig. 9. Corn leaf elongation as influenced by salinity and days after irrigation during an irrigation interval. Numbers labeling the lines indicate the osmotic potentials of the irrigation solutions. Reproduced from ref.[47] by permission of the American Society of Agronomy.

growth stages, so that water of higher salinity would not be used during sensitive growth stages. D. Pasternak and Y. De Malach (personal communication) showed that a relatively short period of supplying nonsaline water at the young seedling stage considerably increased the final yields of corn and tomatoes irrigated with saline water. (b) Mixing of the water in the distribution network to adjust its salt concentration to the crop. (c) Mixing of the water in the soil, i.e. supplying the water from the different sources alternately in a proportion tolerated by the crop. Using potatoes and drip irrigation, Meiri *et al.*[40] compared these three strategies and found that, except for specifically sensitive growth stages, the crop responded to the average water salinity regardless of the mixing procedure.

Increased yield on the field level – stand and row spacing

A salinity level that causes yield reduction per plant does not necessarily reduce yield on the field level to the same extent. The field yield is the product of stand density and yield per plant. Conventional planting density was established under nonsaline conditions. With stunted plants an increase in stand may increase yield on the field level. Cotton was the first crop for which modified plant spacing under saline conditions was studied[17,26] (J.D. Rhoades, personal communication), and the expected economic benefit was calculated[15]. For this row crop, a change in stand can be obtained by changing inter-row and/or intra-row spacing. Data from Keren *et al.*[26] (Table 2) showed that reduced intra-row spacing resulted in strong competition between plants and had no effect on the yield per area. However, reduced inter-row spacing did not cause plant competition and therefore increased the yield per meter^2.

Table 2. Influence of inter-row and intra-row spacing on the yield of seed cotton (Acala SJ-2) sprinkler irrigated with saline water[26]

Inter-row spacing (cm)	Intra-row spacing (cm)	Stand (plants/m^2)	Yield (g/m^2)	Yield (g/plant)
75	12.5	10.7	597	56
75	9.1	14.7	616	42
96.5	12.5	8.3	486	59
95.5	9.1	11.4	503	44

Reduced salinity damage at the plant level

This implies a change in the yield *vs.* salinity curve, in response to modified management or environmental conditions. Four types of modification of this curve are presented in Fig. 10: (1) simultaneously changing the threshold (a) and the slope (b) and maintaining the zero yield salinity (Y_0) constant; (2) simultaneously changing a, b, and Y_0; (3) changing only b and Y_0; and (4) changing a and Y_0.

In case 1 the constant Y_0 indicates an osmotic effect and constant maximum salt tolerance by the crop. That is, Y_0 corresponds to the critical, lower water potential that the plant can resist and thus the maximum salinity level against which it can absorb water when the water supply is ample. This critical value was already considered as the upper limit for root-zone salinization under reduced leaching[3,49]. Changing conditions to influence the water demand and the water stress level in the plant should not change the Y_0 at which there is no water uptake. The a, on the other hand, is the plant stress level at which production starts to be affected and is a function of the rates of water uptake and transpiration. Therefore, the a value should be sensitive to environmental conditions that influence water demand.

In case 2 the changes in a, b and Y_0 indicate a change in the critical maximum tolerance to salt stress. This may indicate a change in tolerance to osmotic stress or to a specific toxic effect.

Cases 3 and 4 may be considered modifications of case 2.

All these types of modifications can be expected as a result of changing management practices. They may be useful in analysis of

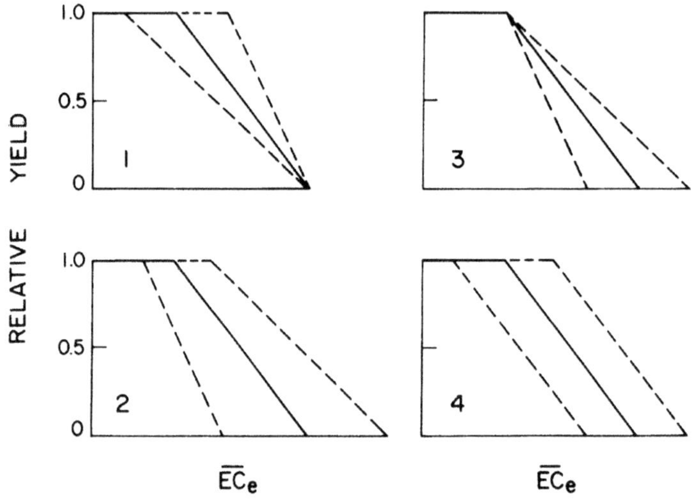

Fig. 10. Schematic presentation of types of crop response to salinity after a change in environmental conditions and management.

the effects of management on the response and should be considered when management decisions are made.

Climate

Greater salinity damage in a warm, dry climate than in a cool, wet one was reported[31]. This may indicate a more severe salinity stress with an increase in the atmospheric demand for water. Temperature- and humidity-controlled glasshouse experiments were used to study these climatic factors' effects on salt tolerance of root crops[23], cereals[19], and beans[22]. The influence of relative air humidity on the responses of corn and barley to salinity (Table 3) is expressed as a higher a and similar Y_0 values at the higher relative humidity for both crops (case 1). A comparison of data from two experiments with muskmelons which differed in many aspects – one with var. Galia grown in a glasshouse in Israel in the spring and irrigated daily[36], the other with varieties grown outdoors in California in the mid-summer and irrigated twice a week[46] – also shows a lower a value under the more severe water stress and only a small difference in Y_0 (case 1)(Table 3). Meiri *et al.*[34] studied the effect of reduced transpiration demand on salt tolerance by shading muskmelons grown in a temperature-controlled glasshouse in the spring. Again, in this experiment, mainly the a value was affected (Table 3). However, contrary to the results reported above, the un-shaded plants had a higher salt tolerance than the shaded ones. Apparently the reduced

Table 3. Influence of environmental and management factors on the response curves of some crops to salinity (salinity as electrical conductivity of the saturated paste extract $\overline{EC_e}$)

Factor modified	Crop	Treatment	a^z (dS m^{-1})	b^z (% yield/dS m^{-1})	Y_0^z (dS m^{-1})	Case[y]	Ref.
Transpiration demand	Barley	High RH[w]	10.6	3.1	42.8	1	19
		Low RH	4.4	2.7	41.9		
	Corn	High RH	4.7	9.7	15.0	1	19
		Low RH	0.6	7.4	14.2		
Many[x]	Muskmelon	Spring glasshouse[x]	6.8	10.0	16.8	1	36
		Summer outdoors[x]	1.3	5.9	18.1		46
Light intensity	Muskmelon	Unshaded	3.7	10.6	13.1	1 + 2	34
		Shaded	0.4	7.8	13.2		
Growth stage	Corn	Constant salinity	5.5	10.0	15.5	2	30
		Increased salinity after tasseling	6.4	6.0	21.6		
Irrigation method	Potato	Sprinkler	1.1	11.7	10.1	2	38, 39, 40
		Drip	2.6	6.3	17.1		
Nutrient level	Clover	N_3P_3	0.2	5.9	17.2	3	41
		N_0P_0	0.6	11.0	9.7		
	Millett	N_3P_3	7.0	8.0	19.6	4	41
		N_0P_0	1.3	8.5	13.1		

[w]RH = relative humidity.
[x]See text for details.
[y]1, 2, 3 and 4 = type of response curve modification in Fig. 10.
[z]a, b, Y_0 = threshold, slope and zero-yield salinity.

rate of photosynthesis (P) in the shaded plants had a stronger effect on the response to salinity than the expected benefit from the reduction in transpiration due to shading (cases 1 and 2). Increased light was also reported to be beneficial by Hellal and Mengel[18].

Increased CO_2 concentration

Increasing the CO_2 concentration above normal may increase the rate of photosynthesis, which in turn may increase the salt tolerance. Enoch *et al.*[13] actually showed a marked change in the response of tomato to salinity at 8000 ppm CO_2 as compared with 350 ppm (Fig. 11). Although the entire salinity range was not covered, the response curve at 8000 ppm CO_2 showed an increase in the Y_0 value. This change may correspond to case 2 and indicates an increased salt tolerance. Higher salt tolerance under higher CO_2 concentration was also found for beans, corn, Xanthium and Atriplex by Schwartz and Gale[42], and for roses by M. Zerony and J. Gale (this volume).

Growth stage

Changes in salt tolerance at different growth stages were recognized long ago. Plants were usually found to be more sensitive at the young seedling stage. Maas *et al.*[30] showed greater salt tolerance of corn after tasseling. The higher tolerance shows up in larger a and Y_0 and smaller b (Table 3). A similar response was found by D. Pasternak

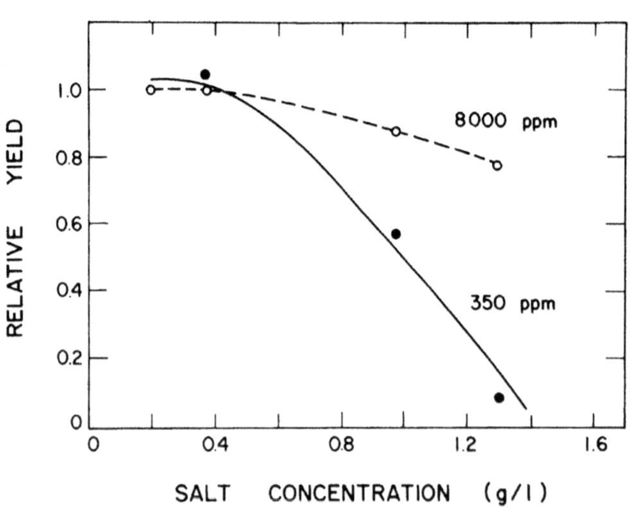

Fig. 11. Influence of CO_2 concentration on salt tolerance by tomatoes[13] (Data recalculated on a relative basis).

and Y. De Malach (personal communication) for corn and processing tomatoes, which exhibit lower salt tolerance at early growth stages resulting from reduced a and increased b. These changes in response to salinity, which indicate a real change in salt tolerance, correspond to case 2 of Fig. 10.

Irrigation method

The specific leaf damage caused by canopy wetting during sprinkler irrigation was documented by Ehlig and Bernstein[12], Bernstein and Francois[2,4], and Maas *et al.*[27,28]. Potatoes are one of the most sensitive crops to such leaf damage. Recently, the quantitative response of potatoes to salinity was studied by Meiri *et al.*[38,39,40] with drip and sprinkler irrigation. The data in Table 3 show that the specific sprinkling damage was expressed as a smaller a and Y_0 and larger b (case 2). This indicates that sprinkler irrigation had a specific toxic effect.

Fertilization

Recent reviews by Feigin[14] and Kafkafi[25] indicate various interactions between fertilization and salinity. A clear increase in salt tolerance under high fertilization was shown by Ravikovitch and Yoles[41] with millet and clover (Table 3). With clover there was no change in the threshold, while the slope decreased and Y_0 increased (case 3). With millet the threshold and Y_0 values increased, with no change in the slope (case 4). The beneficial effect could be an overcoming of nutritional imbalance or a reduction in toxicity, two responses that may result in similar changes in the curves.

Acknowledgements The authors acknowledge the permission given by publishers and authors to reproduce figures and tables in this review.

References

1 Ayers R S and Westcot D W 1976 Water quality for agriculture. FAO Irrigation and Drainage Paper No. 29. Food and Agriculture Organization of the United Nations, Rome.
2 Bernstein L and Francois L E 1973 Comparison of drip, furrow and sprinkler irrigation. Soil Sci. 115, 73–86.
3 Bernstein L and Francois L E 1973 Leaching requirement studies: sensitivity of alfalfa to salinity of irrigation and drainage waters. Proc. Soil Sci. Soc. Am. 37, 931–943.
4 Bernstein L and Francois L E 1975 Effect of frequency of sprinkling with saline waters compared with daily drip irrigation. Agron. J. 67, 185–190.
5 Bernstein, Francois L E and Clark R A 1975 Minimal leaching with varying root depths of alfalfa. Proc. Soil Sci. Soc. Am. 39, 112–115.
6 Bernstein L and Pearson G A 1954 Influence of integrated moisture stress achieved by varying the osmotic pressure of culture solution on growth of tomato and pepper plants. Soil Sci. 77, 355–368.

7 Bower C A, Ogata G and Tucker J M 1969 Rootzone salt profiles and alfalfa growth as influenced by irrigation water salinity and leaching fraction. Agron. J. 61, 783–785.

8 Bresler E and Hoffman G J Irrigation management for salinity control. I. Theories and field tests. (*In press*)

9 Bresler E, McNeal B L and Carter D L 1982 Saline and Sodic Soils: Principles-Dynamics-Modeling. Springer-Verlag, Berlin.

10 Childs S W and Hanks R J 1975 Model for soil salinity effects on crop growth. Proc. Soil Sci. Soc. Am. 39, 617–622.

11 DeWit C T 1958 Transpiration and crop yields. Inst. Biol. Chem. Res. on Field Crops and Herbage. The Netherlands. Verslagen Landbondrez 64,6.

12 Ehlig L F and Bernstein L 1959 Foliar absorption of NaCl as a factor in sprinkler irrigation. Proc. Am. Soc. Hort. Sci. 74, 661–670.

13 Enoch H Z, Zieslin N, Biran Y, Halevy A H, Schwartz M, Kessler B and Shimshi D 1973 Principles of carbon dioxide nutrition research. Acta Hort. 32, 97–110.

14 Feigin A 1985 Fertilization management of crops irrigated with saline water. Plant and Soil 89, 285–299.

15 Feinerman E 1983 Crop density and irrigation with saline water. Western J. Agric. Econ. 8, 134–140.

16 Francois L E 1981 Alfalfa management under saline conditions with zero leaching. Agron. J. 73, 1042–1046.

17 Francois L E 1982 Narrow row cotton (*Gossypium hirsutum* L.) under saline conditions. Irrig. Sci. 3, 149–156.

18 Hellal H M and Mengel K 1981 Interaction between light intensity and NaCl salinity and their effects on growth, CO_2 assimilation, and photosynthate conversion in young broad beans. Plant Physiol. 67, 999–1002.

19 Hoffman G J and Jobes J A 1978 Growth and water relation of cereal crops as influenced by salinity and relative humidity. Agron. J. 70, 765–769.

20 Hoffman G J and Jobes J A 1983 Leaching requirement for salinity control. III Barley, cowpea and celery. Agric. Water Mgmt. 6, 1–14.

21 Hoffman G J, Jobes J A and Alves W J 1983 Response of tall fescue to irrigation water salinity, leaching fraction, and irrigation frequency. Agric. Water Mgmt. 7, 439–456.

22 Hoffman G J, Jobes J A, Houscow Z and Maas E V 1978 Timing of environmental stress affects growth, water relation and salt tolerance of pinto beans. Trans. Am. Soc. Agric. Eng. 21, 713–718.

23 Hoffman G J and Rawlins S L 1971 Growth and water potential of root crops as influenced by salinity and relative humidity. Agron. J. 63, 877–880.

24 Ingvelson R D, Rhoades J D and Page A L 1976 Correlation of alfalfa yield with various indexes of salinity. Soil Sci. 122, 145–153.

25 Kafkafi U 1984 Plant nutrition under saline conditions. *In* Soil Salinity Under Irrigation – Processes and Management. Ecological Studies IV. Eds. I Shainberg and J Shalhevet. Springer Verlag, Berlin. (*In press*).

26 Keren R, Meiri A and Kalo Y 1983 Plant spacing effect on yield of cotton irrigated with saline water. Plant and Soil 74, 461–465.

27 Maas E V, Clark R A and Francois L E 1982 Sprinkling-induced foliar injury to pepper plants: effects of irrigation frequency, duration and water composition. Irrig. Sci. 3, 101–109.

28 Maas E V, Grattan S R and Ogata G 1982 Foliar salt accumulation and injury in crops sprinkled with saline water. Irrig. Sci. 3, 157–168.

29 Maas E V and Hoffman G J 1977 Crop salt tolerance – current assessment. J. Irrig. Drainage Div. ASCE 103 [IR2], 115–134.

30 Maas E V, Hoffman G J, Chaba G D, Poss J A and Shannon M C 1983 Salt sensitivity of corn at various growth stages. Irrig. Sci. 4, 45–57.

31 Magisted O C, Ayers A D, Wadleigh C H and Gauch H G 1943. Effect of salt concentration, kind of salt and climate on plant growth in sand cultures. Plant Physiol. 18, 151–166.

32 Mantell A, Frenkel H and Meiri A Drip irrigation of cotton with saline-sodic water. Irrig. Sci. (*In press*).

33 Meiri A 1984 Plant response to salinity: application of experimental results. *In* Soil Salinity under Irrigation – Processes and Management, Ecological Studies IV. Eds. I Shainberg and J Shalhevet. Springer Verlag, Berlin. (*In press*).

34 Meiri A, Hoffman G J, Shannon M C and Poss J A 1982 Salt tolerance of two muskmelon cultivars under two radiation levels. J. Am. Soc. Hort. Sci. 107, 1168–1172.

35 Meiri A, Kamburov J and Shalhevet J 1977 Transpiration effects on leaching fractions. Agron. J. 69, 779–782.

36 Meiri A, Plaut Z and Pinkas L 1981 Salt tolerance of glasshouse-grown muskmelon (cv. Galia). Soil Sci. 131, 189–193.

37 Meiri A and Shalhevet J 1973 Pepper plant response to irrigation water quality and timing of leaching. *In* Physical Aspects of Soil, Water and Salts in Ecosystems. Eds. A Hadas, D Swartzendruber, P E Rijitema, M Fuchs and B Yaron. Ecol. Studies IV, 421–428. Springer Verlag, Berlin.

38 Meiri A, Shalhevet J, Shimshi D and Tibor M 1980 Irrigation of spring potatoes with saline water. Ann. Report, Inst. Soils and Water, Agric. Res. Org., Bet Dagan, Israel. (*In Hebrew*).

39 Meiri A, Shalhevet J, Shimshi D and Tibor M 1982 Irrigation of spring potatoes with saline water. Ann. Report, Inst. Soils and Water, Agric. Res. Org. (*In Hebrew*).

40 Meiri A, Shalhevet J, Stolzy L H, Sinai G and Steinhardt R 1983 Managing multi-source irrigation water of different salinities for optimum crop production. Ann. Report submitted to U.S.-Israel Binational Agricultural Research and Development Fund, Bet Dagan, Israel.

41 Ravikovitch S and Yoles D 1971 The influence of phosphorus and nitrogen on millet and clover growing in soils affected by salinity. I. Plant development. Plant and Soil 35, 555–567.

42 Schwartz M and Gale J 1984 Growth response to salinity at high levels of carbon dioxide. J. Exp. Bot. 35, 193–196.

43 Shalhevet J, Heuer B and Meiri A 1983 Irrigation interval as a factor in the salt tolerance of eggplant. Irrig. Sci. 4, 83–93.

44 Shalhevet J, Hoffman G J, Meiri A, Heuer B and Francois L E 1983 Salinity tolerance of crops in irrigated agriculture under dynamic conditions. Final Report submitted to U.S.-Israel Binational Agricultural Research and Development Fund, Bet Dagan, Israel.

45 Shalhevet J and Yaron B 1983 Effect of soil and water salinity on tomato growth. Plant and Soil 39, 285–292.

46 Shannon M C and Francois L E 1978 Salt tolerance of three muskmelon cultivars. J. Am. Soc. Hort. Sci. 103, 127–130.

47 Stark J C and Jarrell W M 1980 Salinity-induced modifications in the response of maize to water deficits. Agron. J. 72, 745–748.

48 U.S. Salinity Laboratory 1954 Diagnosis and improvement of saline and alkali soils. Handbook U.S. Dept. Agric. 60.

49 van Schilfgaarde J, Bernstein L, Rhoades J D and Rawlins S L 1974 Irrigation management for salt control. J. Irrig. Drainage Div. ASCE 100 (IR3), 321.

50 Wadleigh C H and Fireman M 1949 Salt distribution under furrow and basin-irrigated cotton and its effect on water removal. Proc. Soil Soc. Am. 13, 527–530.

51 Wadleigh C H, Gouch H G and Magistad O C 1946 Growth and rubber accumulation in guayule as conditioned by salinity and irrigation regime. Tech. Bull. U.S. Dep. Agric. 925, p 34.

52 Yaron B, Shalhevet J and Shimshi D 1973 Patterns of salt distribution under trickle irrigation. *In* Physical Aspects of Soil, Water and Salts in Ecosystems, Eds. A Hadas, D Swartzendruber, P E Rijtema, M Fuchs and B Yaron Ecol. Studies IV, 389–394. Springer-Verlag, Berlin.

Plant and Soil 89, 273–284 (1985). Ms. BR 18

Crop tolerance to saline sprinkling water

E. V. MAAS
USDA, ARS, U.S. Salinity Laboratory, Riverside, CA 92501, USA

Key words Foliar injury Foliar ion absorption Salt tolerance Soil salinity Sprinkler irrigation

Summary Crops sprinkled with saline irrigation water are subject to foliar salt absorption and injury as well as to injury from soil salinity. Yield reductions caused by soil salinity alone are well documented, and data are presented here for 71 agricultural crops. Factors affecting these data and their applicability to sprinkler-irrigated crops are discussed. Although foliar injury has been observed with many sprinkled crops, particularly tree crops, essentially no information is available to predict yield losses as a function of the salt concentration of the irrigation water. Salinity thresholds for sprinkling-induced foliar injury are estimated for some crops; however, climatic conditions greatly affect the onset and degree of injury. Management strategies that minimize sprinkling injury are mentioned.

Introduction

It is well known that crops vary widely in their tolerance to salinity. During the past forty years innumerable experiments have been conducted throughout the world to quantify the tolerance of individual crops[10]. This data base has been invaluable for establishing salt tolerance guidelines for many of the commonly grown agricultural crops[21,24]. These guidelines, however, are expressed in terms of the concentration of salts in the soil or root medium. When the crop foliage is wet by sprinkling with saline water, plants are subject to additional salt damage. Salts are directly accumulated by the leaves, and as a result, some crops may become severely injured and lose their leaves. Of course, sprinkler-irrigated crops are subject to injury from both soil salinity and salt spray. Therefore, this paper begins by discussing crop tolerance to soil salinity and the factors affecting the application of these data to crop production with sprinkler irrigation. The effects and consequences of foliarly-absorbed salts are considered, and management strategies to minimize sprinkling injury are presented.

Crop tolerance to soil salinity

All soils contain a mixture of soluble salts. Many are essential for plant growth, while others are innocuous, at least at low concentrations. However, when the salt concentration is excessive, growth is suppressed. The yield response curve for increasing salinity is typical of the dose-response curve for nonessential elements[4]. For many crops this curve resembles a sigmoidal curve, for others, the curve drops abruptly to zero when lethal levels of salinity are reached. Maas and Hoffman[24] represented this response curve by two intersecting straight lines, one, a plateau with zero slope, and the second, a concentration-dependent line whose slope indicates the yield reduction per unit increase in salinity. The point at which the two lines intersect designates the 'threshold,' *i.e.* the maximum soil salinity that does not significantly reduce yield below that obtained under nonsaline conditions. With these two parameters, the relative yield (Y_r) at any given soil salinity (EC_e) exceeding the threshold can be calculated with the following equation:

$$Y_r = 100 - B(EC_e - A)$$

where A = the salinity threshold and B = the slope. Salt tolerance ratings expressed in terms of the threshold and slope and based on the electrical conductivity of saturated soil extracts (EC_e) are given for 71 agricultural crops in Table 1. The last column provides a qualitative salt tolerance rating that can be useful for categorizing crops in general terms. The limits of these categories are illustrated in Fig. 1.

These data apply to soils where chloride is the predominant anion. Because of the dissolution of $CaSO_4$ in preparing saturated soil extracts, the EC_e values of gypsiferous soils will be 1–3 dS/m higher than those of non-gypsiferous soils having the same soil water conductivity at field capacity. Therefore, plants grown on gypsiferous soils will tolerate EC_e values approximately 2 dS/m higher than those listed in Table 1 (Ref. 1).

Factors affecting salt tolerance

Salt tolerance data provide only relative guidelines to predict yield losses for crops grown under saline conditions. Absolute yield losses are subject to numerous cultural and environmental effects. Interactions between salinity and various soil, water, and climatic conditions change the plant's ability to tolerate salt. Sprinkled crops, like surface-irrigated crops, are subject to these interactions as well as to the effects of other factors.

Climatic conditions probably influence plant response to salinity as much as if not more than any other factor. Most crops can tolerate greater salt stress when the weather is cool and humid than when it is hot and dry. Suboptimal soil conditions can also affect the apparent salt tolerance of crops. For example, plants grown on low fertility soils may seem more salt tolerant than those grown with adequate fertilization. The reason is that soil fertility, not salinity, is the primary factor limiting

Table 1. Tolerance of agricultural crops to soil salinity[a]

Crop		Electrical conductivity of saturated soil extract		
Botanical name[b]	Common name	Threshold[c] (dS/m)	Slope (% per dS/m)	Rating[d]
FIBER, SEED AND SUGAR CROPS				
Arachis hypogaea	Peanut	3.2	29	MS
Beta vulgaris	Sugarbeet[e]	7.0	5.9	T
Glycine max	Soybean	5.0	20	MT
Gossypium hirsutum	Cotton	7.7	5.2	T
Hordeum vulgare	Barley[f]	8.0	5.0	T
Linum usitatissimum	Flax	1.7	12	MS
Oryza sativa	Rice, paddy[h]	3.0[g]	12[g]	S
Phaseolus vulgaris	Bean	1.0	19	S
Saccharum officinarum	Sugarcane	1.7	5.9	MS
Sorghum bicolor	Sorghum	6.8	16	MT
Triticum aestivum	Wheat[f,i]	6.0	7.1	MT
T. turgidum	Durum wheat	5.7	5.4	MT
Vicia faba	Broadbean	1.6	9.6	MS
Vigna unguiculata	Cowpea	4.9	12	MT
Zea mays	Corn	1.7	12	MS
GRASSES AND FORAGE CROPS				
Agropyron cristatum	Wheatgrass, fairway crested	7.5	6.9	T
A. longatum	Wheatgrass, tall	7.5	4.2	T
A. sibiricum	Wheatgrass, standard crested	3.5	4.0	MT
Alopecurus pratensis	Foxtail, meadow	1.5	9.6	MS
Cynodon dactylon	Bermudagrass[j]	6.9	6.4	T
Dactylis glomerata	Orchardgrass	1.5	6.2	MS
Elymus triticoides	Wildrye, beardless	2.7	6.0	MT
Eragrostis sp.	Lovegrass[k]	2.0	8.4	MS
Festuca elatior	Fescue, tall	3.9	6.4	MT
Hordeum vulgare	Barley (forage)[f]	6.0	7.1	MT
Lolium perenne	Ryegrass, perennial	5.6	7.6	MT
Lotus corniculatus tenuifolium	Trefoil, narrow-leaf birdsfoot	5.0	10	MT
L. uliginosus	Trefoil, big	2.3	19	MS
Medicago sativa	Alfalfa	2.0	7.3	MS
Phalaris tuberosa	Hardinggrass	4.6	7.6	MT
Sesbania exaltata	Sesbania[h]	2.3	7.0	MS
Sorghum sudanense	Sudangrass	2.8	4.3	MT
Sphaerophysa salsula	Sphaerophysa	2.2	7.0	MS
Trifolium alexandrinum	Clover, Berseem	1.5	5.7	MS
T. fragiferum	Clover, strawberry	1.5	12	MS
T. pratense	Clover, red	1.5	12	MS
T. hybridum	Clover, alsike	1.5	12	MS
T. repens	Clover, ladino	1.5	12	MS
Vicia angustifolia	Vetch, common	3.0	11	MS
Vigna unguiculata	Cowpea (forage)	2.5	11	MS
Zea mays	Corn (forage)	1.8	7.4	MS

Table 1 (*continued*)

Crop		Electrical conductivity of saturated soil extract		
Botanical name[b]	Common name	Threshold[c] (dS/m)	Slope (% per dS/m)	Rating[d]
VEGETABLE CROPS				
Allium cepa	Onion	1.2	16	S
Apium graveolens	Celery	1.8	6.2	MS
Beta vulgaris	Beet, red[e]	4.0	9.0	MT
Brassica oleracea botrytis	Broccoli	2.8	9.2	MS
B. oleracea capitata	Cabbage	1.8	9.7	MS
B. rapa	Turnip	0.9	9.0	MS
Capsicum annuum	Pepper	1.5	14	MS
Cucumis sativus	Cucumber	2.5	13	MS
Cucurbita pepo melopepo	Squash, scallop	3.2	16	MS
C. pepo melopepo	Squash, zucchini	4.7	9.4	MT
Daucus carota	Carrot	1.0	14	S
Ipomoea batatas	Sweet potato	1.5	11	MS
Lactuca sativa	Lettuce	1.3	13	MS
Lycopersicon lycopersicum	Tomato	2.5	9.9	MS
Phaseolus vulgaris	Bean	1.0	19	S
Raphanus sativus	Radish	1.2	13	MS
Solanum tuberosum	Potato	1.7	12	MS
Spinacia oleracea	Spinach	2.0	7.6	MS
Zea mays	Corn, sweet	1.7	12	MS
FRUIT CROPS[l]				
Citrus paradisi	Grapefruit[m]	1.8	16	S
C. sinensis	Orange	1.7	16	S
Fragaria sp.	Strawberry	1.0	33	S
Phoenix dactylifera	Date palm	4.0	3.6	T
Prunus armeniaca	Apricot[m]	1.6	24	S
P. domestica	Plum; Prune[m]	1.5	18	S
P. dulcis	Almond[m]	1.5	19	S
P. persica	Peach	1.7	21	S
Rubus sp.	Blackberry	1.5	22	S
R. ursinus	Boysenberry	1.5	22	S
Vitus sp.	Grape[m]	1.5	9.6	MS

[a] These data serve only as a guideline to relative tolerances among crops. Absolute tolerances vary depending upon climate, soil conditions, and cultural practices.

[b] Listed alphabetically within groups. Botanical and common names follow the convention of Hortus Third[20] where possible.

[c] In gypsiferous soils plants will tolerate EC_e values about 2 dS/m higher than indicated.

[d] Ratings are defined by the boundaries in Figure 1.

[e] Sensitive during germination. The EC_e should not exceed 3 dS/m.

[f] Less tolerant during emergence and seedling stage. The EC_e at this stage should not exceed 4 or 5 dS/m.

[g] Because paddy rice is grown under flooded conditions, the values refer to the electrical conductivity of the soil water while the plants are submerged.

[h] Less tolerant during emergence and seedling stages.

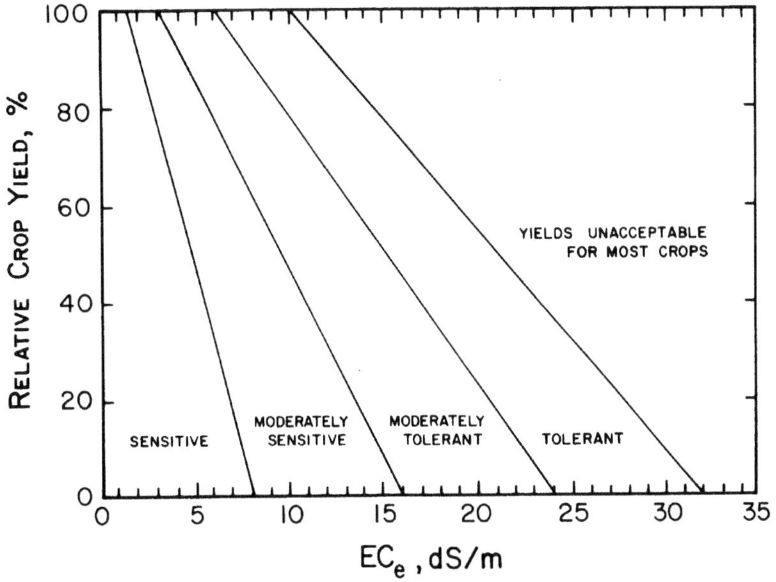

Fig. 1. Divisions for classifying crop tolerance to soil salinity.

plant growth. Proper fertilization would increase yields whether the soil was saline or not, but proportionally more if it were nonsaline. Other soil conditions limiting growth could be expected to have similar effects. For example, soils with poor structure or impermeable layers could restrict root growth as well as influence water and salt distribution in the soil.

Salt tolerance data must also be considered in light of irrigation management practices. As the soil water content decreases between irrigations, the salt concentration increases. Consequently, plants are exposed to increasingly saltier water in the soil as the time between irrigations increases. If water becomes limiting, plants may experience matric stress as well as osmotic stress. One would expect, therefore, that plants can tolerate moderately saline irrigation water better when irrigations are frequent. Sprinkler irrigation offers the capability of

[i] Data may not apply to new semidwarf cultivars.

[j] Average of several varieties. Suwannee and Coastal are about 20% more tolerant, and common and Greenfield are about 20% less tolerant than the average.

[k] Average for Boer, Wilman, Sand, and Weeping cultivars. Lehmann seems about 50% more tolerant.

[l] These data are applicable when rootstocks are used that do not accumulate Na^+ or Cl^- rapidly or when these ions do not predominate in the soil.

[m] Tolerance is based on growth rather than yield.

more frequent as well as more uniform water application than does surface irrigation. Excessive irrigation, on the other hand, can cause poor soil aeration, particularly in heavy, poorly-drained soils. Inadequate drainage that results in a shallow water table could also influence the apparent salt tolerance of crops. Plants could extract water directly from this source and, depending upon the quality of water, they might respond much differently than expected from the soil salinity levels.

In addition to the detrimental effects of soil salinity, sprinkled crops are subject to foliar injury caused by leaf absorption of salts. The degree of injury is related to the salt content in the leaves, but weather conditions and water stress are critical in inducing injury. For instance, leaves may contain potentially toxic levels of Na^+ or Cl^- for several weeks without exhibiting any injury symptoms, but the first hot, dry weather will cause severe leaf necrosis. Consequently, there are no useful guidelines for correlating foliar injury with the salt concentrations in the leaves.

Susceptibility to foliar injury varies considerably among crop species and depends more on leaf characteristics and rate of foliar salt absorption than on tolerance to soil salinity. Numerous factors affect the amount of salt accumulated by leaves, including the properties of the leaf surface, its age, shape, angle, and position on the plant, the type and concentration of salt, the ambient temperature and humidity, and the length of time the leaf remains wet. Some leaves possess a waxy cuticular layer that restricts the absorption of ions. Furthermore, water tends to coalesce into droplets on these leaves which then run off. Generally the permeability of leaves decreases with age. Several reviews cover these factors and the mechanisms of foliar absorption in considerable detail[11,16,18,19].

Na^+, Ca^{2+}, and Cl^- are all readily absorbed by leaves of many plants when they are sprinkled with irrigation water containing these elements[23]. The rate of absorption increases with increasing concentration of the ions. Although absorption is dependent upon metabolic energy, it apparently does not require light. Reports that light stimulates ion uptake have been conflicting[19]. Nevertheless, foliar absorption and injury are reduced if plants are sprinkled at night[5,6]. Since rates of absorption increase with increasing temperature, salt accumulation should be greatest during hot weather.

Frequency of sprinkling also influences the rate of salt accumulation. In greenhouse experiments with pepper (*Capsicum annuum* L.), increasing the frequency from 1 to 8 times per day without changing the total duration of sprinkling doubled the rate of salt absorption[22]. In contrast, 3-fold increases in the duration of once-per-day sprinkling had no measurable effect on the amount of salt taken up. Since salt continues to be absorbed by wet leaves after sprinkling has been stopped, frequent, short irrigations provide a much longer opportunity for absorption than indicated by the total duration of sprinkling. Furthermore, the rate of absorption, being a function of the salt concentration, increases rapidly as the solution evaporates from the leaf and the salt is concentrated. Therefore, the more often the wetting-drying cycle is repeated, the greater the salt accumulation.

Experiments with pepper, soybean, and tomato indicate that salts must remain in solution to be absorbed[14]. Once the salts become dry on the leaves, absorption stops. However, if the salts are redissolved, *e.g.* during periods of high humidity or dew, absorption resumes.

Foliar salt absorption and injury

Leaves of deciduous fruit trees (almond, apricot, and plum) readily absorbed Na^+ and Cl^-, and trees were partially or completely defoliated when they were sprinkled for over 50 hours with either 10 m*M* NaCl or $CaCl_2$ (Ref.[8]). Citrus leaves absorbed these ions more slowly but still fast enough to cause severe leaf burn[15]. Avocado, in contrast, absorbs salt so slowly that foliar absorption is a negligible factor in sprinkling[8]. Similar results were obtained for strawberry, which is very sensitive to soil salinity[7].

A comparative study of eleven herbaceous crops revealed wide differences in the rates of Na^+ and Cl^- absorption when the plants were sprinkled with saline water in the greenhouse[23]. The mean rates of Na^+ and Cl^- absorption among species increased in the order: sorghum $\ll$ cotton = sunflower $<$ cauliflower $<$ sesame = alfalfa = sugarbeet $<$ barley = tomato $<$ potato = safflower (Figs. 2 and 3). Salt was absorbed as a linear function of the weeks of sprinkling initially, but the duration of this linear relationship, like the rates, varied among species. For most crops, the rate of absorption decreased after about 4 to 6 weeks. Susceptibility to leaf injury among species did not vary in strict relation to the rates of salt absorption, however. Foliar injury symptoms increased in the order: sugarbeet $<$ cotton = sunflower = cauliflower $<$ safflower = barley = sorghum $<$ alfalfa $<$ sesame $\ll$ tomato = potato. Both potato and tomato readily absorbed Na^+ and Cl^- and soon developed leaf necrosis. Safflower, with one of the highest rates of absorption, was nonetheless only slightly injured by sprinkling. Barley readily absorbed salt but exhibited only minor injury symptoms; whereas sesame and alfalfa, with lower absorption rates, were somewhat more susceptible to injury. Sugarbeet was uninjured by sprinkling but

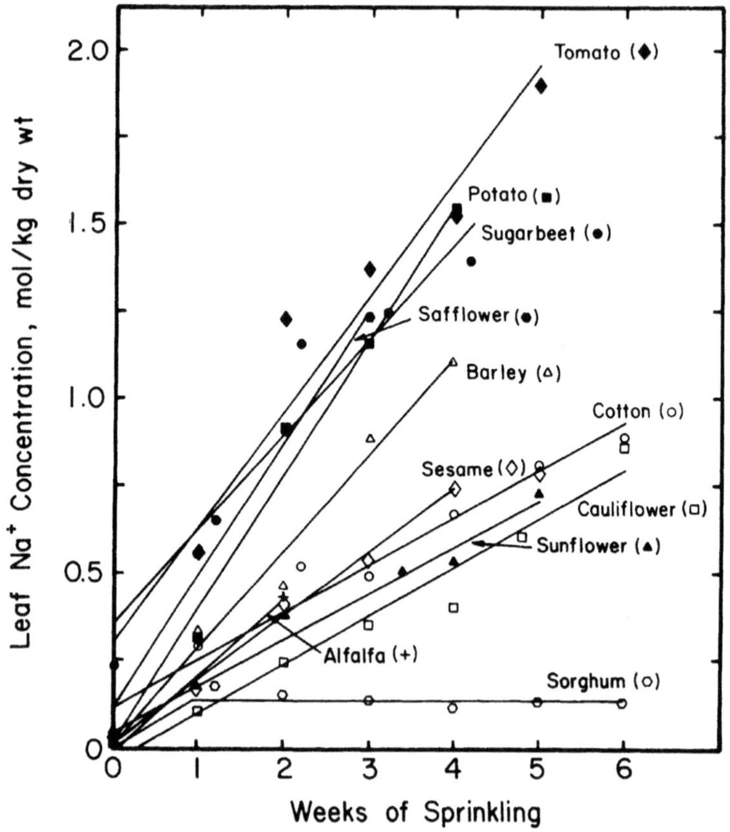

Fig. 2. Foliar absorption of Na^+ by crops sprinkled with 30 meq/l water (9:1 NaCl:$CaSO_4$). Plants were sprinkled 1 h/day, 5 days each week[23].

absorbed appreciable amounts of Na^+, whereas sorghum developed some leaf margin necrosis but absorbed very little salt. Cauliflower, cotton, and sunflower absorbed salt slowly and developed little or no injury.

Salt tolerance of sprinkled crops

Since sprinkled crops are exposed to soil salinity, yield reduction caused by saline irrigation water will be at least as great as for surface-irrigated crops. However, some sprinkled crops are subject to additional yield reduction caused by foliar damage. In a comparative study, Bernstein and Francois[2] found that drip, furrow and sprinkler irrigation with water with an electrical conductivity of 4.5 dS/m decreased yields

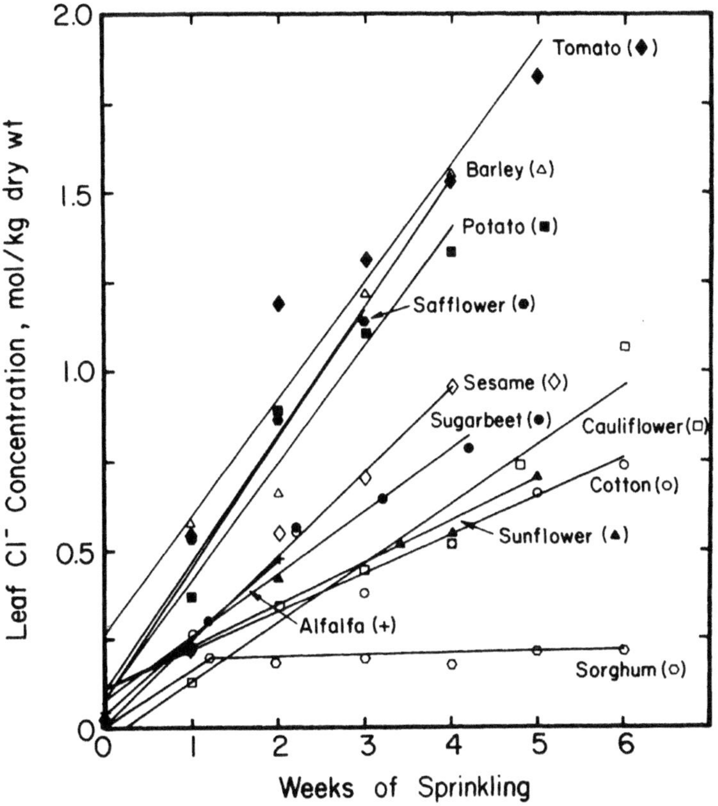

Fig. 3. Foliar absorption of Cl^- by crops sprinkled with 30 meq/l water (9:1 NaCl:$CaSO_4$). Plants were sprinkled 1 h/day, 5 days each week[23].

of pepper 2, 16, and 54%, respectively, compared to similar irrigations with nonsaline water (0.6 dS/m). Furrow- and sprinkler-irrigated plots were irrigated at equal intervals, an average of once every 5 days, while drip-irrigated plots were irrigated daily. A later study indicated that pepper yields were reduced about 16% per dS/m increase in the electrical conductivity of the irrigation water[3]. Sprinkling daily with water of 3.1 dS/m caused similar reductions in the yield of peppers in Israel[12]. Tomatoes sprinkled with water of 3.6 dS/m produced only 38% as much fruit as plants that were drip irrigated with the same water[13]. Even cotton, which is one of the most salt-tolerant crops, produced significantly less (15%) seed cotton when sprinkled during the day than when surface-irrigated with water of 4.4 dS/m (Ref.[5]). It is worth noting that sprinkling at night did not decrease the yield of cotton.

Except for peppers, no information is available to predict yield losses as a function of salt concentration in the irrigation water. Table 2 gives salinity thresholds for both sprinkling-induced foliar injury and yield reduction from soil salinity. Because of a lack of data on foliar injury, the list of crops is relatively short. Those crops that have foliar injury thresholds below the soil salinity threshold have a high likelihood of foliar injury when sprinkled with waters that have salt concentrations equal to or above the soil salinity threshold. At concentrations above both thresholds, both foliar injury and yield reductions can be expected. If the effects are additive or multiplicative, yield losses greater than those indicated in Table 1 are probable.

Minimizing sprinkling injury

Several management strategies should be considered if it is necessary to sprinkle sensitive crops with saline water. First, every effort should

Table 2. Relative tolerance of sprinkled crops to salinity impinging on the leaves or roots. Salinity levels are expressed as the electrical conductivity of the irrigation water (EC_i)

Crop	Salinity threshold: Max. EC_i without injury from foliarly-absorbed salts[a] (dS/m)	Max. EC_i without detrimental soil salinity effects[b] (dS/m)
Almond	< 0.5	1.0
Apricot	< 0.5	1.1
Citrus	< 0.5	1.1
Plum	< 0.5	1.0
Grape	0.5–1.0	1.0
Pepper	0.5–1.0	1.0
Potato	0.5–1.0	1.1
Tomato	0.5–1.0	1.7
Alfalfa	1.0–2.0	1.3
Barley	1.0–2.0	5.3
Corn	1.0–2.0	1.1
Cucumber	1.0–2.0	1.7
Safflower	1.0–2.0	
Sesame	1.0–2.0	
Sorghum	1.0–2.0	4.5
Strawberry	2.0–4.0	0.7
Cauliflower	3.0–6.0	
Cotton	3.0–6.0	5.1
Sugarbeet	3.0–6.0	4.7
Sunflower	3.0–6.0	

[a] Saline waters (primarily NaCl) with EC_i values higher than the threshold are expected to cause foliar injury on crops sprinkled 5 hours or more each week during the irrigation season. The degree of injury is influenced by the cultural and environmental conditions. Data compiled from references [2,7,8,9,12,13,22,23].

[b] Salinity exceeding the threshold is expected to decrease the yield below that of crops irrigated with nonsaline water. The relationship, $EC_e = 1.5\, EC_i$, was used to express the soil salinity threshold in terms of EC_i. Data computed from reference [21].

be made to eliminate or reduce wetting of the foliage. If possible, susceptible crops should be irrigated below the plant canopy. Although not practical, mini-sprinklers used experimentally on field corn successfully limited foliar burn to the first or second leaves when irrigation was with 4 to 8 dS/m water[17]. The non-wetted upper leaves exhibited no evidence of similar foliar injury. Yield reductions for the sprinkled corn were equivalent to those obtained with corn that was subirrigated with ditches. Since injury is related more to the number of sprinklings than to their duration, infrequent heavy irrigations would be preferable to frequent light irrigations. Intermittent wetting by slowly rotating sprinklers that allow drying between cycles should be avoided. Laterally-moving sprinkler systems might be moved downwind, when possible, so that salts accumulated on the leaves from salt drift would be washed off as the sprinkler moves past. Nielson and Cannon[25] observed that damage to alfalfa sprinkled with saline water occurred on plants at the far edges of the sprinkler pattern where water applications were light.

Furthermore, sprinkling should be done when conditions minimize salt absorption by the plants. Since both evaporation and foliar absorption are greater during the day, nighttime sprinkling should be practiced to the extent possible. If daytime sprinkling is necessary, one should avoid hot, dry, windy periods.

References

1 Bernstein L 1962 Salt-affected soils and plants. *In* The Problems of the Arid Zones, Proc. UNESCO Symp. (Paris, France) 18, 139–174.

2 Bernstein L and Francois L E 1973 Comparisons of drip, furrow, and sprinkler irrigation. Soil Sci. 115, 73–86.

3 Bernstein L and Francois L E 1975 Effects of frequency of sprinkling with saline waters compared with daily drip irrigation. Agron. J. 67, 185–190.

4 Berry W L and Wallace A 1981 Toxicity: the concept and relationship to the dose response curve. J. Plant Nutrition 3, 13–19.

5 Busch C D and Turner F Jr 1967 Sprinkler irrigation with high salt-content water. Trans. ASAE. 10, 494–496.

6 Eaton F M and Harding R B 1959 Foliar uptake of salt constituents of water by citrus plants during intermittent sprinkling and immersion. Plant Physiol. 34, 22–26.

7 Ehlig C F 1961 Salt tolerance of strawberries under sprinkler irrigation. Proc. Am. Soc. Hort. Sci. 77, 376–379.

8 Ehlig C F and Bernstein L 1959 Foliar absorption of sodium and chloride as a factor in sprinkler irrigation. Proc. Am. Soc. Hort. Sci. 74, 661–670.

9 Francois L E and Clark R A 1979 Accumulation of sodium and chloride in leaves of sprinkler-irrigated grapes. J. Am. Soc. Hort. Sci. 104, 11–13.

10 Francois L E and Maas E V 1978 Plant responses to salinity: an indexed bibliography. USDA, ARM-W-6, Riverside, CA (Updated Supplement *in press*).

11 Franke W 1967 Mechanisms of foliar penetration of solution. Annu. Rev. Plant Physiol. 18, 281–300.

12 Goldberg D and Shmueli M 1971 Sprinkle and trickle irrigation of green pepper in an arid zone. Hort. Sci. 6, 559–562.

13 Gornat D, Goldberg D, Rimon D and Ben-Asher J 1973 The physiological effect of water quality and method of application on tomato, cucumber, and pepper. J. Am. Soc. Hort. Sci. 98, 202–205.

14 Grattan S R, Maas E V and Ogata G 1981 Foliar uptake and injury from saline aerosol. J. Environ. Qual. 10, 406–409.

15 Harding R B, Miller M P and Fireman M 1958 Absorption of salts by citrus leaves during sprinkling with water suitable for surface irrigation. Am. Soc. Hort. Sci. 71, 248–256.

16 Haynes R J and Goh K M 1977 Review on physiological pathways of foliar absorption. Scientia Horticulturae. 7, 291–302.

17 Hoffman G J, Maas E V, Prichard T and Meyer J L 1983 Salt tolerance of corn in the Sacramento-San Joaquin delta of California. Irrig. Sci. 4, 31–44.

18 Hull H M 1970 Leaf structure as related to absorption of pesticides and other compounds. Residue Rev. 31, 1–155.

19 Jeschke W D 1976 Ionic relations of leaf cells. *In* Encyclopedia of Plant Physiology, New Series, Vol. 2B Transport in Plants II Part B. Tissues and Organs. Eds. U Luttge and M G Pitman. pp 160–194. Springer-Verlag, Berlin.

20 Liberty Hyde Bailey Hortorium Staff 1976 Hortus Third. A Concise Dictionary of Plants Cultivated in the United States and Canada, Macmillan, New York.

21 Maas E V 1985 Salt tolerance of plants. *In* Handbook of Plant Science in Agriculture. Ed. B R Christie, CRC Press Inc, Boca Raton, Florida. (*In press*)

22 Maas E V, Clark R A and Francois L E 1982 Sprinkling-induced foliar injury to pepper plants: effects of irrigation frequency, duration and water composition. Irrig. Sci. 3, 101–109.

23 Maas E V, Grattan S R and Ogata G 1982 Foliar salt accumulation and injury in crops sprinkled with saline water. Irrig. Sci. 3, 157–158.

24 Maas E V and Hoffman G J 1977 Crop salt tolerance – current assessment. J. Irrig. Drainage Div., ASCE 103 (IR2), 115–134.

25 Nielson R F and Cannon O S 1975 Sprinkling with salty well water can cause problems. Utah Sci. 36, 61–63.

Plant and Soil 89, 285–299 (1985). Ms. BR 19

Fertilization management of crops irrigated with saline water*

A. FEIGIN
Institute of Soils and Water, Agricultural Research Organization, The Volcani Center, Bet-Dagan, Israel

Key words Calcium Chloride Ion antagonism Magnesium Nitrate Nitrogen Nutrition disorders Phosphorus Salinity-fertility relationship Sodium

Summary Available data concerning nutrition and fertilization effects on crops irrigated with saline water are presented and discussed. Published data on the salinity-fertility relationship are, at least to some extent, contradictory; both positive and negative effects as well as no effect of fertilization on salinity tolerance have been recorded. However, a great deal of the experimental work supports the view that standard fertilization recommendations for non-saline conditions are also suitable for saline conditions. In addition, available data indicate that the apparent salt tolerance of agricultural crops varies with soil fertility level. Consequently, crops showing exceptionally high apparent salt tolerance at a low fertility level become more sensitive when adequately fertilized, although the absolute yield may be greatly increased. On the other hand, some data seem to show a real increase in salinity tolerance under improved fertility conditions.

Important information concerning the responses of plants to salinity under various fertility levels was obtained by tissue analysis. It should be kept in mind that this information may be influenced also by the plant species and by environmental conditions.

Introduction

With the increasing use of saline water for agriculture, fertilizer application under saline conditions has been the subject of considerable interest. The salt tolerance of crops is usually presented as a function of the electrical conductivity of the saturated soil extract (EC_e)[3,20,36]. However, it seems that the situation is more complex, in that other soil factors, including fertility[20], can influence salt tolerance.

Salinity-fertility relationships are of great economic importance and have been the subject of many greenhouse and field studies (*e.g.*[4,13,14,15,23,25,28,29,32]). These studies were conducted to evaluate improved fertilization management as a means of alleviating growth inhibition due to salinity. The effects of fertilization on nutrient uptake by plants, on the chemical composition of plant tissues, and on crop yield were studied under various salinity conditions and soil-crop combinations. The research work resulted in different and even contradictory conclusions. Positive, negative and no effects of fertilization on crop salinity tolerance were found.

* Contribution from The Agricultural Research Organization, The Volcani Center, Bet Dagan, Israel. No. 1085-E, 1984 series.

The present paper aims at assessing improved fertilization management as a means of increasing the salinity tolerance of agricultural crops. In other words: Is it possible to alleviate the effects of salinity on agricultural crops by changing the fertilization regime (type and quantity of fertilizers and the method of their application) from that considered appropriate for nonsaline conditions?

Soil properties affecting salinity-fertility relationship

Both salinity and soil fertility are determined by the concentration of various ions in the root zone of plants. Salinity develops when the concentrations of Na, Mg, Ca, Cl, HCO_3, SO_4 ions and B in the soil solution is high enough to affect plant growth[21]. As the salt concentration exceeds a threshold level, the growth rate and final size of plants progressively decrease; consequently, the crop yield also declines[20]. It was concluded[1,21] that the effect of salinity on plants is a result of two factors: restriction of water uptake by high osmotic potential in the soil, and high concentration of specific ions that may cause physiological disorders in plant tissues.

The level of soil fertility, as far as plant nutrition is concerned, is determined by the level of available nutrient elements (N, P, K, Ca, Mg, S and micronutrients). The availability of N, P and K in soil is often too low for economic yields of agricultural crops. Addition of fertilizers to maintain an adequate fertility level is therefore a standard practice in agricultural soils. The concentration of the other nutrient elements is rarely below the desirable level, although these materials may be applied in standard fertilizers or under specific situations.

Many publications have reported yield reduction due to high osmotic pressure in the root zone of crops (*e.g.* [1,3,4,36]). Salinity-induced specific effects – nutritional disorders and various damage symptoms in plant tissue – have also been recorded[1,20].

The desired concentration of nutrient elements in soil is usually well below the level needed to cause a salinity effect. However, excessive application of fertilizers, particularly of N and K, due to their high solubility, may result in a salinity buildup in the soil. This is typically reflected by a decline in yield when the fertilizer levels exceed the optimum values[30].

The simultaneous presence of salts and nutrient elements in the root zone can influence ion uptake by plants and affect their chemical composition. Synergistic and antagonistic effects can increase or decrease the intensity of these processes. Specific ion effects, *e.g.* nutritional disorders such as Ca deficiency, may be created or corrected, depending on the specific conditions. The cation − anion (C − A) balance may also be affected. The level of soil fertility may affect salinity tolerance through its effect on root growth and activity[15]. Finally, since both salinity and low fertility are important growth-limiting factors, moderation or elimination of one factor may affect the crop's response to the other, as presented in the following section.

Interactive effects of salinity and fertility on the growth and yield of agricultural crops

Many greenhouse and field experiments have been conducted to study the interactive effects of salinity and fertility on the yields of agricultural crops. Most of the work was related to N and P fertilization, although other nutrient elements were also considered.

Possible types of interaction of salinity with other variables, including soil fertility level, were described by Bernstein *et al.*[4] (Fig. 1). The curves given in this figure are idealized, showing only the range of yield that is sensitive to salinity, which is represented as a straight line. The upper line in each part of the figure shows the yield obtained under optimum nutrition conditions while the lower line represents the yield obtained when the fertility level is inadequate. One can calculate the relative yields by considering the yields in the nonsaline treatments for each level of nutrition as 100%. In the first case (type A)

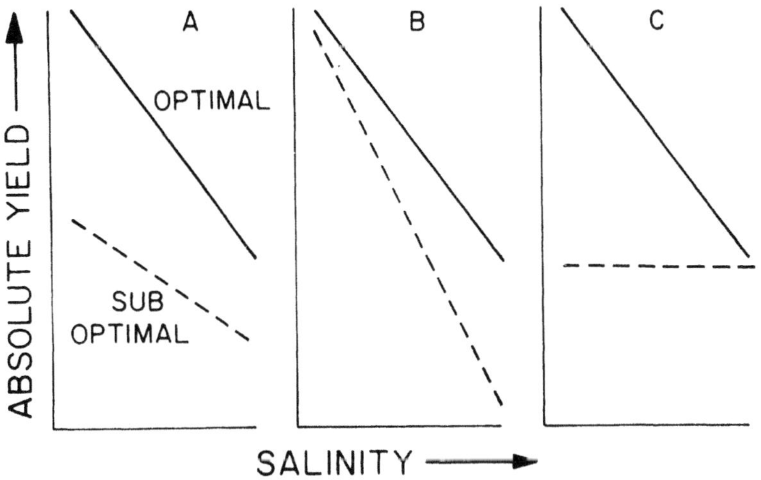

Fig. 1. Types of salinity-fertility relationships (Reproduced from Bernstein *et al.*[4] by permission of the American Society of Agronomy). See text for explanation.

(Fig. 1A), the salinity and fertility effects at the optimum and suboptimum fertility levels are independant, since the relative yield is the same for the two conditions. The second case (type B)(Fig. 1B) represents decreased salt tolerance at the suboptimal fertility level, while the third case (type C)(Fig. 1C) represents increased salt tolerance.

Bernstein *et al.*[4] presented data supporting the validity of the proposed types of salinity-fertility interactions. For instance, type A response occurs in lettuce (*Lactuca sativa* L.), onion (*Allium capa* L.), carrot (*Daucus carota* L.) and beets (*Beta vulgaris* L.) grown in sand at different N and P levels, each combined with three different osmotic potentials (obtained by the addition of NaCl and $CaCl_2$): 0, -0.2 and -0.4 MPa. Type B response occurred in corn plants (*Zea mays* L.) suffering from salinity-induced Ca deficiency[1]. Type C was detected in cabbage (*Brassica oleracea* L. var. 'capitata') and broccoli (*Brassica oleracea* var. 'botrytis'). The latter two crops were relatively insensitive to salinity when the levels of N and P in the nutrient solution were inadequate. Other published data on interactive effects of salinity and fertility on the yield of crops fit well into the scheme suggested by Bernstein *et al.*[4].

Some of the published data on salinity-fertility relationships indicate a positive effect of fertility on crop salinity tolerance. Among the most striking data are those published by Ravikovitch and Yoles[29]. The yields of clover (*Trifolium alexandrinum*) and millet (*Setaria italica*) growing in pots in a greenhouse experiment were greatly increased by the addition of N and particularly P to saline soil. The yields of the fertilized plants were greater both absolutely and relatively. Champagnol[7] reviewed 17 papers dealing with phosphate nutrition of plants in relation to salt toxicity. Thirty-four crops out of 37 mentioned in the paper responded positively to phosphate fertilization, including cases of positive P x salinity interaction. Positive effects of N and P application on the yield of wheat (*Triticum aestivum* L.) growing on saline soils under dryland conditions were detected by Luken[18]. The addition of P improved the yield from tomato (*Lycopersicon esculentum* L.) growing under saline conditions[6]. Kafkafi[13] concluded that the use of saline water for irrigation should be combined with a continuous supply of nutrients in the proper concentration. The yield of beans (*Phaseolus vulgaris* L.) was increased by P fertilization for the salinity range studied by Lunin and Gallatin[19]. The addition of P fertilizer (15 and 40 mg P l^{-1}) to pepper plants (*Capsicum annum* L.) during the first 55 days of growth increased the percent germination of seeds obtained from these plants in the presence of 0.5% NaCl solution, in comparison with untreated plants[8]. Langdale and Thomas[17] found that application of N fertilizer to moderately saline soils resulted in an increase in the yields of forage and protein obtained from coastal Bermuda grass (*Cynodon dactylon* L.).

In contradiction to the positive effect of increased levels of nutrient elements on crop salinity tolerance recorded in the above-mentioned papers, other sources present data indicating a decrease in salinity tolerance. Khalil *et al.*[15] measured increased sensitivity of corn and cotton (*Gossypium hirsutum* L.) to salinity with increased level of nutrition. Ogo and Morikawai[24] found an apparent decrease in salt tolerance when excessive N fertilizer was given to rice (*Oryza sativa* L.). The response of barley (*Hordeum vulgare* L.) to N fertilization was rapidly reduced as the salinity level was increased, and in one of the field trials a tendency toward reduced crop tolerance to salinity was detected[27]. A considerable deterioration in the salt tolerance of corn plants caused by P treatment was reported by Bernstein *et al.*[4]. Increasing the level of P from 0.1 to 2 m*M* in sand culture resulted in a striking decrease in the yield of husked ears. Similar results were published by Nieman and Clark[23]: the yield of fresh corn shoots was 2259, 1101, 549 and 252 g per two plants for control high P, control low P, saline low P, and saline high P, respectively (the low P concentration was 0.1 m*M* and the high one was 2 m*M*; the osmotic potential of the control and saline solutions was −0.04 and −0.44 MPa, respectively).

A contrasting response by tomato and wheat plants to an increase of P concentration in a saline nutrient solution was measured by Cerda and Bingham[6]. While tomato yield increased as P levels were increased (from 0.5 to 50 mg P l^{-1}), the yield of wheat was reduced at P levels greater than 5 mg l^{-1}.

Unlike the above cases of positive or negative effects of fertilization on crop sensitivity to salinity, other publications indicate no significant change in the relative salt tolerance, some in the same plants as above. Such results were obtained in beans[19] and corn[29] when N levels were increased. Increasing N levels in nutrient solutions from 8 to 64 and 120 mg l^{-1} resulted in a positive yield response from tomato plants growing in low-salinity soil (1 dS m^{-1}); however, no yield increase due to N application was found when the salinity level was 5 dS m^{-1} or more (9 dS m^{-1})[25]. Another example of the absence of a positive fertilizer effect on salinity tolerance was reported by Selassie and Wagenet[32], who studied the interactive effects of salinity (4 levels), irrigation (6 levels), and N fertilization (5 levels) in a 'line source' field experiment on corn. Yield was not improved by N application.

It seems that the available information on crop response to fertilizer application under saline conditions is not clear, with some of the published data even contradictory. Chemical analyses of plant tissue and studies of the physiological processes involved in the salinity-fertility relationship were employed in the search for a valid explanation for these phenomena.

Interactive effects of salinity and fertility on the chemical composition of plants

Crops can be characterized by the typical chemical composition of growing or developed tissues. Chemical analysis of plant parts is often used for diagnostic purposes in determining fertilizer needs. Poor fertility level, excessive concentration of available nutrients, or high salinity in the root zone is reflected in lower or higher concentrations of certain elements in plant tissues in comparison with the optimum range. Some of the available information on the interactive effects of salinity and fertility on the chemical composition of crops will be presented and discussed in the following sections.

Cations

Some cations influence the uptake of other cations by plants[21], *e.g.* antagonism occurs between K and Ca, or Mg and Na. Such effects may be involved in the occurrence of nutritional disorders in plant tissues. The addition of Ca ions (up to 10 m*M*) to bean plants (*P. vulgaris* L.) grown in a 50 m*M* NaCl solution resulted in decreased Na uptake and an increased yield of dry matter[16]. High levels of $CaCl_2$ in the nutrient medium resulted in a great increase in Ca concentration and a reduction of K and Mg levels in the tissue of bean plants; on the other hand, K addition (as KCl) increased K and decreased Ca and Mg contents in plant tissue[1]. Corn plants (*Z. mays*) tested in the same study responded differently to different Cl salts. Plant growth was better in the presence of $CaCl_2$ than with any combination of NaCl, $MgCl_2$ or KCl at comparable osmotic pressures. However, high Ca content depressed the K and Mg levels, while a mixed salt solution (Ca + K, Ca + Na) corrected the imbalance. Bernstein[1] concludes that 'the likelihood of unbalanced nutrition is greatly lessened by the mixed-salt solutions normally prevalent in saline soils.' Nassery *et al.*[22] studied the effect of various ion solutions on the growth and chemical composition of sesame (*Sesamum indicum* L.). Calcium chloride suppressed top growth at the early fruiting stage much more than did NaCl, causing severe necrosis in the lower leaves of the plants. The detrimental effect of Ca resulted from the presence of an insufficient level of Mg rather than from the accumulation of toxic levels of Ca in plant tissues. The authors concluded that 'at extreme Na-Ca ratios, ionic imbalance predominates and growth suppression exceeds that due to osmotic stress.'

A simultaneous increase in the concentrations of Na and Ca, as the chlorides, in the nutrient solution resulted in a considerable increase in the levels of these elements in melon (*Cucumis melo* L.) shoots and a concomitant decrease in the K and Mg levels in the plants[9] (Fig. 2). The concentration of Ca was almost doubled and that of Na was almost tripled in certain treatments, and the reduction in the K and Mg contents in the melon shoots was of about the same magnitude.

The steady state of Na ion fluxes as well as the cytoplasmic and vascular Na ion contents were lower in the presence of K ions than in their absence in wheat, sunflower (*Helianthus annuum* L.), and onion (*A. cepa*); moreover, the addition of K to the roots induced a significant net Na extrusion from wheat roots[12]. These authors suggested that an efficient K/Na exchange system at the plasmalemma increases crop tolerance to salinity.

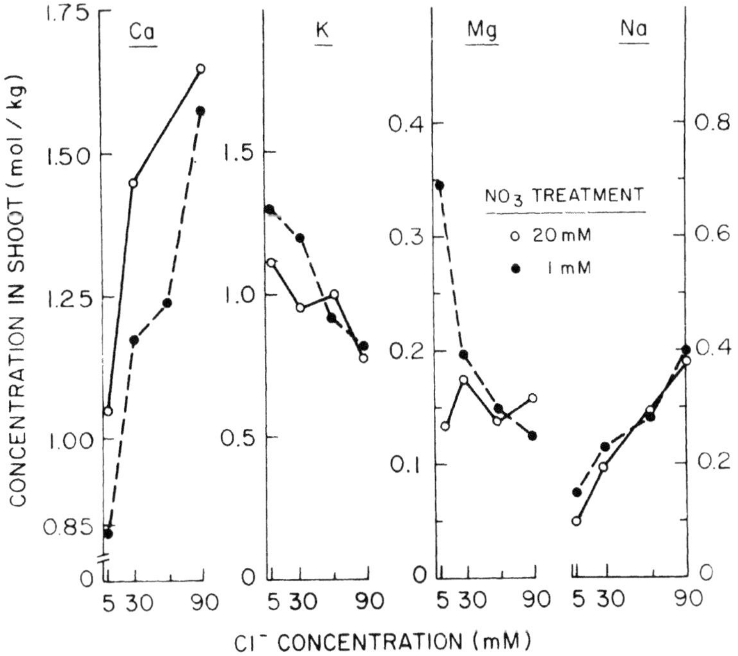

Fig. 2. Effect of the chemical composition of the nutrient-salt solution on the concentrations of Ca, Mg, K and Na in dry matter in the tops of melon plants (Feigin *et al.*[9]).

Anions

The clear effect of NO_3 concentration in the nutrient solution on the level of Cl in plant tissue and *vice versa* illustrates another case of antagonism between ions. Kafkafi *et al.*[14] showed that an increase in Cl concentration in the nutrient solution led to a reduction in the NO_3 content of the tissue of tomato plants. Increasing the concentration of NO_3 in the nutrient solution from 7.5 to 20 meq l^{-1} in the absence of Cl had no effect on the NO_3 concentration in plant tissue. Feigin *et al.*[9] found that the concentration of Cl in the tissue of melon shoots increased linearly as the level of Cl in the nutrient-salt solution was elevated from 5 to 60 m*M* (Fig. 3). A dramatic increase in the concentration of Cl in the tissue of melon shoots was detected when the amount of Cl added (in all cases as 2/3 NaCl and 1/3 $CaCl_2$) was increased from 60 to 90 m*M*. It is possible that the ability of cell membranes to control salt uptake was reduced by the high levels of Cl. Figure 3

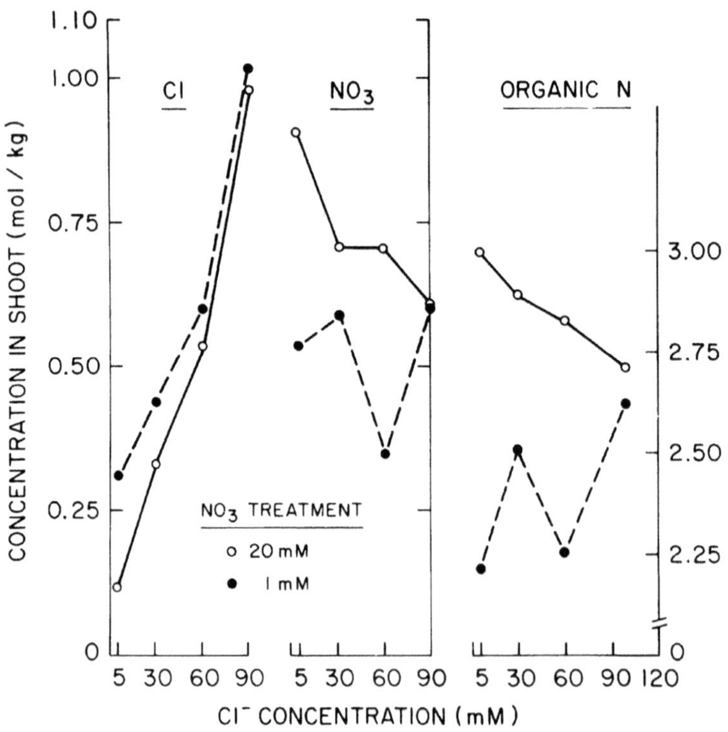

Fig. 3. Effect of the chemical composition of the nutrient-salt solution on the concentration of Cl, NO_3 and organic N in dry matter in the tops of melon plants (Feigin *et al.*[9]).

also shows that the Cl concentration in melon shoots was reduced when the level of NO_3 in the nutrient-salt solution was increased. However, this effect was significant only for the lower Cl levels tested, mainly between 5 and 30 m*M*. As the level of Cl was raised, the moderating effect of NO_3 on the Cl concentration in the shoots disappeared. A clear effect of the Cl content of the nutrient-salt solution on the NO_3 level in melon shoots was also found in this study (Fig. 3). The concentration of NO_3 in the tissue of melon tops was not clearly affected by the level of Cl in the nutrient-salt solution when the level of NO_3 was low (1 m*M*). On the other hand, a clear Cl-induced decrease in NO_3 in plant tissue was found when the NO_3 concentration in the nutrient salt-solution was higher[9]. A similar effect of Cl on organic N content in the melon shoot was also detected (Fig. 3). Antagonistic effects between Cl and NO_3 in potato plants (*Solanum tuberosum* L.) were reported by Saffigna and Keeney[31]. Torres and Bingham[35] presented data suggesting that Cl-induced NO_3 deficiency was responsible for growth retardation in wheat plants exposed to high levels of NaCl in the substrate.

Papadopoulos and Rendig[25] published data indicating depression by Cl of the concentration of P in tomato plants. Champagnol[7] concluded that it seems unlikely that competition between H_2PO_4 and Cl ions is important, due to the great differences in the physical and physiological properties of these ions. Champagnol[7] has shown, in his review on the relationship between the phosphate nutrition of plants and salt toxicity, that there are three types of reaction to salinity, with respect to the P content of plant tissues. The level of P was increased in some crops [*e.g.* sorghum (*Sorghum bicolor* L.), corn and sesame], reduced in others (tomato, onion and barley), and unchanged in millet, cabbage, Brussels sprouts (*Brassica oleracea* L. var. 'gemmifera'), carrot and wheat. Bernstein *et al.*[4] attributed the growth retardation measured in wheat plants by Torres and Bingham[35] to the higher levels of P (2 or 3 m*M*) in the nutrient solution used in their experiment. Aggravated salt injury in corn plants due to increased P levels in the solution used in sand culture (from 0.1 to 2 m*M*)[4] was explained by Nieman and Clark[23] as a result of salinity damage to the mechanism regulating the internal levels of orthophosphate, which resulted in excessive accumulation of P and toxicity. It was also found that a combination of salinity and a high level of orthophosphate (2 m*M*) resulted in a considerable increase in the level of sugar phosphates in the leaves, showing that translocation rather than photosynthesis was the main factor limiting growth.

Cation-anion balance

The ionic composition of the nutrient-salt solution may affect the balance between the total concentration of cations and anions (C − A) in plant tissues. Larger C − A values are correlated with greater production of dry matter, since the 'extra' cations are balanced by a greater concentration of organic anions in plant tissue[21]. The C − A value declines when the Cl concentration is increased (Table 1)[14]. Table 1 shows that the C − A balance of young tomato plants was 2090 meq kg^{-1} dry matter when no Cl was added to the nutrient solution, but its value decreased greatly (to 1420 meq kg^{-1}) when the Cl concentration in the solution was raised to 35 meq l^{-1}. A reduction in the quantity of organic

Table 1. The effect of Cl concentration in the nutrient-salt solution on the cation and anion composition and C − A balance in tissues of young tomato plants (from Kafkafi *et al.*[14])

Cl concentration	Cations (meq/kg)			Anions (meq/kg)				
(m*M*)	K	Ca	Mg	Cl	NO_3	SO_4	P*	Cations − Anions
0	1430	1860	740	160	1260	450	70	2090
35	1280	1960	720	1190	960	320	70	1420

* total P concentration.

salts in gloxinia (*Sinningia speciosa*) and chrysanthemum (*Chrysanthemum morifolium*) was found by Bik[5] when the level of Cl in the growth medium was raised. Thomas and Langdale[34] reported that nitrogen fertilization increased the C − A values in Bermuda grass (*C. dactylon*) at all salinity levels, but Cl addition decreased it. The greater yield obtained with N fertilization on moderately saline soils was attributed to a change in the C − A balance, which was affected by substitution of NO_3 for Cl.

Visible injury to plant tissues

Nutritional disorders can adversely affect plant growth and yield and its quality. Specific symptoms detected under saline conditions were: Ca-deficiency-induced blossom end rot in tomatoes[10] and Na and Cl injuries in various fruit crops; scorch symptoms when the levels of Cl in leaves exceeded 1% of dry matter; and specific Na scorch effects[1]. Calcium-induced necrosis in sesame plants, possibly due to Mg deficiency, was reported by Nassery *et al.*[22]. The combination of salinity and high P levels caused severe leaf injury in corn plants[23]. Other cases of plant injury caused by salinity-induced nutritional imbalances or deficiencies have been mentioned in various publications (*e.g.*[20]).

Physiological disorders

Published evidence points toward interference of salinity with various metabolic processes related to plant nutrition. Helal and Mengel[11] found that incorporation of labelled inorganic N into the protein fraction was impaired by salinity, even though its concentration in plant tissue was unaffected. The addition of K enhanced N uptake and its incorporation into protein. The authors conclude that impairment of protein metabolism by salinity is an important aspect of salt stress and is probably induced by disturbance of the K/Na balance. Shomer-Ilan *et al.*[33] found that Rhodes grass (*Chloris gayana* Kunth.) grown under high atmospheric humidity responded positively to a combination of low salt (50 m*M* NaCl) and 19 m*M* N. Higher levels of salinity cancelled the positive effect of N nutrition. The authors suggested that the major negative effect of salinity is on protein synthesis.

General discussion and conclusions

Maas and Hoffman[20] concluded that the apparent salt tolerance of crops may vary with soil fertility; thus, due to the limiting effect of inadequate fertility, crops grown under low fertility levels may show exceptionally high relative salt tolerance. Adequate fertilizer application will increase the absolute yield, but will also aggravate the apparent salt sensitivity (Fig. 4). When the absolute yield of dry matter in melon plants was limited by a low level of N (1 m*M*) in a circulating nutrient-salt solution, an increase in salinity from about 1.8 dS m^{-1} to about 9 dS m^{-1} (from 5 to 90 m*M* Cl) resulted in a relatively small

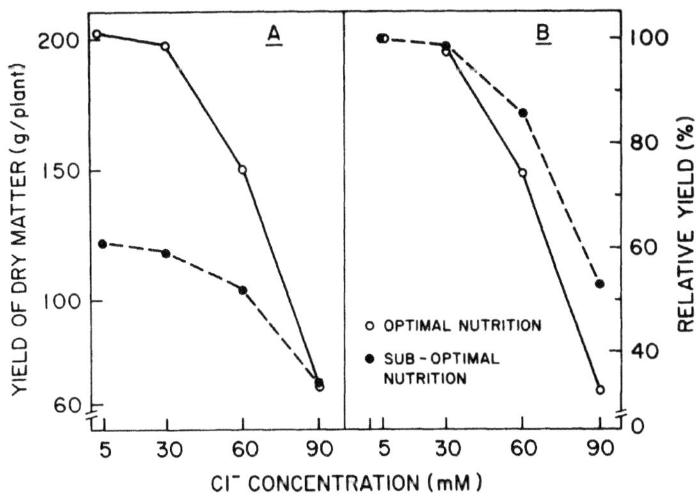

Fig. 4. Effect of Cl (salinity) and NO_3 (optimal and suboptimal nutrition) levels in solution on the (**A**) absolute and (**B**) relative yields of dry matter in the shoots of melon plants (Feigin *et al.*[9]). See text for explanation.

decrease in the yield of dry matter[9]. The addition of NO_3 (from 5 to 20 mM) increased the absolute yield of dry matter (Fig. 4A) but at the same time decreased the apparent salt tolerance (Fig. 4B). Other similar cases were mentioned by Maas and Hoffman[20], who also concluded that 'unless salinity causes specific nutritional imbalances, additional fertilization has little effect or reduces salt tolerance.' Bernstein *et al.*[4] concluded that the correction of either salinity or fertility limits is more effective when only one of these factors is the main limiting factor. When the limiting effect of both of the above-mentioned factors is severe, the response obtained by the elimination of the other factor will be relatively small; maximal effects of salinity correction and fertilization will be achieved when both factors become nonlimiting.

Salt tolerance is usually studied under optimum fertility conditions[2,20]; consequently, the published lists of crop tolerance to salinity were obtained from research carried out at the U.S. Salinity Laboratory[2,3,20,36] at fertility levels considered optimum for nonsaline conditions. Bernstein *et al.*[4] pointed out that salinity tolerance under suboptimal conditions is important only under dryland conditions where high levels of fertilization are not economical or the availability of fertilizers is limited. In addition, these authors conclude that salinity *per se* rarely causes or aggravates nutrient deficiency. The above-mentioned cases of a large reduction in salinity tolerance in the presence of a high level of available P in the root zone of crops occurred in solution or sand cultures. Such high levels of available P are not likely to exist in soils. Consequently, the idea that the standard fertilization recommendations for nonsaline conditions may also be followed under most saline conditions is well accepted. Much of the available data concerning salinity-fertility relationships support this view. However, the occurrence of contradictory results, some of which were presented and discussed in this paper, indicate that the picture is not yet clear.

The disagreement between some of the publications dealing with salinity-fertility relationships could stem from the use of different salinization techniques, including the use of different chemicals to change the level of salinity in the root zone of plants. For example, Nassery *et al.*[22] have shown different responses of sesame plants to $CaCl_2$ and NaCl. Moreover, differences in species, varieties, duration of experiment and growth conditions could also explain the variations in crop response. Standardization of methodology, where feasible, could reduce the variation between experiments.

Different plant parts are not equally affected by salinity[11,37], the growth of tops often being more suppressed than that of roots. The effect of specific nutrient elements on different plant organs also varies[21]. The positive effect of P fertilization on crops growing under saline conditions was attributed to its effect on root growth[15]. Although it is not easy to evaluate the salinity-fertility effect on the growth of particular plant organs under field conditions, this aspect should not be ignored.

Although tissue analysis is a valuable tool in research on phenomena and processes related to plant nutrition, interpretation of the data obtained is often complex. The effect of certain ions or salinity conditions on the content of other elements in plant tissue often differs between crops and varies with growth conditions. For instance, Bernstein *et al.*[4] found that deficiencies in N and P had variable effects on tissue mineral concentrations. For instance, N deficiency increased the Cl level in corn, barley and some vegetable crops, but decreased it in wheat. Champagnol[7] presented data showing that the effect of salinity on P content in the tissue of different crops varied greatly. Other data in the literature support these findings.

The effect of salinity on the concentration of nutrient elements in plant tissue can also be explained, at least partially, by its effect on plant or population size. Patel *et al.*[26] pointed out that reduced growth of sorghum plants decreased the total amounts of N and P needed by each plant. Similarly, Selassie and Wagenet[32] found that the increase in nitrogen content of corn plants affected by salinity was probably due to the reduction in the plant population, and consequently to greater quantities of available N per plant.

References

1 Bernstein L 1964 Effects of salinity on mineral composition and growth of plants. Plant Anal. Fert. Probl. 4, 25–45.

2 Bernstein L 1964 Salt tolerance of plants. USDA Inf. Bull. 283, 23 p.

3 Bernstein L, Francois L E and Clark R A 1972 Salt tolerance of ornamental shrubs and ground covers. J. Am. Soc. Hort. Sci. 97, 550–556.

4 Bernstein L, Francois L E and Clark R A 1974 Interactive effects of salinity and fertility on yields of grains and vegetables. Agron. J. 66, 412–421.

5 Bik R A 1970 Nitrogen, salinity, substrates and growth of gloxinia and chrysanthemum. Centre for Agric. Publ. and Documentation, Wageningen. Agric. Res. Report No. 739.

6 Cerda A and Bingham F T 1978 Yield, mineral composition, and salt tolerance of tomato and wheat as affected by NaCl and P nutrition. Agrochimica 12, 140–149.

7 Champagnol F 1979 Relationship between phosphate nutrition of plants and salt toxicity. Phosphorus in Agriculture 76, 35–44.

8 Feder Z, Markovitch T, Bar-Yosef B, Perl M, Kremer S and Adler I 1984 Effect of fertilization on quality of pepper seeds. Hassadeh 64, 1574–1575. (*In Hebrew*)

9 Feigin A, Rylski I and Shalhevet J 1984 Response of melon plants to different combinations of NO_3 and Cl in saline nutrient solutions. Paper presented at Soil Salinity under

Irrigation-Processes and Management, Inter. Conference, Inter. Soc. Soil Sci. and Israel Soc. Soil Sci., Bet Dagan, Israel.

10 Geraldson C M 1957 Control of blossom-end-rot of tomatoes. Proc. Am. Soc. Hort. Sci. 69, 309–317.

11 Helal H M and Mengel K 1979 Nitrogen metabolism of young barley plants as affected by NaCl-salinity and potassium. Plant and Soil 51, 547–562.

12 Jeschke W D and Nassery H 1981 K^+-Na^+ selectivity in roots of *Triticum*, *Helianthus* and *Allium*. Physiol Plant. 52, 217–224.

13 Kafkafi U 1984 Plant nutrition under saline conditions. *In* Soil Salinity under Irrigation – Processes and Management. Eds. I Shainberg and J Shalhevet. Springer-Verlag, Berlin. pp 319–338.

14 Kafkafi U, Letey J and Valoras N 1982 Chloride interaction with nitrate and phosphate nutrition in tomato. J. Plant Nutr. 5, 1369–1385.

15 Khalil M A, Amer F and Elgabali M M 1967 A salinity-fertility interactions study on corn and cotton. Soil Sci. Soc. Am. Proc. 31, 683–686.

16 Lahaye P A and Epstein E 1971 Calcium and salt toleration by bean plants. Physiol. Plant 25, 213–218.

17 Langdale G W and Thomas J R 1971 Soil salinity effects on absorption of nitrogen, phosphorus and protein synthesis by coastal Bermuda grass. Agron. J. 63, 708–711.

18 Lüken H 1962 Saline soils under dryland agriculture in south-eastern Saskatchewan (Canada) and possibilities for their improvement. II: Evaluation of effects of various treatments on soil salinity and crop yield. Plant and Soil 17, 26–48.

19 Lunin J and Gallatin M H 1965 Salinity-fertility interactions in relation to the growth and composition of beans. I: Effect of N, P and K. Agron. J. 57, 339–342.

20 Maas E V and Hoffman G J 1977 Crop salt tolerance – current assessment. J. Irrigation Drainage Div. ASEC 103 (IR2), 115–134.

21 Mengel K and Kirkby E A 1979 Principles of Plant Nutrition. Inter. Potash Inst., Bern, Switzerland.

22 Nassery H, Ogata G and Maas E V 1979 Sensitivity of sesame to various salts. Agron. J. 71, 595–597.

23 Nieman R H and Clark R A 1976 Interactive effects of salinity and phosphorus nutrition on the concentration of phosphate and phosphate esters in mature photosynthesizing corn leaves. Plant Physiol. 57, 157–161.

24 Ogo T and Morikawai S 1965 Relationship between certain nitrogen fractions in leaf blade of crops and salt tolerance (I). Shimane Agr. Coll. Bull. 13A, 5–9.

25 Papadopoulos I and Rendig V V 1983 Interactive effects of salinity and nitrogen on growth and yield of tomato plants. Plant and Soil 73, 47–57.

26 Patel P M, Wallace A and Wallihan E F 1975 Influence of salinity and N-P fertility levels on mineral content and growth of sorghum in sand culture. Agron. J. 67, 622–625.

27 Peters J R 1983 The effects of phosphorus and nitrogen fertilizer on the relationship between soil salinity levels and the grain yield and protein content of barley grown on stubble land. Can. J. Soil Sci. 63, 705–718.

28 Ravikovitch S and Porath A 1967 The effect of nutrients on the salt tolerance of crops. Plant and Soil 26, 49–71.

29 Ravikovitch S and Yoles D 1971 The influence of phosphorus and nitrogen on millet and clover growing in soils affected by salinity. I. Plant development; II. Plant composition. Plant and Soil 35, 555–567; 569–588.

30 Russel E W 1968 Soil Conditions and Plant Growth. Longmans, Green and Co., London.

31 Saffigna P G and Keeney D R 1977 Nitrogen and chloride uptake by irrigated Russet Burbank potatoes. Agron. J. 69, 258–264.

32 Selassie T G and Wagenet R J 1981 Interactive effects of soil salinity, fertility and irrigation on field corn. Irrigation Science 2, 67–78.

33 Shomer-Ilan A, Samish Y B, Kipnis T, Elmar D and Waisel Y 1979 Effects of salinity, N-nutrition and humidity on photosynthesis and protein metabolism of *Chloris gayana* Kunth. Plant and Soil 53, 477–486.

34 Thomas J R and Langdale G W 1980 Ionic balance in coastal Bermuda grass as influenced by nitrogen fertilization and soil salinity. Agron. J. 72, 449–452.

35 Torres B C and Bingham F E 1973 Salt tolerance of Mexican wheat. I: Effect of NO_3^- and NaCl on mineral nutrition, growth and grain production of four wheats. Soil Sci. Soc. Am. Proc. 37, 711–715.

36 United States Salinity Laboratory Staff 1954 Diagnosis and Improvement of Saline and Alkaline Soils, USDA Handbook 60, Washington, D.C.

37 Waisel Y 1971 Biology of Halophytes, Academic Press, NY.

Plant and Soil 89, 301–307 (1985). Ms. BR 20

Effect of salinity on quality of various agricultural crops

YOSEF MIZRAHI[1,2] and DOV PASTERNAK[1]
[1] *The Boyko Institute for Agriculture and Applied Biology, The Institutes for Applied Research, and* [2] *Department of Biology; Ben-Gurion University of the Negev, P.O. Box 1025, Beer-Sheva 84110, Israel*

Key words Chinese cabbage Crop quality Iceberg lettuce Melon Peanuts Saline irrigation Tomato

Summary The effect of salinity on the quality of various agricultural crops has not yet been explored much. This information is very important to Israel due to the increasing use of saline water for irrigation. This paper reports the effect of saline irrigation water on the quality, especially the taste, of several crops. Fruits from a processing tomato cultivar exposed to various degrees of salinity had higher values for total soluble solids (TSS) and acidity than their controls. The yield of fruit after saline water irrigation is lower, but this is offset by the higher fruit quality and its consequent higher value. Melon fruits from plants subjected to saline water scored higher in taste than their controls when the fruits were analyzed fresh. After 3–4 weeks of storage at room temperature, there was no longer any difference in taste. Even though salinity slightly increased the TSS content, this did not correlate with the taste scores. Iceberg lettuce grown with saline water did not significantly differ in taste from its control, even when the sensitive triangle taste test was used. The same was true for peanuts. Thus, for these two crops no advantage of better quality would compensate for possible lower yields. Salinity had little effect on the yield of two varieties of Chinese cabbage, but increased the frequency of tipburn.

Introduction

Israel uses 95% of its known fresh water resources. Thus, any additional agricultural development must make use of alternative water resources, *e.g.* the saline water recently found under the Negev Desert. The question of the quality of agricultural products irrigated with saline water is crucial for the future of Israeli agriculture.

The effect of salinity on various processes in plants has been studied intensively over a long period of time. However, very limited research has been done on the effect of salinity on the quality of crops[3]. One study on fresh market tomatoes demonstrated that the ripening process, as measured by the CO_2 and C_2H_4 evolution rates, pericarp pigment concentrations, and polygalacturonase activity, was enhanced in fruits from plants irrigated with saline water. In addition, the amount of total soluble solids (TSS), content of reducing sugars, total acidity, and NaCl concentration were higher in fruits from salinized plants. Their taste was much better than that of their proper controls, but their shelf life was shorter[3].

This paper presents a study on the effects of saline water on the quality (especially the taste) of five agricultural crops grown in the Negev Desert of Israel.

Materials and methods

Plant material

Processing tomatoes, *Lycopersicon esculentum* Mill cv M-82-1-8; melons, *Cucumis melo*, cultivars Noy Tsahov, Makdimon, and Hemed; iceberg lettuce, *Lactuca sativa* cv Vangard; peanuts, *Arachis hypogaea* cv Shulamit; and two species of Chinese cabbage, *Brassica pekinensis* cv Michili and *B. chinensis* cv Spring A, were grown in the Negev Plateau (Ramat HaNegev) Experimental Station according to the agricultural practices recommended by the extension service. The control treatment consisted of irrigation with water from the National Water Carrier, with an electrical conductivity (EC_i) of 1.2 dS m^{-1}, and the saline treatments, with saline water from local wells (EC_i = 5 dS m^{-1}). The processing tomatoes were irrigated with water of four levels of salinity, which were achieved by addition of NaCl and $CaCl_2$ (3 : 1 ratio) to the control water for electrical conductivities of 1.2 (control), 4.5, 7.5, and 10.0 dS m^{-1}. The Chinese cabbage was irrigated with water of 1.2, 4.0, 7.0, and 10.0 dS m^{-1}. All the crops were harvested according to the agricultural practices used in the area.

Chemical analyses of tomato fruit

Tomato slices (one quarter from each of 20 fruits) were homogenized in a blender. The total acidity (as percent citric acid, determined by titration with NaOH), pH, electrical conductivity, and percent total soluble solids (TSS) of the juice were determined by conventional methods.

Organoleptic taste tests

The yields of several plants of the same treatment were pooled for taste tests. Melon fruits were sliced longitudinally and then the flesh (mesocarp) was cut into cubes (2 × 2 × 2 cm) and piled on a plate. Iceberg lettuce heads were sliced longitudinally and then transversely (approximately 3 × 3 × 3 cm) so that each piece consisted of several leaf layers. The pieces were piled on a plate. Peanuts were shelled and roasted in a home oven until the cotyledons started to separate. Then roasted peanuts from the same treatment were piled on the same plate.

The taste testers were usually employees of our Institute. The same people participated in all tests. Each tester was required to taste each sample and to grade it according to the following scale: 1, worst; 2, intermediate; and 3, best. In the preference test the tester was required to taste one sample from the salinity and one from the control treatment and to judge which sample tasted better. A plus sign indicates that the product from the saline treatment tasted better than the control, a minus sign that the control was better than the treated, and a zero that the tester did not prefer either treatment.

Lettuce was also tested by the more sensitive triangle taste test[2], as follows. Each head from each treatment (8–10 heads per plate) was cut in half. Each half was cut into slices as above and piled on a separate plate. Two plates from the same treatment were tested along with only one from the other treatment. The tester was required to taste the three samples and to decide which were identical. A reciprocal triangle test was conducted afterwards, in which two samples of the other treatment and a single sample from the first treatment were used.

Results

Processing tomatoes

The juice of ripe processing tomatoes from plants exposed to increasing degrees of salinity showed increases in TSS content, total acidity, and electrical conductivity. The pH of the juice was not changed significantly, except at the highest salinity, where it was reduced (Table 1).

Table 1. Effect of salinity on quality parameters of juice of processing tomatoes (cv M-82-1-8)

Electrical conductivity of irrigation water (dS m^{-1})	Quality parameters			
	TSS (%)	pH	Acidity (% citric acid)	Electrical conductivity (dS m^{-1})
1.2	4.76 ± 0.15	4.26 ± 0.02	0.36 ± 0.01	6.3 ± 0.2
4.5	5.11 ± 0.12	4.31 ± 0.03	0.37 ± 0.01	7.4 ± 0.3
7.5	5.76 ± 0.19	4.30 ± 0.02	0.40 ± 0.02	7.6 ± 0.3
10.0	6.10 ± 0.14	4.16 ± 0.03	0.53 ± 0.02	8.9 ± 0.6

Melons

Fruits were harvested and taste-tested one day after harvesting (storage time = 0). The rest of the fruits were stored at room temperature (20–25°C) for various times according to their shelf lives. Before storage, the fruits from the salt-treated plants were tastier than their controls in all three cultivars (Table 2). After 3 weeks in storage the short-shelf-life cultivar Hemed had already deteriorated. The taste of fruits from the salinized plants of the other two cultivars was somewhat better than that of the controls, but not as significantly as in the zero-time test. After 4 weeks in storage the Noy Tsahov cultivar (the only one with such a long shelf life) showed no difference in taste score between the treatments. The results of the preference tests supported those of the taste tests (Table 2). However, when the TSS levels were plotted against the taste scores of the same samples (Fig. 1), no correlation was found between these two parameters.

Peanuts and lettuce

In peanuts no difference was found between saline and control treatments either in taste score or in the preference test (Table 3). The same was true for iceberg lettuce. Thus, lettuce was retested in triangle taste tests as described under Materials and Methods. The two reciprocal tests yielded similar results, and the testers were unable to distinguish between the treatments. The results of a typical test are given in Table 4.

Chinese cabbage

When the two Chinese cabbage varieties were harvested, one hundred randomly selected heads per plot from each salinity treatment were weighed and then cut longitudinally, and the yield and occurrence

Table 2. Effect of salinity on the taste of melon fruits[a]

Cultivar	Treatment	Storage time, weeks: 0 — Average taste score[b]	0 — Frequency of preference score[c] +	0 — −	0 — 0	3 — Average taste score	3 — Frequency of preference score +	3 — −	3 — 0	4 — Average taste score	4 — Frequency of preference score +	4 — −	4 — 0
Noy Tzahov	Control	1.58	17	5	7	1.90	14	11	0	2.06	11	15	0
	Salinity	2.20				1.96				2.06			
Makdimon	Control	1.79	22	4	3	2.03	18	11	4	already deteriorated			
	Salinity	2.58				2.33							
Hemed	Control	1.96	21	4	4	already deteriorated							
	Salinity	2.55											

[a] For each test (same date, same cultivar), slices taken from at least 8 fruits were combined in separate plates for each treatment.
[b] Taste score: best = 3, medium = 2, worst = 1.
[c] Preference score: +: testers preferring fruits from the salinated treatments; —: testers preferring the control fruits; 0: no preference.

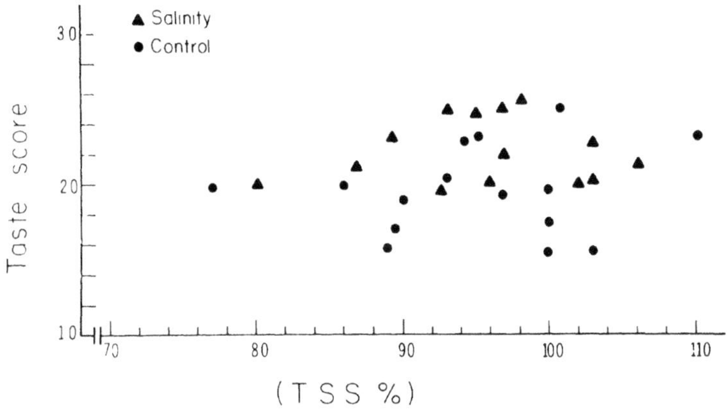

Fig. 1. Relationship between TSS and taste in melon fruits taken from saline and control treatments. These measurements were made with the fruits used in the experiments shown in Table 2.

of tipburn were assessed (Table 5). Increasing salinity very slightly decreased the yield in both varieties. In cv Spring A the frequency of tipburn increased from 5% in the controls to 90% in the plants irrigated with 4.0 dS m^{-1} water. The other variety (Michili) appeared to be less sensitive to tipburn under increasing salinity of irrigation water, even though it was more sensitive to yield reduction.

Discussion

Many studies have been done on the effects of salinity on yield in agricultural crops. However, very little has been done on the effects of saline irrigation water on the quality of the products. A study of fresh market tomatoes[3] showed that fruits produced under saline conditions were smaller than the controls, but they developed a better color, due to higher pigment concentrations in the pericarp, and had a much better taste. However, their shelf-life tended to be shorter, due to higher polygalacturonase activity. A chemical analysis of the fruit demonstrated higher values for TSS, reducing sugars, and total acidity[3]. Because of their better taste, the fruits from the salt-treated plants might obtain better prices on the market, which could compensate for the lower yields. This study showed that processing tomatoes also exhibit higher values for TSS and acidity in the juice of fruits from salinized plants (Table 1). This finding means that the quality of the products is better. Since most processing tomatoes are of the determinant type, their ripening periods are more or less synchronized[6].

Table 3. Effect of salinity on the taste of peanuts

Treatment	Frequency of taste score best (3)	medium (2)	worst (1)	Average taste score	Frequency of preference score[a] +	–	0
Control	10	10	0	2.5	9	7	4
Salinity	12	8	0	2.6			

[a] Preference score as in Table 2.

Table 4. Effect of salinity on the taste of iceberg lettuce cv Vangard[a]

	Frequency of samples judged identical 1 = 2	1 = 3	2 = 3	1 = 2 = 3
Number of testers	8	5	7	4
Percent	33	21	29	17

[a] Triangle taste test as described under Materials and Methods.

Table 5. Effect of salinity on yield and on occurrence of tipburn in two cultivars of Chinese cabbage

EC of irrigation water ($dS\ m^{-1}$)	cv Michili Yield ($kg/10\ m^2$)	Head weight (kg)	Tipburn (%)	cv Spring A Yield ($kg/10\ m^2$)	Head weight (kg)	Tipburn (%)
1.2	13.5^a	2.21^a	0	5.8^a	1.65^a	5
4.0	11.8^b	2.01^b	20	5.8^a	1.56^a	90
7.0	10.8^b	1.86^{bc}	60	5.5^a	1.59^a	90
10.0	10.9^b	1.78^c	80	6.5^a	1.40^b	95

[a, b, c] Values in column designated by the same letter do not differ significantly at $p = 0.05$. Each value is a mean of six replicates.

Our unpublished data show that irrigating such tomato plants with saline water for two weeks before the expected time of ripening will increase the TSS and acid contents of the fruits. This late treatment should prevent the yield reduction due to smaller fruits obtained under saline conditions, because by this stage the fruits have almost completed their growth[4]. In effect, we are proposing sodium chloride as a special late-season fertilizer for increasing the quality of processing tomatoes. Late application of salt to indeterminant cultivars would reduce the yield more, but still can be considered, if high-enough prices would be obtained for better-tasting fruits.

In melons, the fresh fruits from salinized plants were tastier than their controls. However, during storage at room temperature, this advantage was reduced, probably due to post-harvest processes leading to a loss of chemicals important for taste. These chemicals are probably not the total sugars measured in the fruit as TSS, because no correlation was found between TSS and the taste score (Fig. 1). A loss of volatile chemicals can probably account for this phenomenon[5], and storage in the cold rather than at room temperature might help preserve the taste after harvest.

In iceberg lettuce as well as in peanuts, the salinity of the irrigation water had no detectable effect on the taste (Tables 3 and 4). These differences in results might be due to the fact that melons and tomatoes are fruits, while lettuce heads are leaves, and peanuts are seeds. However, to verify this hypothesis, various plant commodities constituting various organs will have to be analyzed for taste after saline and non-saline irrigation.

At least in the one case of the tomatoes, the quality of the product is better under saline conditions. This might result in a higher price, which could compensate for any yield reduction. Salinity can also have a detrimental effect on the product quality, as was shown for Chinese cabbage (Table 5). Tipburn results from a deficiency of Ca^{2+} (Ref.[1]), which apparently was aggravated by the high concentration of sodium ions in the soil solution.

The effects of salinity on the quality of agricultural products needs further exploration, but these preliminary experiments should encourage researchers and farmers to think of saline water, under certain circumstances, as a positive factor in agricultural production.

References

1 Bangerth F 1979 Calcium-related physiological disorders of plants. Annu. Rev. Phytopathol. 17, 97–122.
2 Kopeliovitch, E, Mizrahi Y, Rabinowitch H D and Kedar N 1982 Effect of the fruit-ripening mutant genes *rin* and *nor* on the flavor of tomato fruit. J. Am. Soc. Hort. Sci. 107, 361–364.
3 Mizrahi Y 1982 Effect of salinity on tomato fruit ripening. Plant Physiol. 69, 966–970.
4 Monselise S P, Varga A and Bruinsma J 1978 Growth analysis of tomato fruit, *Lycopersicon esculentum* Mill. Ann. Bot. 42, 1245–1247.
5 Rhodes M J C 1980 The maturation and ripening of fruits. *In* Senescence in Plants. Ed. K V Thimann. pp 178–180. CRC Press, Boca Raton, Florida.
6 Rick C M 1978 The tomato. Sci. Am. 239, 67–76.

Section 4

Seawater agriculture

Plant and Soil 89, 311–321 (1985). Ms. BR 21

Agricultural production of halophytes irrigated with seawater

J. W. O'LEARY, E. P. GLENN and M. C. WATSON
Environmental Research Laboratory, University of Arizona, Tucson International Airport, Tucson, AZ 85706, USA

Key words Atriplex Brackish water Halophytes Irrigation Salinity Seawater

Summary Growing agricultural crops with direct seawater irrigation has progressed within the past few years from the conceptual to the experimental phase. This has been accomplished by selecting halophytes with inherently high salinity tolerance for use as crop plants rather than by increasing the ability of traditional crop plants to tolerate seawater. Some of the halophytes being investigated for use as crops in seawater irrigation scenarios have high nutritional value as forage or fodder crops. Most of them also have high digestibility. The limiting factor in such use is their high salt content, but this limitation can be moderated. However, since seeds of halophytes do not accumulate salt any more than do those of glycophytes, the greatest promise for seawater-irrigated halophytes probably will be as seed crops. The seeds of many halophytes have high protein and oil contents and compare favorably with traditional oilseed crops. Sustained high yields of seed and biomass already have been obtained from some halophytes irrigated with seawater, and within the next few years seawater agriculture should proceed from the experimental to the operational phase.

Introduction

The notion of using seawater to grow crops probably occurred to at least one person sometime in the distant past. It would be surprising if it had not, given the proximity of such a large source of water to vast expanses of unused desert land. The first person to formally conceptualize and actively promote such a notion was Boyko[1]; however, his zeal received mixed reactions. Even though several people subscribed to his views, those views were not well received by the scientific community in general. Much of the reason lies in the faulty reasoning used by Boyko to explain why seawater irrigation of crops grown on dune sand would succeed. He coined such 'principles' as partial root contact, subterranean dew, *etc.* and assigned unreasonably beneficial roles to those 'principles.' In spite of these problems of interpretation, the use of dune sand was an important part of the reason for the limited successes reported by Boyko and his supporters[1]. The good drainage associated with sand allows for a high volume and/or high frequency of irrigation with seawater, thus preventing salt accumulation.

Nevertheless, the fact remains that no conventional agricultural crops are yet grown with undiluted seawater, even on dune sand. Successful seawater irrigation depends on the use of halophytes. In fact, in more recent times, several people have promoted the use of halophytes to solve the seawater irrigation problem[3, 17, 26, 27], even though the roles prescribed for halophytes differed considerably among the authors. For example, even though Mudie[17] was optimistic about the significant potential contributions of halophytes to seawater-based agriculture, she thought the benefits would accrue via transfer of germplasm from halophytes to glycophytic crops. The conclusion was that 'prospects appear to be rather low for large scale use of halophytes as novel foods for human consumption'[17]. Even Chapman[3], who cited several instances of human consumption of halophytes and was fairly optimistic, concluded that 'it must be admitted that the outlook for new crops is not promising.' Others, though, such as Somers[26], have actually promoted cultivation of wild halophytes and directly irrigating them with seawater. Interest in this area has grown considerably in recent years, as attested to by the growing number of publications on the subject.

Thus, the movement from the conceptual phase to the experimental phase has been rapid. The next step, the transition from the experimental to the operational phase, *i.e.*, actual cultivation of halophytic crops, depends upon satisfactory demonstration of high yields under reasonably representative agricultural conditions and of the usefulness of the products. For forage or fodder crops, that means high yields must be obtained with halophytic crops grown in the field strictly with seawater irrigation, and utilization of the crop by animals must be demonstrated.

Within the past few years we have made substantial progress toward accomplishing these objectives. Those results will be reviewed here and used as the basis for critically assessing the future prospects of seawater-based agriculture.

Biomass yields

Several potential forage or fodder crops have been grown on the Sonoran Desert seacoast, where they were irrigated solely with hypersaline seawater[8] that had first been used in a shrimp aquaculture facility[25]. As a result of passing through the aquaculture facility, the salinity was increased to 40 parts per thousand (ppt) total dissolved solids (TDS), and the water contained approximately 20 ppm nitrate-nitrogen, 0.3 ppm ammonia-nitrogen, and 0.5 ppm phosphate-phosphorus. This constant supply of nutrient-enriched hypersaline water allowed us to grow several crops on beach sand with high-volume, high-frequency irrigation. Much of the early work was associated with evaluation of candidate species for their survival, nutritional value, and relative productivities. Preliminary results from that work have been reported[5, 6, 10, 20]. We were also able to obtain annual productivity values for several potential halophytic crops grown under those field conditions[8].

Productivity varied widely among species (Table 1). The interesting observation that can be made is that the highest productivities were obtained from native species. The highest productivity from an exotic

Table 1. Annual productivity of halophytes irrigated with 40 ppt seawater at Puerto Peñasco, Sonora, Mexico. (N = native, E = exotic)

Plant species		Productivity ($gDW\,m^{-2}\,y^{-1}$)
Atriplex lentiformis	(N)	1794
Batis maritima	(N)	1738
Atriplex canescens subsp. *linearis*	(N)	1723
Salicornia europaea	(N)	1539
Atriplex barclayana	(N)	863
Atriplex nummularia	(E)	801
Atriplex glauca	(E)	514
Cressa truxillensis	(N)	394
Atriplex polycarpa	(N)	312
Atriplex canescens subsp. *canescens*	(N)	303
Atriplex repanda	(E)	61
Atriplex patula	(E)	8

species was that of *Atriplex nummularia*, but it was less than half of the productivity of the best-producing species. There were also some native species that had low productivities, but it is important to note that no exotic species had high productivity. It will be interesting to see if this result will be found in other locations, because it could be a serious constraint on widespread development of seawater-based halophyte agriculture. For example, *Atriplex nummularia* long has been promoted as one of the most promising Atriplex species for extensive use as a forage crop on marginal lands[9, 13], primarily because of its relatively high protein content. It even has been grown successfully far from its native environment[12], but usually not under severe stress conditions, such as with highly saline water, and no annual productivity information was obtained. The only other attempt to grow *A. nummularia* with seawater of which we are aware is that of Pasternak[22] in Israel, which is at approximately the same latitude as our site. The yield was 429.5 $g\,m^{-2}\,y^{-1}$, even lower than ours.

The most productive halophytes yielded the equivalent of 8–17 tons of dry matter per hectare (Table 1). This compares favorably with a conventional forage crop such as alfalfa grown on fresh water, which yields 5 to 20 tons of dry matter per hectare annually[30]. Thus, at first glance it seems like very high productivity can be achieved from halophytes irrigated solely with seawater. However, it is not fair to compare total dry weights since one of the characteristics of most halophytes is a high ash content[7]. Thus, ash-free dry weight (AFDW) really should be the basis for comparison and would more accurately reflect the productivity of the crops, insofar as total organic matter production is concerned. Since typical ash contents of plants such as Atriplex grown on seawater are on the order of 20 to 30%, that would lower the actual productivities of the top performers to approximately 12 tons of AFDW per hectare annually, which still is a respectable value, albeit not outstanding. However, if the caloric or nutritional value of the dry matter is higher than that of alfalfa, then the economic productivity of the halophytes might actually be greater (see below).

Since forage or fodder crops such as alfalfa usually are harvested or cropped several times per year, it was of interest to investigate the ability of these potential halophyte crops to tolerate multiple cuttings per year and to determine the effect on annual yield. Four species of Atriplex were planted in April, irrigated solely with seawater, and clipped to a height of 30 cm three times during the first 16 months after planting[8]. There was a considerable loss of plants with *A. lentiformis* and *A. canescens* sbsp. *linearis* when they were cut back, and the annual productivity was reduced substantially as a result (Table 2).

Table 2. Annual productivity and mortality of *Atriplex* species planted in April 1980 and clipped to a height of 30 cm the succeeding October, January and August

Plant species	Productivity ($gDW\,m^{-2}\,y^{-1}$)	Percent mortality
Atriplex barclayana	2336	3
Atriplex nummularia	2080	0
Atriplex lentiformis	1456	20
Atriplex canescens sbsp. *linearis*	1104	40

On the other hand, there was little or no loss in *A. barclayana* and *A. nummularia*, and their annual productivities were increased considerably, by 170% and 160%, respectively. That puts them at the upper end of the annual yield range for alfalfa grown on fresh water, even when corrected for ash content. In other studies of response to defoliation[14] *A. nummularia* was found to be superior to other Atriplex species. *A. nummularia* was capable of producing new leaves along the main stems and branches after complete defoliation and even from the base of the plant after being cut almost to ground level.

Nutritional value of forage

In contrast to the paucity of information on the annual productivity of cultivated halophytes, there is an abundance of reports giving values for protein content of potential forages such as Atriplex. Much of that information has been reviewed by Goodin[9], and the crude protein levels typically range from about 11 to 24% of the dry weight. We also have determined the crude protein content of numerous halophytes, and, depending on the age of the plant tissue at harvest and the salinity at which the plant was grown, have found that the crude protein content of most halophytes examined so far ranges from 10 to 20% of the dry weight, with a few species having higher values. For example, *Atriplex barclayana* and *A. lentiformis* grown on seawater had protein contents of 11.6% and 16.7%, respectively[8]. Salicornia, in contrast, had only 5.7% protein when grown under the same conditions (Table 3). For comparison, the protein content of alfalfa ranges from 12% to 22%, with an average of 16.9%[30]. Fat and fiber contents of the Atriplexes also compare favorably with those of alfalfa. As mentioned above, the ash content of the foliage typically is high, and this can be a serious drawback to the potential forage value of these plants. For example, *A. lentiformis* and *A. barclayana* grown on seawater had ash contents

Table 3. Nutritional analysis of halophytes irrigated with 40 ppt seawater at Puerto Peñasco, Sonora, Mexico, as percent of dry weight

	Plant species		
Constituent	*Atriplex lentiformis*	*Atriplex barclayana*	*Salicornia europaea*
Protein	16.7	11.6	5.7
Fat	1.3	1.3	0.4
Fiber	14.1	14.2	17.0
Ash	26.8	33.1	41.3
Oxalate	3.6	5.9	1.5

of 27% and 33%, respectively. As previously stated, this reduces the productivity values if yield is expressed on the basis of AFDW, but if the total protein yield is calculated, the picture is not so bad. The most productive halophytes (in Table 1) yielded the equivalent of 0.6 to 2.6 tons of protein per hectare when grown on seawater, compared to typical yields of alfalfa grown on fresh water of 0.5 to 3.0 tons of protein per hectare[30].

We recently completed a study in which *A. barclayana* and *A. nummularia* were grown in lysimeters at five different salinities and harvested at three different times. The protein levels in the leaves were determined at three different times during the growing season and averaged 18% for *A. barclayana* and 24% for *A. nummularia* at the highest salinity (30 ppt). Thus, it might make more sense to harvest only leaves rather than the entire tops of the plants. The lower annual productivity of *A. nummularia* might be more than compensated for by the higher protein content in the leaves and its great ability to regenerate new leaves[14]. The significance of this is that amenability to harvesting has to be an integral part of the evaluation of any potential halophytic forage, and as described above, there is considerable variability within a genus such as Atriplex in the ability to tolerate clipping.

Forage digestibility

To survey as many species of halophytes as possible in the shortest possible time as a prelude to conducting large-scale animal feeding trials, we conducted *in vitro* organic matter digestibility studies of 45 different halophytes[16]. Foliage samples were collected from plants growing in our halophyte germplasm collection, incubated for 48 h in a rumen fluid-buffer inoculum followed by 48-h digestion in acid pepsin. A standard alfalfa hay sample was included also. At the end of this time, the amount of *in vitro* organic matter disappearance (IVOMD) was measured. The IVOMD for the 45 halophytes ranged from 50.1 to 87.2%. The standard alfalfa hay sample had an IVOMD of 64.3%. Thirty-five of the 45 halophytes had values in excess of the standard alfalfa sample. When samples from plants grown on fresh water were compared with those grown on water with a salt content up to 20 ppt, no detrimental effect of salinity was apparent. In fact, in some cases the IVOMD was actually higher at the higher salinities. The data from the *Atriplex* species discussed here are presented in Table 4. Both the digestibility and the calculated digestible organic matter content (DOMC) compare reasonably well with alfalfa.

The high salt content of the foliage is a serious constraint to consumption by animals. Thus, halophytic forages probably will be limited to use as components in feed mixes rather than as sole sources of food. Limited feeding trials with goats were conducted using *Atriplex barclayana* and *Atriplex lentiformis* that had been irrigated with seawater[32]. When the halophytes were included as 25% of the total diet, their acceptability and digestion were high. *A. lentiformis* seemed superior to *A. barclayana* in acceptability by the animals. Some of the plant forage was washed to leach out salts. Approximately 10 kg of dried Atriplex was placed in a revolving drum mixer and rinsed with fresh running water for 2 to 3 h and then air-dried. That treatment

Table 4. Percent *in vitro* organic matter disappearance (IVOMD) and calculated digestible organic matter content (DOMC) of halophytes grown with 0, 10 or 20 ppt total dissolved salts in the irrigation water

Plant species	Salinity level (ppt)	IVOMD	DOMC
Atriplex barclayana	0	59.8	51.0
	10	61.7	46.5
	20	67.6	47.0
Atriplex lentiformis	0	58.8	53.3
	10	55.7	47.6
	20	63.1	50.0
Atriplex nummularia	0	77.0	59.7
Standard alfalfa hay	0	64.3	58.4

reduced the ash content to approximately two-thirds of the original amount but did not improve the acceptability or digestibility.

Nutritional content of the seeds

Salt content of the tissue is not a problem with seeds. Even though the vegetative tissues of plants may have extremely high salt contents, the seeds of those plants normally have low salt levels, even when irrigated with highly saline water (Table 5). Some halophytes, such as Spartina and Distichlis, might become crop plants analogous to rice, as suggested by Somers[26], but our observations lead us to give serious consideration to the use of halophytes as potential oilseed crops. Our own seed analyses, as well as those of others, indicate that there is a wide range of protein and oil contents among halophyte seeds; however, several species compare very favorably with the major oilseed crops of the world (Table 5). Thus, we propose that the greatest promise for halophytes may be as oilseed crops. The plants could be grown with seawater with no serious salt accumulation in the portion of the plant of economic interest. The seeds could be harvested and processed with extant equipment, yielding a vegetable oil for food or industrial use and a protein-rich seed meal that could be incorporated into animal feeds, much like soybeans are presently utilized. We have just begun to obtain seed yield data, and so far the results are promising. Details of this work are still being prepared for publication, but we can report that the annual seed yield of Salicornia irrigated with seawater is consistently greater than 250 g m^{-2}, which is equivalent to 2.5 tons per hectare. Thus, not only does the seed compare well with soybeans qualitatively (Table 5), but it has yields that are comparable[31].

Table 5. Protein, oil, and ash contents of seeds, as percents of dry weight

Plant	Protein	Oil	Ash
Present oilseed crops[a]			
Soybean	40.0	18.8	4.8
Safflower	14.3	30.4	2.5
Sesame	18.6	49.1	5.3
Sunflower	17.5	36.0	3.6
Crambe	25.0	47.0	5.1
Halophytes[b]			
Mesembryanthemum crystallinum	31.2	7.8	5.0
Atriplex canescens	5.4	1.0	6.5
Atriplex lentiformis	12.5	2.1	13.6
Chenopodium quinoa	12.1	7.5	3.1
Salsola pestifer	16.9	1.3	17.2
Halophytes[c]			
Atriplex triangularis	16.4	9.4	3.5
Cakile edentula	28.6	51.2	5.2
Cakile maritima	21.5	47.1	5.0
Crithmum maritimum	21.5	41.4	8.0
Kosteletzkya virginica	23.8	18.1	5.0
Salicornia europaea	30.2	28.0	7.5

[a] from Weiss[31].
[b] from Earl and Jones[4].
[c] original data by authors this paper.

Discussion

It has been demonstrated that halophytes can be cultivated and irrigated solely with seawater and have productivities within the range of conventional crops. Several questions require answers, however, before the actual seawater farming of halophytes becomes a reality. One of these questions is the geographic range within which high productivities can be achieved with a given species. If, as the meager evidence so far obtained suggests, productivity is less when the plant is grown out of its native range, then ecotypic selection should be intensively investigated. This should not be a major obstacle to the domestication of a given halophyte, since many of our contemporary crops are grown in areas well beyond the native range of their wild progenitors, but it will limit the areas within which successful cultivation of that plant can be conducted at first. We have established test plantings within and outside of the native range of Salicornia with several ecotypes collected from widely separated areas to make comparative productivity measurements. The same should be done for other halophytes, such as Atriplex.

The protein content of the leading halophyte candidates should be studied in greater detail. The standard procedure for calculating crude protein content (6.25 × total N) probably is not valid for these plants. This conversion factor was based on the assumptions that all of the Kjeldahl-N is protein-N and that the proteins contain 16% N. Even though this factor has been found to be invalid for many plant tissues[2,21], it still is used in virtually all proximate analyses[11], probably since the percent error is not unusually high in most cases, and it is, in fact, usually emphasized that it is 'crude' protein that is being reported. However, for halophytes the error may be considerable, since the nitrate levels can be very high. Even in sugar beet (a crop with some halophytic ancestry), nitrate-N has been reported to represent 88% and 38% of the total N in petioles and blades, respectively[15]. The levels of other non-protein nitrogen compounds, such as quaternary ammonium compounds and proline, also are often very high in halophytes[28,29] and would bias the calculated value as well. This will not be an easy question to resolve, since all methods for determining the protein content of plant tissues have their shortcomings; yet we must attempt to do a better job of reporting the protein content of halophyte forage than we have been doing.

The question of what to do about the inevitable high salt content of seawater-irrigated halophyte forage is one that also must be addressed concurrent with the productivity question in any evaluation. We have emphasized the utilization of this forage as one component of a multi-component feed mix, but there are other alternatives that have not yet been subjected to serious investigation. One alternative is to use the foliage for extraction of leaf protein (LP). The technology is improving, and interest is increasing rapidly enough now that LP will soon become an important part of the world food budget. It has been emphasized before[23,24] that LP should be extracted from those plant sources that are not now consumed, even those that might contain undesirable or toxic substances (since those substances should be lost in the extraction process), in order to make a significant impact on increasing the world protein supply. Halophytes meet these criteria pretty well. The high salt content and even high levels of oxalic acid or other undesirable constituents would be no constraint to their utilization as source material for extraction of LP. The halophytes could even serve as sources of more purified leaf protein concentrate (LPC), which has great promise for use as human food because it is odorless, colorless, and tasteless and also has high, uniform nutritional quality (Fraction I protein). We have been able to extract relatively pure Fraction I protein from Salicornia tissue (Bourque, unpublished results) and will be scaling up that work as well as extending it to other halophyte species.

In conclusion, it looks like halophytes irrigated solely with seawater have great potential as crop plants, in spite of some problems such as high salt content of tissues. In particular, it looks like high yields of good quality protein can be produced. There is a lot of work yet to be done before the story is complete, particularly, determination of the acceptability and utilization of the material by animals, but there is good reason to be highly optimistic, even for those of us who were originally very pessimistic[18, 19], and the transition from the experimental phase to the operational phase should occur within the present decade.

References

1 Boyko H Ed. 1966 Salinity and Aridity. Dr. W. Junk, The Hague.
2 Byers M 1983 Extracted leaf proteins: their amino acid composition and nutritional quality. *In* Leaf Protein Concentrates. Eds. L Telek and H D Graham. pp 135–175. AVI Publishing Co., Westport, Connecticut.
3 Chapman V J 1975 Terrestrial halophytes as potential food plants. *In* Seed-Bearing Halophytes as Food Plants. Ed. G F Somers. pp 75–90. DEL-SG-75, College of Marine Studies, Univ. of Delaware, Newark, Delaware.
4 Earl F R and Jones Q 1962 Analyses of seed samples from 113 plant families. Econ. Bot. 16, 221–250.
5 Glenn E P, Fontes M, Katzen S and Colvin L B 1982 Nutritional value of halophytes grown on hypersaline seawater. *In* Biosaline Research. Ed. A San Pietro. pp 485–490. Plenum Press, New York.
6 Glenn E P, Fontes M and Yensen N 1982 Productivity of halophytes irrigated with hypersaline seawater in the Sonoran Desert. *In* Biosaline Research. Ed. A San Pietro. pp 491–494. Plenum Press, New York.
7 Glenn E P and O'Leary J W 1984 Relationship between salt accumulation and water content of dicotyledonous halophytes. Plant Cell Environ. 7, 253–261.
8 Glenn E P and O'Leary J W 1984 Productivity and irrigation requirements of halophytes grown with seawater in the Sonoran Desert. J. Arid Environments (*In press*).
9 Goodin J R 1979 Atriplex as a forage crop for arid lands. *In* New Agricultural Crops. Ed. G A Ritchie. pp 133–148. Westview Press, Boulder, Colorado.
10 Hodges C N, Fontes M R, Glenn E P, Katzen S and Colvin L B 1981 Seawater-based agriculture as a food production defense against climate variability. *In* Food-Climate Interactions. Eds. W Bach, J Pankrath and S H Schneider. pp 81–99. Reidel Publishing, New York.
11 Hulse J H, Rachie K O and Billingsley L W 1977 Nutritional Standards and Methods of Evaluation for Food Legume Breeders. IDRC-TS7e, International Development Research Center, Ottawa, Canada.
12 Hyder S Z 1981 Preliminary observations on the performance of some exotic species of Atriplex in Saudi Arabia. J. Range Management 34, 208–210.
13 Jones R, Hodgkinson K C and Rixon A J 1970 Growth and productivity in rangeland species of *Atriplex*. *In* The Biology of Atriplex. Ed. R Jones. pp 31–42. CSIRO, Canberra.
14 Leigh J H and Wilson A D 1970 Utilization of *Atriplex* species by sheep. *In* The Biology of Atriplex. Ed. R Jones. pp 97–104. CSIRO, Canberra.
15 Lorenz O A 1978 Potential nitrate levels in edible plant parts. *In* Nitrogen for the Environment, Vol. 2. Eds. D R Nielsen and J G MacDonald. pp 201–219. Academic Press, New York.
16 Moore J A, Swingle R S, O'Leary J W, Glenn E P and Colvin L B 1982 *In vitro* organic matter digestibility of salt tolerant plants. Proc. Western Sec. Am. Soc. Anim. Sci. 33, 319–322.

17 Mudie P J 1974 The potential economic uses of halophytes. *In* Ecology of Halophytes. Eds. R J Reimold and W H Queen. pp 565–596. Academic Press, New York.
18 O'Leary J W 1975 Potential for adapting present crops to saline habitats. *In* Seed-Bearing Halophytes as Food Plants. Ed. G F Somers. pp 91–119. DEL-SG-3-75, College of Marine Studies, Univ. of Delaware, Newark, Delaware.
19 O'Leary J W 1979 Yield potential of halophytes and xerophytes. *In* Arid Land Plant Resources. Eds. J R Goodin and D K Northington. pp 574–581. ICASALS, Texas Tech Univ., Lubbock, Texas.
20 O'Leary J W 1984 The role of halophytes in irrigated agriculture. *In* Salinity Tolerance in Plants. Eds. R C Staples and G A Toenniessen. pp 397–414. John Wiley & Sons, New York.
21 Orr M L and Watt B K 1957 Amino Acid Content of Foods. Home Economics Research Report No. 4, US Government Printing Office, Washington, DC.
22 Pasternak D 1982 Biosaline research in Israel: alternative solutions to a limited fresh water supply. *In* Biosaline Research. Ed. A San Pietro. pp 39–58. Plenum Press, New York.
23 Pirie N W 1982 Food protein from leaves. *In* Food Proteins. Eds. P F Fox and J J Condon. pp 191–210. Applied Science Publishers, London.
24 Pirie N W 1983 Introduction on the knowledge needed for successful leaf protein production. *In* Leaf Protein Concentrates. Eds. L Telek and H D Graham. pp 1–6. AVI Publishing Co., Westport, Connecticut.
25 Salser B, Mahler L, Lightner D, Ure J, Danald D, Brand C, Stamp N, Moore D and Colvin B 1978 Controlled environment aquaculture of penaeids. *In* Drugs and Food From the Sea. Eds. P N Kaul and C J Sindermann. pp 345–355. Univ. of Oklahoma Press, Norman, Oklahoma.
26 Somers G F 1975 Seed-bearing Halophytes as Food Plants. DEL-SG-3-75, College of Marine Studies, Univ. of Delaware, Newark, Delaware.
27 Somers G F 1982 Food and economic plants: general review. *In* Biosaline Research. Ed. A San Pietro. pp 127–148. Plenum Press, New York.
28 Stacy R, Ahmad N and Wyn Jones R G 1977 Taxonomic and ecological aspects of the distribution of glycinebetaine and related compounds in plants. Oecologia 27, 319–332.
29 Stewart G R and Larher F 1980 Accumulation of amino acids and related compounds in relation to environmental stress. *In* The Biochemistry of Plants – A Comprehensive Treatise, Vol. 5. Eds. P K Stumpf and E E Conn. pp 609–635. Academic Press, New York.
30 USDA 1979 Agricultural Statistics 1979 US Government Printing Office, Washington, DC.
31 Weiss E A 1983 Oilseed Crops, Longman Group Limited, London.
32 Wiley S T, Swingle R S, Brown W H, Glenn E P, O'Leary J W and Colvin L B 1982 Evaluation of two *Atriplex* species grown with hypersaline water and the effect of water leaching on their digestibility by goats. Proc. Western Sec. Amer. Soc. Anim. Sci. 33, 12–15.

Plant and Soil 89, 323–336 (1985). Ms. BR 22

Halophytic crops for cultivation at seawater salinity

JOHN L. GALLAGHER
College of Marine Studies, University of Delaware, Lewes, DE 19958, USA

Key words *Atriplex triangularis* *Distichlis spicata* Edible oil Grain Halophytes Hay Pasture Seawater *Sporobolus virginicus* *Spartina alterniflora* *Spartina patens*

Summary Several hundred halophytes from salt marshes and salt deserts of the world have been evaluated in our laboratory at various degrees of intensity, and a few have been selected for development as crops. The development of the cultivars and the basic biology of the plants is being studied in Delaware in the United States. Agronomic testing, feeding trials, and development of the best agronomic practices are taking place in the saline desert at the American University in Cairo research station in Sadat City. Our present efforts focus primarily on three forages, one grain, and one vegetable. *Sporobolus virginicus* cultivars for both hay and pasture are being tested. A *Distichlis spicata* cultivar for hay has been identified, and *Spartina patens* is being evaluated as a hay as well. Although we do not yet have the data for a full year's growth in Egypt, forage yields of these various cultivars, when harvested as hay crops, range to 6.9 or more tons per acre, depending on the salinity and other environmental conditions, and the crude protein content as indicated by the nitrogen content ranges from 6 to 10%. Cultivars having the most useful agronomic qualities have been identified and are being increased in quantity. The grain crop *Kosteletzkya virginica* is a perennial, producing a seed which resembles millet; its whole seeds contain approximately 25% protein and 15% oil. The yields of one of our better cultivars are about 1460 kg/ha (1300 lb/acre) when grown under 25‰ salinity. The vegetable *Atriplex triangularis* (similar to spinach) has been under mass selection for four years; a cultivar has been identified and seed is now being increased for this species.

Introduction

The need for salt-tolerant crops increases each year as a growing world population seeks to feed itself with ever-decreasing soil resources and dwindling freshwater supplies. Heightened expectations with respect to the quality of life are increasing this demand, while salinization of soils in many parts of the world is reducing the acreage available for conventional agriculture. However, in addition to humanitarian reasons, there are economic reasons for developing halophytic crops. Soils that are otherwise fertile, but that have become salinized because of inadequate irrigation practices, are low in economic value but would have much of that value restored if they could be used for halophytic crops. Furthermore, there are naturally saline areas, such as parts of the Western Desert in Egypt, that have never been used agriculturally, but which could be brought into production as irrigation projects are developed, if salt-tolerant crops were readily available.

Three approaches are being used by researchers in various parts of the world to develop salt-tolerant crops. In the first approach, searches have been initiated within traditional crops for salt-tolerant cultivars. This effort has met with some success, especially with rice, barley, and wheat[12,14]. Selection of salt-tolerant lines developed through tissue culture techniques also holds promise[13].

A second approach that has proven successful has been to cross commercial species low in salt tolerance with wild salt-tolerant relatives. Sacher *et al.*[17] as well as Rush and Epstein[16] have pursued this avenue of research, producing salt-tolerant tomato plants with moderate-sized fruit.

The third approach, and the one initiated in our laboratory by G. F. Somers and also used by a number of investigators in other laboratories around the world (University of Arizona, Ben-Gurion University of the Negev, University of Western Australia), has been to begin with wild species that have evolved high salt tolerance in their native habitats. These species are first evaluated for their potential agronomic value, and desirable cultivars are then developed using both traditional and modern biotechnological methods. This paper describes the plants we have chosen for development into a series of halophytic crops, presents the work we have done in developing them as crops, and identifies their major deficiencies which need to be addressed in further research.

Methods

Growth chamber growing conditions

Plants grown in chambers were placed in Sherer cabinets with a mixture of fluorescent and incandescent light which gave a maximum intensity of approximately 25% that of full sunlight. Temperature and light-cycle conditions were varied, depending upon the experiment. In some cases nutrients were supplied by half-strength Hoagland's solution in hydroponics or in washed quartz-sand culture. In other cases, we used Rumford loamy sand from our field plots. When the latter was used, we supplemented the natural soil fertility with the slow-release fertilizer Osmocote (14% N, 14% P, 14% K).

Greenhouse growing conditions

The greenhouse used in all experiments was heated in the winter but was not cooled in the summer. The temperatures in the winter were a minimum of 25°C during the day and a minimum of 18°C at night. In the summer the temperatures typically ranged from 20°C to 25°C at night, and during the day they often rose to the mid-thirties. The light intensity was approximately 50% of full sunlight. Plants were grown in the same media and nutrients supplied in the same fashion as in the growth chamber experiments. Saline water was used directly from our running seawater system (30‰) or was diluted to a lower salinity with fresh well-water. Day-length cycles were those of the latitude (38° 47′ N) or were controlled with clocks and 1000-watt metal halide lamps.

Field growing conditions

Soils in the Delaware field plots were either Rumford loamy sand or a variable, coarse-textured dredged material. The plots were leveled areas with low dikes (~ 20 cm). Grasses were grown in swards on this leveled ground (Fig. 1). *Atriplex triangularis* and *Kosteletzkya virginica* were grown on the sides of longitudinal mounds approximately 15 cm high. The plots were flood-irrigated with water from either a canal or a salt marsh stream, three days a week during the growing season. Salinities during most of the season were between 25 and $30‰$ at the canal site where the soil was coarse, and between 15 and $25‰$ at the stream site where the soil was a Rumford loamy sand. Because it was thought that sprinkler irrigation would be the most economical method of watering in commercial production, some of the grass swards (Fig. 2) at the marsh-stream site utilized such a system. The soils at Sadat City are very variable, with deep sands overlying heavy-textured cobbly subsoils. In the swales the surface soils are sand, while on the ridges the surface layer has been blown away, exposing an often saline subsoil. Interstitial water salinities vary from near zero adjacent to an opening in the biwall irrigation line, where irrigation water with a salinity of less than $1‰$ flows, to $25‰$ and higher 30 cm away. Seasonal data are not yet available. Soil reaction ranges from approximately pH 6.5 to 8.5 at one site.

Results and discussion

Forages

Spartina alternifora *Loisel.* This species at first appeared to be promising for use as a forage because it grows vigorously in full-strength seawater, often producing over 3000 g dry weight/m^2 under natural conditions[6]. Multiple harvesting strategies similar to those often used for hay crops were tested, but the results were disappointing, since regrowth was minimal and the stand itself was nearly killed by the midsummer harvests. The problem appears to be that the species uses virtually all of its underground reserves to produce an initial burst of

Fig. 1. Flood-irrigated plot of *Spartina patens* and *Kosteletzkya virginica* in an intercropping study in Delaware.

Fig. 2. Swards of halophytic forage in Delaware.

spring growth; reserves in May were only 12 g/m^2 compared to 60 g/m^2 for a short-grass stand in Delaware[4]. Harvesting a vigorous experimental stand in one of our flood-irrigated plots in June, August, and October resulted in the plants going into the winter with only 12 g/m^2 reserves, while uncut plots averaged 120 g/m^2 (Ref.[4]). Apparently, there has been little selection pressure for this species to retain reserves below ground for the repair of a damaged canopy. Because the substrates are soft and typically will not support large herbivores, the few grazers present in most of the natural stands of this plant are insects that eat only about 8% of the annual productivity[21]. Most of the *S. alterniflora* biomass enters an estuarine detritus-based food web after the plants die at the end of the growing season. The management pattern for carbohydrate reserves evolved by this species would appear to make most populations of these plants of little use for perennial forage development since they will not give good regrowth after mid-season harvesting. We will be testing this species as a winter annual forage in the saline desert in Egypt in 1984–85 (Fig. 3).

Two of my colleagues, G. F. Somers and D. Grant[19], have selected a strain of *S. alterniflora* that germinates at 30‰ salinity and has high seedling vigor. Although this species has not yet proven to be useful as a forage for domesticated herbivores, we have used these plants for

Fig. 3. Stand of *Spartina alterniflora* on the Rumford loamy sand in Delaware.

mitigation projects involving the restoration of damaged, and the creation of new, wetlands for wildlife habitats.

Distichlis spicata (L.) *Green*. This plant has been suggested as a native forage for the southern United States by Leithead *et al.*[10]. Wintertime grazing by horses and cattle would be facilitated if walkways were built on the marshes to enable the animals to traverse the soft substrate. Native *D. spicata* has been used to vegetate saline soils near Mexico City, where the plant served as the total dietary forage for cattle[20].

We have completed a series of tests of harvesting strategies with *D. spicata* in our flood-irrigated plots in Delaware. Indications are that June harvests produce almost as much forage as late harvests, whereas mid-season cutting does not produce as high a yield (Table 1). A similar

Table 1. Yields of *Distichlis spicata* harvested at different times during 1983 from a field plot in Lewes, Delaware

	Yields (dry weight)	
Harvest date	kg/ha*	tons/acre*
June	8420	3.8
August	5200	2.3
September	6800	3.0
October	9480	4.2

* Yields are the means of four samples.

mid-season decline was observed in natural stands of *S. alterniflora* in Georgia[6]. Experiments on harvesting strategies are now in progress in Egypt and the results may differ. We are currently concentrating on selecting for the hay, rather than the pasture, potential of *D. spicata.* To date we have collected a wide variety of genetic material of this species and have selected a cultivar that produces a good rhizome mat yet has a relatively high yield of forage. Table 2 shows the yields after 10 months of several lines of *D. spicata.* The Seabrook cultivar has the highest yield and the least tendency to exhibit dormancy and is our currently recommended line for saltgrass hay in warm climates.

In choosing marsh plants for development into upland forage plants, one must consider that in the managed setting they will probably be subjected to rather widely fluctuating salinities, depending on the phase of the irrigation cycle. Plants that have evolved in the low elevations of the salt marsh are 'irrigated' frequently by the tides and have not been subjected to fluctuating salinity during their evolution, while those plants native to higher elevations, where flooding with saline tidal waters is less frequent, have been exposed to such environmental pressures. An experiment in our laboratories by Rumer[15] with *D. spicata* involved subjecting hydroponically grown plants to constant low (10‰) and constant high (30‰) salinities and to one-week high-low and two-day high-low fluctuations. Growth at the constant high salinity and in the two-day fluctuating treatment was approximately 1/3 that at the constant low salinity, while the yield of the one-week fluctuating treatment was 2/3 that of the low salinity. Examination of proline levels in the plant tissue indicated that they were twice as high in both the constant high salinity and the one-week fluctuating treatments as in either the constant low salinity or the two-day fluctuating treatments (20 *vs.* 38 μmol/g ash-free dry weight of leaf tissue). These results are compatible with the concept that an inability to make rapid biochemical adjustments to the short-term salinity fluctuations resulted in poor growth.

Table 2. Yields of halophytes harvested on 27 March 1984 from field plots at Sadat City, Egypt

	Yields (dry weight)	
Species	kg/ha*	tons/acre*
Sporobolus virginicus – Dixie	15500	6.9
Sporobolus virginicus – Smyrna	7850	3.5
Spartina patens – North Carolina	7170	3.2
Distichlis spicata – Seabrook	5600	2.5
Distichlis spicata – Lewes	1790	0.8
Distichlis spicata – Utah	2240	1.0

* Based on % dry weight determined in summer of 1983. Yields represent net growth since May 1983.

One of the problems with this species is that, even though it has salt glands, its leaves tend to have a higher salt content than traditional forage plants. The ash contents of several genotypes grown in a common plot are shown in Table 3: the range was from 7 to 17% and indicates that lines can be developed that accumulate relatively little salt.

Sporobus virginicus *(L.) Kunth.* While doing productivity studies, Gallagher and Plumley[5] and Linthurst and Reimold[11] found that feral cattle in the Georgia marshes were selectively grazing *Sporobolus virginicus* in preference to *Spartina alterniflora* and *Distichlis spicata.* Leithead *et al.*[10] also mention the forage value of this marsh plant when it is accessible to herbivores. For these reasons we thought palatability would not be a problem in the development of this plant as a forage. We have collected numerous populations of the species from the southeastern United States and have selected individuals which are particularly vigorous and have desirable agronomic qualities.

There are generally two forms of the plant, which are listed by Hitchcock[8] as normal and robust forms. We have two cultivars of this plant. The first cultivar, Smyrna, is intended for use as a pasture plant, while the second, Dixie, has a canopy structure more appropriate for use as a hay. The pasture cultivar was selected from plants of the normal, or type specimen, form. We have found that form growing primarily on the upper fringes of marshes, where salinity and aeration fluctuate widely, depending on both the phase of the lunar tide cycle and the amount of rainfall. Our Smyrna cultivar of the normal form has fine stems that are not easily broken; in contrast, the hay cultivar, Dixie, was selected from the robust form which is found primarily on coastal dunes and has comparatively brittle stems. Donovan and Gallagher[2] found that the shoot/root ratio of the robust form was relatively high (10.7) when compared to the normal form (2.1).

Table 3. Ash content (percent of dry weight) of the leaves of different selections of *Distichlis spicata* growing under the same salinity conditions in field plots in Lewes, Delaware, 1981

Plant source	Percent ash
Delaware, USA	8.0
Georgia, USA	10.0
Peru	17.4
Utah, USA	8.0
Florida, USA	8.9
Oregon, USA	7.1

This species is very salt-tolerant, and in a hydroponic study 80% of the plants exposed to 80‰ NaCl salinity survived, although growth was minimal[3]. The response, with respect to vigor and morphology, of the normal form to various salinities is shown in Fig. 4. Survival and growth of both cultivars have been excellent in the saline desert in Egypt. Single-harvest yields after 10 months growth at Sadat City were 15 500 kg/ha (6.9 tons/acre) for Dixie and 7850 kg/ha (3.5 tons/acre) for Smyrna (Table 2).

Anaerobic conditions as well as seawater salinity stress caused a reduction in *S. virginicus* growth, but combining these stresses did not have an additive effect. The adjustments made in response to salinity stress also served to relieve the anaerobic stress[1].

The major problem with this species in our plots in Delaware is that seed production is very poor. We have had better success in Egypt, and some seeds are produced in native stands in the southern United States. We are working on solving this important problem, since vegetative propagation is too labor-intensive to be of economic value for commercial production in industrial societies.

Spartina patens (*Ait.*) *Muhl.* The common name of this plant (salt marsh hay) comes from the historical use of the plant by colonists in New England and in the mid-Atlantic states as hay and bedding for cattle and horses. We have done some work with this species and have found yields of a line from Georgia to be 9740 kg/ha (4.3 tons/acre) for our growing season, approximately April to September. In a Delaware study of harvesting strategy, we evaluated five single-date and two multiple harvest programs (Table 4). There seemed to be no advantage either to harvesting after July or to multiple harvesting. We are still examining the effect of the various harvesting regimes on the underground reserves and, consequently, on the over-wintering ability and the next season's regrowth of the plants. A line from North Carolina yielded 7174 kg/ha (3.2 tons/acre) during the 10-month growth period.

We conducted a growth chamber study comparing the responses to several temperature cycles of two lines of *S. patens*. It was somewhat

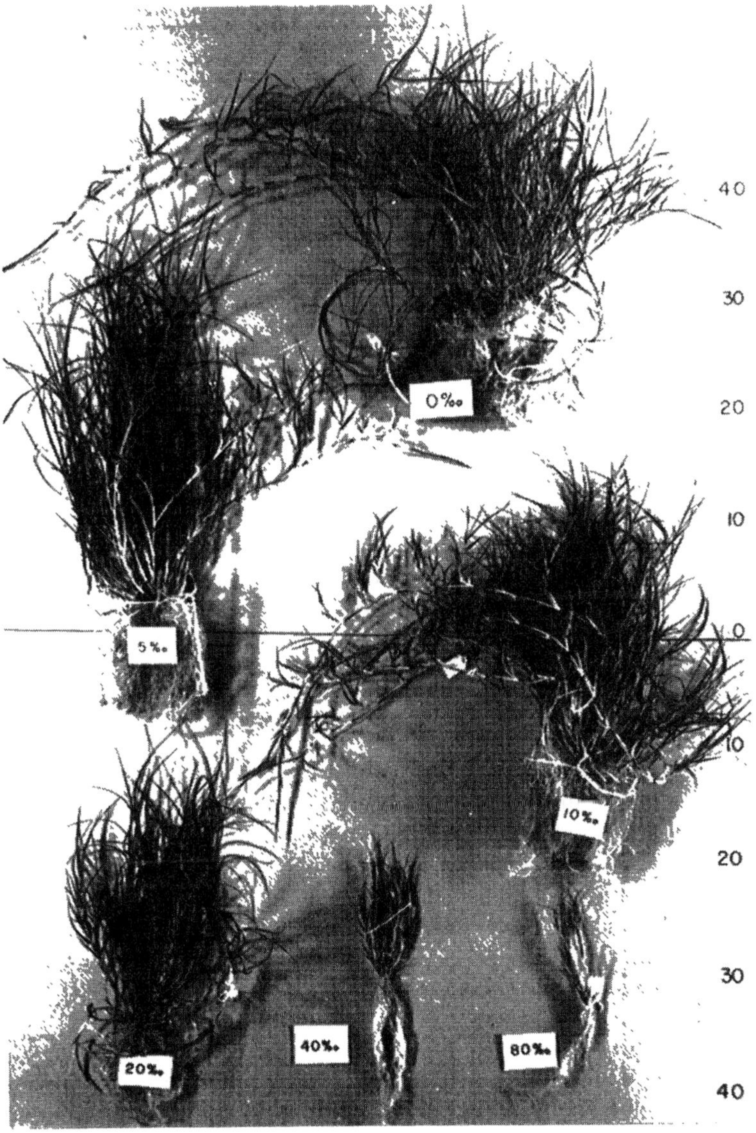

Fig. 4. Response of *Sporobolus virginicus* grown in hydroponic solution to salinity.

Table 4. Yields of *Spartina patens* harvested in 1983 from a field plot in Lewes, Delaware

	Yields (dry weight)	
Harvest dates	kg/ha	tons/acre
Single harvest		
June	8350	3.7
July	14400	6.4
August	11500	5.1
September	14100	6.3
October	7760	3.5
Repeated harvests		
June-August-October (total)	12100	5.4
June	9700	4.3
August	2000	0.9
October	380	0.2
July-September (total)	11600	5.2
July	10900	4.9
September	650	0.3

surprising to find that a line from Connecticut was relatively less inhibited by a regime of 38°C days and 28°C nights than by 28°C days and 18°C nights when compared with a line from Georgia.

At this point we have no results from the feeding trials with sheep and cattle in Egypt. Seed production by the *S. patens* lines is good and germination is high.

Vegetable

Atriplex triangularis *Willd.* This member of the Chenopodiaceae has been used for a number of years as a vegetable[7]. We have collected a large number of ecotypic lines from across the United States and have made selections from between and within these lines for four years. Our selection criteria have been the following:

1. survival and vigorous growth under conditions of high salinity (~ 30‰ seawater salinity)
2. high vitamin A and C contents
3. dark green, large leaves
4. upright growth habit for ease of mechanical harvesting.

We are now increasing seed stocks of the cultivar Seaside Greens (Fig. 5) which had a beta-carotene content of 92 mg/g and an ascorbic acid content of 43 mg/100 g under Delaware growing conditions in 1983 (personal communication from M. Islam). This ascorbic acid content was one of the highest of the lines we have tested, but the beta-carotene content ranked near the bottom. Yield estimates for the fresh market crop are 21300 kg fresh weight/ha (19 000 lb fresh weight/acre) per harvest for growth under 30‰ seawater salinity.

Fig. 5. *Atriplex triangularis*, cv Seaside Greens, grown for seed increase in Delaware.

Multiple cropping is possible under a variety of climatic conditions.

Because seeds will not germinate in highly saline water, it is necessary to use freshwater irrigation for the initial growth period. Pot experiments indicate that one week is a sufficient amount of time to establish the seedlings. A possible alternative, if no freshwater is available, is the fluid-drilling of germinated seeds.

Grain

Kosteletzkya virginica (*L.*) *Presl.* The plant seashore mallow has a rather narrow geographic range along the Atlantic and Gulf coasts of the United States. Since this perennial lives at least five years in cultivation and produces more stems per plant during the first few years, the energy costs for tillage are low. A four-year-old plant with 20 stems is shown in Figure 6. *Kosteletzkya virginica* was identified as a plant with grain potential by my colleague G. Fred Somers[19] who found that it would grow in managed ecosystems if it were germinated in low-salinity water and then grown on mounds. At our latitude (38° 47′N) this plant is not winter-hardy unless grown with a straw mulch. No grain yield data is yet available from Egypt, but plant survival of transplanted fleshy roots was 92%.

In growth chambers *K. virginica* does not survive salinities of approximately 15‰ (personal communication from G. F. Somers and

Fig. 6. Four-year-old *Kosteletzkya* plant with 20 stems growing from the perennial fleshy root in a Delaware field plot.

D. M. Grant). However, in hydroponic studies in the greenhouse my student S. Rumer[15] could grow the plants successfully at salinities of 20‰; if the salinity were kept constant, but when the salinity fluctuated between 0‰ and 20‰, the plants grew less well. At one of our field sites *K. virginica* plants are regularly exposed to as much as 25‰ and do well, but at another site where the irrigation water salinity rises above 30‰, their survival is poor and growth of the survivors is low. These results indicate that plant age and environmental conditions other than salinity (possibly light) may be important factors in the survival and growth of this plant. In Egypt, seashore mallow plants are growing well on the saline soils at Sadat City.

My colleague, Mir Islam, and his associates have evaluated the nutritional quality of the *K. virginica* seeds. Hulled seeds contained 32% protein, with valine as the limiting amino acid, and 22% lipid. Protein scores for both whole and hulled seeds were 64 and 72, respectively. The seeds were high in calcium (> 200 mg/100 g) and potassium (1250 mg/100 g), while the sodium content was only 12–15 mg/g (Ref.[9]).

Grain yields of mass plantings under 25‰ seawater salinity irrigation have ranged from 720 to 1300 lb/acre, and we are now examining 14 seed lines for high yield. In rat-feeding studies, removing the seed hulls improved the protein efficiency ratios (PER's) and also increased the digestibility. The PER's for hulled seeds were higher than those for corn and wheat, but somewhat lower than those for rice, soybean, and cottonseed[9]. The thick, and therefore difficult to remove, seed coat may be responsible for the poor performance of chicks fed *K. virginica* as 20% of their diet (personal communication from M. Islam).

We now have cultivars of three forages (two for hay and one for pasture). All of these cultivars have been tested in Delaware and in Egypt under saline growth conditions. Feeding tests with ruminants are just beginning, but anecdotal evidence indicates that the forages will be quite palatable. A cultivar of the green leafy vegetable is ready for distribution. However, the grain is still under development because of the seed coat problem, and we are not ready to release a cultivar.

Acknowledgements Thanks are extended to all who have worked at or are still working at the Halophytes Biology Laboratory at the University of Delaware (G F Somers, D Grant, D Decker, M Mills, M Siegel, L Donovan, S Rumer, B Dubinski, K Philipp, P Straub, and others who have only worked here a short time), to those in the Department of Food Science and Human Nutrition at the University of Delaware (M Islam, C Wilson and T Watkins), and to my colleagues with whom I work in Egypt (A Bishay, S Seif, and D Seliskar). Financial assistance in this research has come from the Delaware Sea Grant College Program, The Jessie Smith Noyes Foundation, The Crystal Trust, the University of Delaware, AID Title XII Strengthening Grant Program, and John Goelet.

References

1 Donovan L A 1982 Anaerobic substrate tolerance in *Sporobolus virginicus* (L.) Kunth. Masters thesis. Univ. of Del., College of Marine Studies, Lewes, DE.
2 Donovan L A and Gallagher J L 1984 Anaerobic substrate tolerance in *Sporobolus virginicas* (L.) Kunth. Am. J. Bot. (*In press*).
3 Gallagher J L 1979 Growth and element compositional responses of *Sporobolus virginicus* (L.) Kunth. to substrate salinity and nitrogen. Am. Mid. Nat. 102, 68–75.
4 Gallagher J L 1983 Seasonal patterns in recoverable underground reserves in *Spartina alterniflora* Loisel. Am. J. Bot. 70, 212–215.
5 Gallagher J L and Plumley F G 1979 Underground biomass profiles and productivity in Atlantic coastal marshes. Am. J. Bot. 66, 156–161.
6 Gallagher J L, Reimold R J, Linthurst R A and Pfeiffer W J 1980 Aerial production, mortality and mineral accumulation – export dynamics in *Spartina alterniflora* and *Juncus roemerianus* plant stands in a Georgia salt marsh. Ecology 61, 303–312.
7 Gibbons E 1962 Stalking the Wild Asparagus. David McKay Co., NY.
8 Hitchcock A S 1971 Manual of Grasses of the United States. Dover Publ., New York.
9 Islam M N, Wilson C A and Watkins T R 1982 Nutritional evaluation of seashore mallow seed, *Kosteletzkya virginica*. J. Agric. Food Chem. 30, 1197–1198.
10 Leithead H L, Yarlett L L and Shiflet T N 1971 100 Native forage grasses in 11 southern states. U.S. Dept. Agric. Handbook 389. Washington, D.C.
11 Linthurst R A and Reimold R J 1978 Estimated net aerial primary productivity of selected estuarine angiosperms in Maine, Delaware and Georgia. Ecology 59, 945–55.
12 Norlyn J D and Epstein E 1982 Barley production: irrigation with seawater on coastal soil. *In* Biosaline Research: A Look to the Future. Ed. A San Pietro. pp 525–529. Plenum Press, New York.

13 Nabors M W, Daniels A, Nadolny L and Brown C 1975 Sodium chloride tolerant lines of tobacco cells. Plant Sci. Let. 4, 155–159.

14 Richards R A, Dennett C, Schaller C W, Qualset C O and Epstein E 1982 Selection for yield in cereals for salt-affected croplands. *In* Biosaline Research: A Look to the Future. Ed. A San Pietro. pp 535–537. Plenum Press, New York.

15 Rumer S L 1982 A comparison of fluctuating vs constant salinity stress on plants. Masters thesis. University of Delaware, College of Marine Studies, Lewes, DE.

16 Rúsh D W and Epstein E 1976 Genotypic responses to salinity: differences between salt-tolerant genotypes of the tomato. Plant Physiol. 57, 162–166.

17 Sacher R F, Staples R C and Robinson R W 1982 Saline tolerance in hybrids of *Lycopersicon esculentum* × *Solanum pennellii* and selected breeding lines. *In* Biosaline Research: A Look to the Future. Ed. A San Pietro. pp 325–336. Plenum Press, New York.

18 Somers G F 1975 Seed-bearing halophytes as food plants. DEL-SG-75, College of Marine Studies, Univ. of Del., Newark.

19 Somers G F 1979 Natural halophytes as a potential resource for new salt tolerant crops: some progress and prospects. *In* The Biosaline Concept: An Approach to the Utilization of Underexploited Resources. Ed. A Hollaender *et al.* pp 101–115. Plenum Press, New York.

20 Urbina J C M 1980 Determinacion del rango de tolerancia al ensaletramento per el pasto salado (*Distichlis spicata* (L.)) Greene, en suelos del ex Lago de Texcoco. Ciencia Forestal 4, 22, 21–44.

21 Weigart R G and Evans F C 1967 Investigation of secondary productivity in grasslands. *In* Secondary Productivity in Terrestrial Ecosystems. Ed. K Petruservicz. pp 499–518. Pol. Akad. Nauk.

Plant and Soil 89, 337–348 (1985). Ms. BR 23

Developing the seawater agriculture concept

D. PASTERNAK, A. DANON, J. A. ARONSON
The Boyko Institute for Agriculture and Applied Biology, The Institutes for Applied Research, Ben-Gurion University of the Negev, Beer-Sheva, Israel

and R. W. BENJAMIN
The Agricultural Research Organization, Gilat Station, The Negev, Israel

Key words Ash content *Atriplex* species Fodder Protein Seawater irrigation Sheep Yields

Summary The experimental work that is being carried out at Ben-Gurion University of the Negev on the development of seawater agriculture based on salt-resistant halophytes is described. An experimental field, located 100 meters from the Mediterranean seashore, was prepared and equipped for experimentation with seawater irrigation. A detailed description of this system is given.

The work on halophytic fodder plants is basically divided into three components: 1) introduction of a large number of halophytic species and their selection for resistance to seawater salinity; 2) irrigation and management studies of selected fodder halophytes under seawater irrigation; and 3) feeding tests with the halophytic fodder.

One hundred and twenty plant species have been tested since 1982. Of these, 26 have performed at least as well under irrigation with 100% seawater as under irrigation with 15% seawater. The outstanding introductions were *Atriplex lentiformis*, *A. barclayana*, *A. atacamensis*, *A. undulata* and an Atriplex species from Camarones in Argentina.

Atriplex nummularia was irrigated with 100%, 75% and 15% seawater. The corresponding annual yields of dry matter were 1.53, 2.12 and 2.89 kg m^{-2}. The ash content was very high, from 25 to 40% of the dry weight, depending on the treatment and season. The crude protein content was 15–21%.

A first largescale feeding trial was carried out to test the value of *A. nummularia* as a protein supplement for sheep grazing on wheat aftermaths. The daily intake of dry matter was 400 g per head, which was effective only with the daily addition of 400 g corn meal per head. It appears that the low feed intake is at present the main limiting factor in the development of seawater-irrigated fodder.

Introduction

'Salinity and Aridity – New Approaches to Old Problems' was a book edited by H. Boyko that appeared in 1966. In it[2], Boyko drew attention to the possibility of crop production using seawater irrigation and highly salt-resistant plants called halophytes. Consideration of this new approach has raised three major questions:

1) Since the Na^+/Ca^{2+} ratio in seawater is very high, would this cause soil dispersion and a consequent decreased water permeability and poor aeration?

2) Can halophytes actually grow with seawater irrigation? If so, can they be as productive as other plants irrigated with conventional water?

3) With the exception of sugarbeet (*Beta vulgaris*), all the major crop plants are nonhalophytes. Therefore, new crops will have to be developed from wild halophytic plant material. Can this be accomplished? And is it worth accomplishing?

The answer to the first question is well known to soil scientists. The high electrolyte concentration in the seawater should totally prevent soil dispersion, in spite of the high Na^{+}/Ca^{2+} ratio[11]. The experimental project described here was designed to obtain answers to the second and third questions.

In the mid-seventies three centers of halophyte research were established, one at Lewes, Delaware, USA[12], another in Tucson, Arizona, USA[4], and the third in Israel[10]. The initial results in all three centers have indicated that the productivity of halophytes under seawater irrigation is no less than that of conventional crops irrigated with fresh water. These findings complement the observations of ecologists who have measured very high rates of primary productivity (2.3–4.0 kg DW m^{-2} yr^{-1}) in some halophytes growing in salt marshes by the sea[7]. They were substantiated by basic physiological studies summarized in this book by Gorham *et al.*[5], Flowers[3], and others, which throw light on the mechanisms by which halophytes can cope with high salt concentrations in the soil solution and in the plants' tissues, with little sacrifice of primary productivity. The research in Israel on seawater irrigation has gone through two phases so far. In the first phase, which started in 1976 and lasted four years, we carried out small-scale preliminary introduction trials on a 400-m^2 plot by the Red Sea[10]. These trials showed that halophytes can grow with seawater irrigation. The next step was to establish an experimental field on the Mediterranean coast south of the town of Ashkelon. The field was planted during 1981 and 1982, and the experimental work was started in 1983. In this article we will describe in detail the approaches taken in the Research and Development work carried out near Ashkelon and the experimental set-up and will discuss the results of the first year's trials.

Seawater irrigation – the basic approach

Our approach is a long-term maximalistic one. We want to develop, from wild halophytic plants, a range of new crop species that will give economically worthwhile yields with seawater irrigation, or in other words, a seawater-based agriculture. This work is therefore based on the continuous introduction and selection of halophytes for high productivity with Mediterranean Sea water under the particular climatic and edaphic conditions of the experimental site. The selected plants will then be further developed for their ultimate uses as fodder crops, grain crops, firewood, industrial crops, *etc.*

Plant introduction

The approach taken in the plant introduction project is of a very broad nature, because we are dealing mainly with wild plant species whose economic attributes and performance under seawater irrigation are not yet known. There is a need to test as many halophytic species as possible under controlled, reproducible conditions and with continuous evaluation of the potential of each particular species.

The introduction project involves four phases:

Literature survey and data base

Species are being sought in various types of literature according to two criteria:

a) Adaptability of the halophytic species to variable water, climate, and soil conditions

b) Their known or potential economic uses.

Our halophyte list now includes approximately 1250 species from 95 botanical families and is continually being expanded. Information is sought on the taxonomy, geographical distribution, known and potential uses, and maximum salt tolerance that have been reported under any conditions. The list is updated periodically with details on the performance of the individual plant species in the Ashkelon introduction plots and will serve as a data bank for all future halophyte development work. For further details see Aronson[1].

Collection of propagation material

Seeds are being collected through exchanges with botanical gardens and various research institutions, from private seed collectors, through seed-collection expeditions, and through continuing seed-exchange arrangements with the Universities of Arizona and Delaware. In addition, seeds are routinely collected from species grown in the Ashkelon field. The nucleus of a halophyte seed-bank is now being established in Beer-Sheva.

Primary field evaluation

The seedlings are prepared in the Beer-Sheva nursery with freshwater irrigation. They are grown in 1-kg plastic bags filled with potting mixture for 3–8 months to obtain sturdy plants. Before planting in the field they are irrigated for 7–14 days with half-strength seawater

Table 1. Most promising halophytes in Ashkelon introduction field, 1983

Species	Family	Origin	Life form[a]	Aver. height[b] (cm)	Aver. width[b] (cm)	Potential uses and comments
Arthrocnemum fruticosum	Chenopodiaceae	Israel	CH	60	200[c]	Fodder plant? ornamental?
Arthrocnemum halocnemoides	Chenopodiaceae	Israel	CH	60	200	Attractive, compact with blue color
Atriplex barclayana	Chenopodiaceae	Baja California	SH	150	200	Highly promising for fodder and forage, fast growing, self-sowing
Atriplex atacamensis	Chenopodiaceae	Chile	SH	150	150	Fodder, forage, very productive
Atriplex lentiformis	Chenopodiaceae	USA	SH	230	200	Fuelwood, fodder, large biomass
Atriplex canescens var. *linearis*	Chenopodiaceae	North America	SH	150	200	Fodder, forage
Atriplex nummularia	Chenopodiaceae	Australia	SH	150	200	Fodder, forage
Atriplex repanda	Chenopodiaceae	South America	SH	90	100	Fodder, forage
Atriplex sagitifolia	Chenopodiaceae	California	SH	70	100	Fodder, forage
Atriplex undulata	Chenopodiaceae	South America	SH	50	120	Fodder, ornamental
Atriplex vesicaria	Chenopodiaceae	Australia	SH	50	100	Fodder, forage, very productive
Atriplex sp. 'Camarones'	Chenopodiaceae	Chile	SH	120	200	Fodder, forage
Atriplex 'Cin. Wide leaf'	Chenopodiaceae	Australia	SH	60	200	Fodder (spreading by runners), very productive
Batis maritima	Batidaceae	West Indies	HP	25	200	Ornamental ground cover, fodder, minor food
Borrichea frutescens	Asteraceae	Florida	CH	20	150	Ornamental, spreading by runners
Conocarpus erectus	Combretaceae	Central America	T4	70 (recently planted)	70	Fuelwood, ornamental
Chenopodium auricomum	Chenopodiaceae	Australia	CH	(recently planted)		Fodder, forage; 20% protein in leaves
Disphyma clavellatum	Aizoaceae	Australia	HP	20	100	Attractive succulent ground cover, purple flowers
Limonium axillare	Plumbaginaceae	Sinai, Egypt	HP	40	60	Ornamental, cut flowers for export
Limonium perezii	Plumbaginaceae	Canary Islands	HP	70	30	Ornamental, cut flowers for export
Maireana sedifolia	Chenopodiaceae	Australia	SH	60	60	Ornamental, cut branches for export
Melaleuca halmaturorum	Myrtaceae	Australia	T2	80	50	Ornamental, fuelwood

Table 1 (*continued*)

Species	Family	Origin	Life form[a]	Aver. height[b] (cm)	Aver. width[b] (cm)	Potential uses and comments
Nitraria retusa	Zygophyllaceae	Israel	SH	150	200	Ornamental, fuelwood
Sesuvium sp. (green)	Aizoaceae	West Indies	HP	10	200	Ornamental ground cover
Suaeda fruticosa	Chenopodiaceae	India	SH	50	100	Edible seed oil? biomass
Suaeda torreyana	Chenopodiaceae	USA	SH	120	200	Edible seed oil, biomass

[a] Key: HP – Herbaceous perennial; CH – Chamaephyte; SH – Shrub; T2 – Tree to 2 meters; T4 – Tree to 4 meters.
[b] Values given are after 1–2 years growth in undiluted seawater. In all cases, growth rate and size are comparable to or better than those of the control group for the same species (data not shown) grown with 15% seawater in an adjacent plot.
[c] Maximum width allowed = 200 cm.

to adjust them to high salinity. The halophytes are planted in two adjacent plots. One plot (test) is irrigated with full-strength seawater (37 000 ppm dissolved salts), the second (control) with 15% seawater (5500 ppm dissolved salts), which is in the optimum range of salinity for maximum productivity in most halophytes[6]. Fifteen plants per species are planted in three 5-m rows in each of the two plots, at a spacing of 1 × 1 m. The plots are irrigated daily by a drip system. Two 4-l/h emitters are located 10 cm to either side of each plant. More details on the irrigation system are included in the following sections.

The plants in the introduction plots are evaluated every 4 months according to the following criteria: life-form, scale of performance from 0 to 4, height, width, stem circumferance (in woody plants), percent field cover, phenological parameters, *etc.* Table 1 shows results of an evaluation made on 29 November, 1983, for 26 of the best-performing out of 120 species planted so far. Table 2 gives the dry matter production and the crude protein (CP), ash, and fiber contents of six *Atriplex* species evaluated in 1983.

Based on the observations reported in Tables 1 and 2, we planted in 1984 a 500-m^2 plot of *Atriplex barclayana* to study the potential of this promising species as a fodder crop.

Atriplex lentiformis seems to be the most productive species so far tested, both in the experimental and the control plots. The high resistance of this species to salinity has been mentioned by Osmond *et al.*[8]. In one year these plants attained a height of 2.5 m in both treatments. The ash content of the leaves is relatively low and the CP content relatively high (Table 2). The stems are thick and woody. We are therefore investigating the use of this species as a dual-purpose plant for producing both fodder and firewood.

Secondary introduction and selection

Selected introductions are planted in larger plots and evaluated for their potential as crops. If and when proven promising, additional accessions of these particular species are introduced, and a selection of desirable cultivars is begun (*i.e.* in 1985 about 25 accessions of *A. nummularia* collected in various localities in Australia will be planted, from which selections for acceptability to ruminants will be made).

In the next part of this article we will describe the work being carried out on the development of halophytic fodder crops.

Development of fodder crops

Fodder crops were the first seawater crops chosen to be developed, for the following reasons:

Table 2. First evaluation of yields and composition of several Atriplex species irrigated with 15% and 100% seawater

	Dry weight ($kg\ m^{-2}\ yr^{-1}$)		Ash (%)		Fiber (%)		Crude protein (%)	
Species	15%	100%	15%	100%	15%	100%	15%	100%
A. atacamensis	–	2.4	–	31.8	–	23.2	–	13.3
A. barclayana	0.7	1.2	22.6	30.4	26.2	21.0	12.6	14.5
A. sp. 'Camarones'	1.4	0.7	22.8	35.7	36.8	24.3	11.3	13.7
A. cinerea[*]	1.2	1.2	22.3	30.4	32.6	23.1	10.1	15.1
A. lentiformis (leaves only)	–	–	24.0	24.1	18.2	15.6	18.7	23.7
A. sagitifolia	–	–	25.7	28.2	27.7	24.2	16.5	17.8

* This plant has not been conclusively identified as *A. cinerea*.

a) Many halophytes, particularly some members of the Poaceae (grasses) and the Chenopodiaceae (Atriplex) are potentially good fodder crops.

b) In Israel, seawater is expected to be cheap and abundant, relative to other sources of water, and therefore may be economically used for production of fodder over large stretches of land. This will reduce the dependence of Israel on imported animal feeds[9].

c) In arid zones, many of which border on oceans, livestock production is the major if not the only traditional means of agriculture.

The fodder crops to be developed are selections from wild populations. They will have to be domesticated, and specific management practices will have to be developed based on specific agronomical and physiological studies (Table 3). Studies of water requirement, irrigation frequency, fertilization requirements, *etc.* and screening and breeding for specific morphological, chemical, physiological, *etc.* traits will have to be carried out.

No less important are *in vivo* and *in vitro* studies on the value of the fodder as a feed (Table 4). The fodder must be accepted by the animals. A major limiting factor may be the high ash content of the halophytes. To date, most fodder candidates are from the Chenopodiaceae. These plants normally have a high crude protein content but a relatively low energy content. It therefore appears that their main use will be as a cheap protein source, and that the animals' diet will have to be supplemented with an energy-rich feed.

The following section describes in detail the experimental set-up designed specifically for the seawater irrigation studies and gives the first-year yield results for *Atriplex nummularia* and the results of a first feeding trial with *A. nummularia* fodder.

Table 3. Fodder production with seawater: Research and Development methodology. I. Agronomical studies

Agromanagement
Planting technology
Water requirements
Irrigation management
Plant nutrition
Specific crop management
Weed, pest and disease control
Breeding and selection
Introduction and selection from wild populations
Crossing of selected genotypes, if necessary
Physiological studies
General physiology
Productivity studies
Responses to environment and to agromanagement
Specific physiological problems
Salt tolerance
Tolerance studies on whole plants
Mechanisms of salt tolerance
Interaction with environment and agromanagement

Table 4. Fodder production with seawater: Research and Development methodology. II. Nutritional feed value and the value as fodder

NUTRITIONAL VALUE
Chemical analyses
Crude protein
Ash content and composition
Fiber content
Specific analyses (oxalate, tannins, etc.)
In vitro *studies*
Total digestibility
Protein digestibility
Fodder preparations
Green, semidry
Dry (hay), silage, etc.
In vivo *studies*
Total digestibility
Protein balance
Salt balance
Effects on fertility
VALUE AS FODDER
Halophyte fodder as a protein substitute
Body maintenance trials
Production studies (liveweight gain)
Animal adjustment to halophyte fodder
Development of screening methods for preference of animals for halophyte fodder

Experimental set-up

The seawater experimental station is 8 km south of Ashkelon, about 100 m east of the water line and 10 m above sea level. Seawater is pumped by a 10-hp semisubmerged cast-iron pump to the water distribution network (Fig. 1). At the distribution center there is also a source

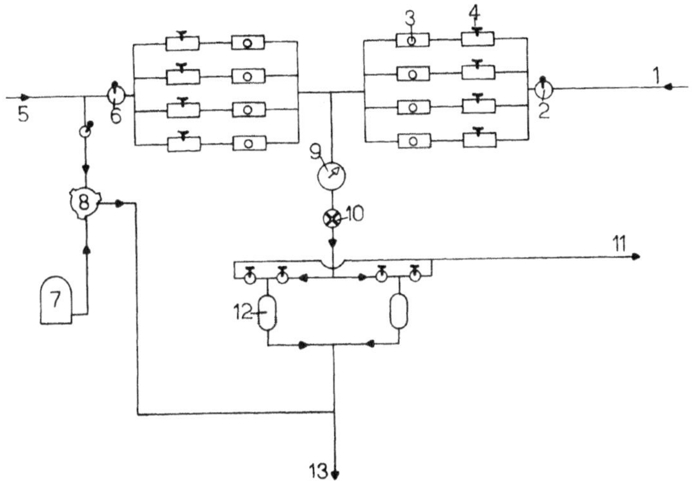

Fig. 1. The water distribution network in the halophyte introduction field at Ashkelon. **1**, seawater pipe; **2**, main seawater valve; **3**, 'Unitrol' discharge regulator; **4**, hydraulic valve; **5**, fresh water pipe; **6**, main fresh water valve; **7**, fertilizer tank; **8**, fertilizer pump; **9**, main water meter and electrical output to irrigation computer; **10**, automatic control of filter flushing; **11**, spill-over of filter flush water; **12**, 2-inch filters (140 mesh); **13**, main pipe to irrigated plots.

of fresh water. The two types of water are distributed through two sets of manifolds which house water-discharge regulators. Each of the manifold branches has a hydraulic valve controlled by a 'Metarol' irrigation computer. This arrangement facilitates the production of a range of mixtures of seawater and fresh water – whatever is required for the experimental purposes.

Both water and fertilizer were applied daily, in excess of the daily requirement (irrigation was 1.2 times the mean daily class A pan evaporation), to ensure that there would be no water deficit and to keep the salt concentration at a minimum. Water was applied through a sprinkler system. The amounts of N, P, and K applied daily per hectare were 3500, 500 and 500 grams, respectively. Microelements were also supplied daily in an approximately half-strength Hoagland solution. Fertilizers were applied towards the end of each irrigation cycle.

Experimental field and treatments

The experimental field is subdivided into twelve 6 × 8 m plots with wide borders between them. Each plot is irrigated by six 'Dan Pal' plastic minisprinklers. Each of the 12 plots is controlled separately by a hydraulic valve. Locally selected *Atriplex nummularia* was planted in February, 1982. A year later the field was divided into three blocks with four treatments in each block. The treatments consisted of irrigation with 15, 50, 75 and 100% seawater.

The plants were harvested by cutting down to a height of 50 cm above ground level when the plants in the 100% seawater treatment reached a height of 80 cm. There were four harvests, the first on 17 May, 1983, and the last on 16 October, 1983. Due to a technical failure, the plots in the 50% seawater treatment received full-strength seawater for some time. Data for this treatment will therefore not be presented.

The fresh and dry weight yields are shown in Table 5. The dry weight was normalized to 12% ash content to enable comparison with the yield of alfalfa. The achievement of a normalized dry weight yield of 1.1 $kg\,m^{-2}$ in the first year of trials is very promising. This fodder is also rich in crude protein (Table 6). Seawater irrigation, however, results in an increase in the ash content from *ca.* 30% in the control to *ca.* 40% of the dry weight in the 100% seawater-irrigated plants.

Table 5. Effect of irrigation with seawater and two dilutions of seawater on yearly production of fresh weight, dry weight, and normalized dry weight (based on 12% ash content) in *Atriplex nummularia*

Seawater dilution	Fresh weight* ($kg\,m^{-2}$)	Dry weight ($kg\,m^{-2}$)	Normalized D W ($kg\,m^{-2}$)
15%	16.8a	2.89a	2.36a
75%	9.4b	2.12b	1.12b
100%	5.4b	1.53c	1.08b

* a, b, c: Values in columns denoted by the same letters do not differ at $p = 5\%$.

First feeding trials

The high crude protein content of *A. nummularia* (Table 6) and its high digestability[13] indicate that its probable value is as a protein supplement rather than as an energy source for animals. Our first feeding trials with *A. nummularia* were designed to check the value of this forage as a protein supplement for sheep grazing on wheat aftermaths during the summer. *A. nummularia* was compared with poultry litter – a cheap protein supplement usually given to sheep in the summer in Israel. There were six feeding rations (treatments) and six ewes per treatment. The crude protein and ash contents of the feeds are shown in Table 7A.

The study was divided into two periods: a preliminary adaptation period of 20 days and an experimental period of 33 days. Liveweight gains per head after 33 days are presented in Table 7B. *A. nummularia* produced a significant weight gain only when supplied with 400 g of corn meal per head. The daily intake of dry *A. nummularia* in treatments

Table 6. Effect of irrigation of *Atriplex nummularia* with seawater and two dilutions of seawater on fodder quality parameters at three harvests during the year

A. Crude protein, ash, and fiber contents (% of dry weight)

Seawater concentr.	Crude protein (%)			Ash content (%)			Fiber content (%)		
	1st	2nd	4th	1st	2nd	4th	1st	2nd	4th
15%	21.1	16.4	19.3	25.3	26.8	30.5	22.8	27	22.0
75%	19.6	17.8	18.8	33.9	32.6	34.4	21.9	20.6	18.1
100%	18.4	15.1	15.6	33.7	37.9	40.7	22.1	20.5	16.7

B. Ionic composition of ash in the last harvest (% of ash weight)

Seawater concentr.	Na^+	K^+	Mg^{2+}	Ca^{2+}	Cl^-	Total
15%	29.7	9.1	10.3	4.3	41.4	94.8
75%	28.9	6.5	7.8	2.4	40.4	86.0
100%	30.0	3.7	6.5	2.3	44.8	87.3
Seawater composition (% of salt content (wt))	30.4	1.1	3.7	1.1	55.2	91.5

Table 7. Trial of *Atriplex nummularia* as a substitute for poultry litter as a protein source for sheep*

A. Crude protein and ash content of feeds

Feed	Crude protein % (N × 6.25)	Ash content (%)
Poultry litter	25	20
A. nummularia	14	42
Wheat straw	3	8.2
Corn meal	10	1.7

B. Performance of sheep (6 ewes per treatment) on these feeds

Treatment	Body weight change after 33 days (kg)
1. *A. nummularia* + 200 g corn + wheat straw	− 0.1
2. *A. nummularia* + 400 g corn + wheat straw	+ 2.3
3. Poultry litter + 200 g corn + wheat straw	+ 0.3
4. Poultry litter + 400 g corn + wheat straw	+ 4.0
5. Poultry litter + wheat straw	− 1.1
6. Grazing on wheat aftermaths	+ 0.7

* Daily intake of *A. nummularia* ~ 0.4 kg/head. *A. nummularia* was provided ad lib in the form of ground, air-dried hay. The daily ration of dry poultry litter was 600 g $head^{-1}$.

1 and 2 was about 400 g $head^{-1}$ day^{-1}. This is very low in relation to the potential dry hay intake per day (> 2000 g) by adult ewes. The low intake could be a result of the high ash content of the feed (Table 7A) or due to a repellant taste of the saltbush.

These results indicate that the acceptability of the feed to animals is now the major bottleneck in the process of developing seawater-irrigated fodder plants. In the near future, efforts will be devoted to the solution of this problem.

References

1 Aronson J A 1984 Economic Halophytes – A Global Review. *In* Proceedings of the Kew International Conference on Economic Plants for Arid and Semi-arid Regions. Kew, England (*In press*).

2 Boyko H, Ed. 1966 Salinity and Aridity – New Approaches to Old Problems. Dr. Junk, The Hague, The Netherlands.

3 Flowers T J 1985 Physiology of halophytes. Plant and Soil 89, 41–56.

4 Glenn E, Fontes M and Jensen N 1982 Productivity of halophytes irrigated with hypersaline seawater in the Sonoran Desert. *In* Biosaline Research: A Look to the Future. Ed. A San Pietro. pp 491–494. Plenum, New York.

5 Gorham J, Wyn Jones R G and McDonnell E 1985 Some mechanisms of salt tolerance in crop plants. Plant and Soil 89, 15–40.

6 Greenway H and Munns R A 1980 Mechanisms of salt tolerance in non-halophytes. Annu. Rev. Plant Physiol. 31, 149–190.

7 Odum E O 1974 Halophytes, energetics and ecosystems. *In* Ecology of Halophytes. Eds. R J Reimold and W H Queen. pp 599–602. Academic Press, New York.

8 Osmond C B, Bjorkman O and Anderson D J 1980 Physiological Processes in Plant Ecology: Toward a Synthesis with Atriplex. Springer-Verlag, Berlin.

9 Pasternak D 1982 Biosaline research in Israel: alternative solutions to a limited water supply. *In* Biosaline Research: A Look to the Future. Ed. A San Pietro. pp 39–58. Plenum, New York.

10 Pasternak D, Ben-Dov Y and Forti M 1979 Is seawater irrigation possible? Hassadeh 59, 2369–2373. (*In Hebrew*).

11 Quirk J P 1971 Chemistry of saline soils and their physical properties. *In* Salinity and Water Use. Eds. T Talsma and J R Philip. pp 79–91. Macmillan, London.

12 Somers G F 1979 Production of food plants in areas supplied with highly saline water: problems and prospects. *In* Stress Physiology in Crop Plants. Eds. H Mussel and R C Staples. pp 127–142. John Wiley and Sons, New York.

13 Wilson A D 1977 The digestibility and voluntary intake of the leaves of trees and shrubs by sheep and goats. Aust. J. Agric. Res. 28, 501–508.

Section 5

Mariculture

Plant and Soil 89, 351–369 (1985). Ms. BR 24

The present situation and problems of mariculture in Japan

T. WATANABE
Department of Aquaculture, Tokyo University of Fisheries, Tokyo, Japan

Key words Artemia Broodstock Essential fatty acids Flounder Live food Mariculture Red sea bream Rotifers Self-pollution Yellowtail

Summary Recently, rearing techniques for various kinds of fish have advanced markedly, and the number of fish species in commercial production increases every year. The establishment of methods for stable, reproducible mass culture of live foods that are highly nutritious is still necessary to improve the survival and growth rates of larval fish. Since, however, the mass propagation of live foods requires costly equipment and depends upon weather conditions, the development of artificial larval diets to replace live foods will be essential. In mariculture local trash fish are commonly used as a feed for juvenile fish because of their low cost and high acceptability to the cultured fish. However, this frequently results in deterioration of water environments, leading to the appearance of fish diseases and pollution. The development of artificial diets such as moist pellets will also improve these conditions.

Mass-cultured fish seed are mainly used for the culture of commercial-sized fish, even though they are generally poorer in taste than wild fish. They are also used for release into coastal waters to promote inshore fishery, but it is difficult to evaluate the effect of stocking on the total catch. Another type of mariculture depends upon raising wild juveniles, though there are clearly too few caught to supply enough fish seed to satisfy the ever-growing demands of fish breeders. Thus, the cultivation of broodstock to produce high-quality eggs is important.

Introduction

Recently, rearing techniques for various kinds of fish and methods for the mass production of living feeds have advanced markedly and, partly as a consequence, the number of fish species in commercial production increases every year. The main species under production in 1982 in Japan are listed in Table 1. Aquacultural production of both freshwater and seawater fish in recent years has amounted to about 10% of the total fisheries production in Japan, but about 20% of the total sales. Yellowtail and sea bream are the major finfish species cultured in net cages. Although the production of yellowtail has been gradually decreasing since 1979 because of excess production, that of sea bream has been increasing (Fig. 1). More than 80% of the yellowtail and sea bream have been cultured in southern Japan, which includes 18 prefectures facing the southern Pacific Ocean, the Seto Inland Sea, and the East China Sea, effectively influenced by the warm Kuroshio and Tsushima currents, except for the central part of the Seto Inland Sea.

For the mass culture of yellowtail and sea bream, trash fish such as anchovy, sand lance, and mackeral have been used as the main foods,

Table 1. Number of seed of main species by length produced in 18 prefectures of southern Japan, 1982

Species		Size range (cm)					Total
Common name	Latin name	1–2	2–4	4–6	6–8	8–10	(× 10^4)
Red sea bream	*Pagrus major*	330	849	413	329	9	1930
Black sea bream	*Acanthopagrus schlegeli*	490	39	47		6	582
Ayu fish	*Plecoglossus altivelis*			452			452
Japanese flounder	*Paralichthys olivaceus*	10	427	164	5		606
Mud dab	*Limanda yokohamas*	105	5				110
Puffer	*Takifugu rubripes*		168	25	510	225	1631
Rockfish	*Sebastes schlegeli*	2	4	8			14
Marbled rockfish	*Sebastiscus marmoratus*		10	2			12
Japanese sea bass	*Lateolabrax japonicus*		12				12
Rabbit fish	*Siganus fuscescens*			10			10
Yellowtail	*Seriola quinqueradiata*		5			1	6

rather than artificial dry pellets, because of the low cost of the trash fish. The use of trash fish as feed results in limited development of yellowtail and in pollution of the environmental water. Artificial diets for marine fish are now under development to improve these conditions.

At present more than 20 fish species are mass-produced in governmental farming centers and private hatcheries in all the prefectures of Japan for the purpose of release into coastal waters and of cultivation to commercial size. Planting of migratory fishes like red sea bream is mainly performed by the farming center of the central government, the Japan Farming Fisheries Association, together with each prefectural governmental farming center. The latter, including the cooperative fisheries associations, have preferably planted species that inhabit closed areas of the prefectural coastline to allow harvesting in local waters. Among the species mass-produced in both central and prefectural government fish farming centers, the red sea bream *Pagrus major* is the most popular and important species in both the northern and southern areas.

Mariculture in Japan

Mariculture can be roughly classified into two types. One is mass propagation of commercially important juvenile fish, shellfish and prawns for the purposes of release to promote coastal fishery and of culture to commercial size in net cages. The other is net-cage culture of wild juvenile fish to commercial size. Yellowtail is typically cultured by the latter method. However, as long as wild juveniles or young fish are needed as culture seed, mariculture is fairly unstable due to limitation by the number of available species and to large annual fluctuations

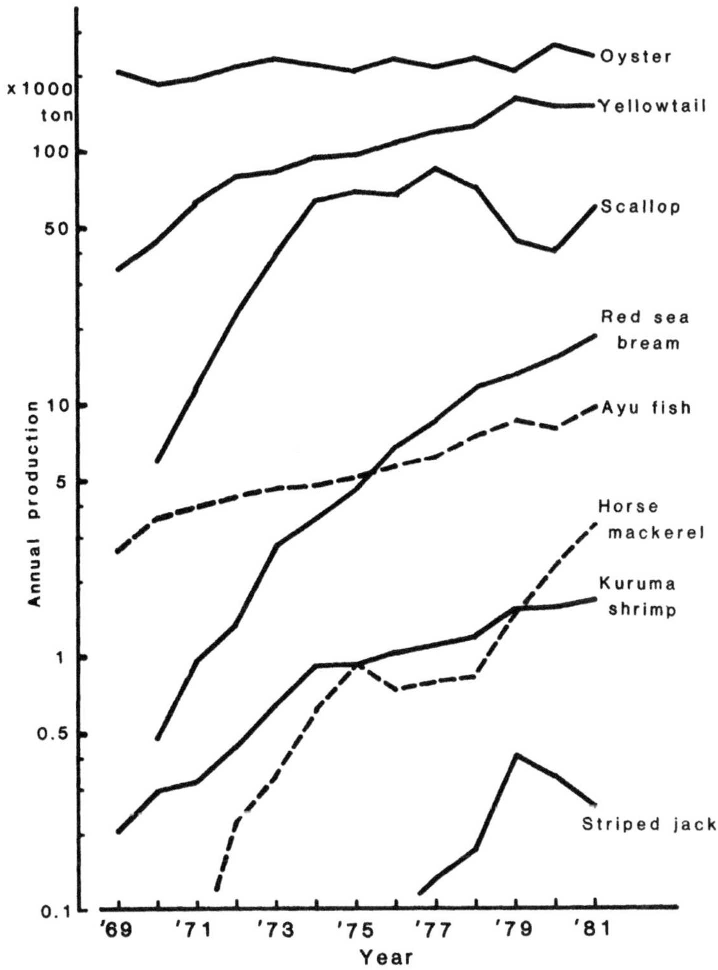

Fig. 1. Trends in the production of representative species cultured in Japan during 12 recent years.

in their catch. Thus the demand for a stable supply of hatchery-produced seed has gradually increased. In addition, experimental projects to release important species for increasing natural stock have been carried out since 1970.

In the procedures for the mass propagation of various marine fish the most important fundamental approaches are:

1. obtaining high-quality eggs that will provide a high percentage of normal larvae for fish seed;

2. cultivation of broodstock to produce high-quality eggs;

3. stable mass-cultivation of living organisms with a high dietary value as feed for the fish larvae.

Taking red sea bream as a typical example of a mass-propagated marine fish, the rearing techniques of the government Fish Farming Center and of hatchery farms can be outlined as follows.

1) Maintenance of broodstock

Broodstock are usually kept in floating net cages in the sea for economic reasons, except in the spawning season. They are fed minced fish meat or an artificial diet. The broodstock are generally transported to a spawning tank on the land from the floating net cages about 4–6 weeks before spawning. A tank of approximately 50–100 m^3 capacity is frequently used for spawning. Equal numbers of male and female fish 3–7 years old are maintained at a density of 0.7–1.5 individuals/m^3. During the spawning season a high density of stock and an unequal sex-ratio of spawners decrease successful spawning[19]. Small fresh fishes such as anchovy, sand lance and mackeral are preferred as feedstuffs. The spawning period extends from 50 to 70 days during March to June. The onset of the spawning season is earlier in the Pacific region than in the Seto Inland Sea and the Japan Sea. In some districts the warm waste water from atomic power stations is used to induce early spawning.

2) Egg supply

Fertilized eggs can be obtained from eggs and sperm of several different origins: a) those stripped from broodstock of both cultured and wild fish without hormone treatment, b) those from hormone-treated broodstock, and c) those spawned or emitted naturally by broodstock reared in a tank. Fertilized red sea bream eggs were obtained by the first or the second method in the early days, but these methods provided few chances to obtain fully mature female and male fish at the same time, and are clearly inadequate to supply enough high-quality eggs. After techniques for natural spawning in captivity[41] were developed, viable eggs have been collected from broodstock maintained in tanks.

Each procedure in the mass propagation of red sea bream is varied technically and in different localities. Pelagic eggs of red sea bream are collected by fine-meshed nets from the overflow water from the spawning tanks, transferred to incubation nets, and then kept in a flow-through system until they hatch. Alternatively, they are introduced directly into a rearing tank at a density of 30 000–90 000 eggs/m^3.

Eggs in the initial stage of the spawning season are, in general, low in percent hatching, and those obtained in the middle phase of spawning are most suitable as seed. About $100–400 \times 10^4$ eggs can be collected from one female broodfish 3–6 years old during one spawning season (50–70 days) in captivity. When a spawning tank of 50–100 tons capacity holds 40–80 fish (equal numbers of males and females), several million eggs can be obtained every day for a month or more[30].

3) Larval rearing

The tank used for larval rearing has a capacity from 10 to 200 m^3, usually 20–100 m^3, and is aerated. Hatched larvae are maintained at a density of 5000 to 50 000/m^3. An optimum stocking density for experimental rearing is recommended to be 4000–5000 larvae/500 l tank[22]. Outdoor tanks are covered with shade nets to reduce the light intensity at the water surface to about 5000 lux.

The most suitable diets for various developmental stages of some fish, according to present knowledge, are outlined below[9,10]. For hatched fish of body length greater than 2.3 mm, rotifers (*Brachionus plicatilis*) are given as the initial diet, which is continued for about 30 days after hatching. When the fish reach 7 mm or more in body length, marine copepods such as Tigriopus, Acartia, Oithona and Paracalanus or, in their absence, freshwater Moina and Daphnia are fed to the larvae together with rotifers, since rotifers are somewhat small for larvae of 7 mm. Brine shrimp (*Artemia salina*), which is distributed commercially, is frequently used as a food for larvae of many marine fishes when there is a shortage of marine copepods. Larvae larger than 10–11 mm are fed on minced fish, shelfish and shrimp or an artificial diet. When juveniles attain 30–50 mm in total body length, larval production is considered to be finished.

The amount of living feed consumed by a single larva per day at each stage of growth has been studied in several species such as red sea bream[27,28], black sea bream[47], rock sea bream[16], flounder[74], common sea bass *Lateolabrax japonicus*, etc., and is roughly estimated to be around 60% of the larval body weight on a wet basis. In red sea bream the relationship between the number of rotifers (R) or *Tigriopus japonicus* (T) taken in by a larva per day and its total body length (L) is reported to be indicated by the following equations:

$$R = 0.3927\,L^{3.676}, \qquad T = 0.06728\,L^{3.466}$$

The average body weight of a rotifer is about 0.003 mg and that of *T. japonicus* 0.0034 mg.

The density of rotifers in a rearing tank must be kept at more than

5 individuals/ml to avoid larval starvation. The amount of living feeds required to supply a rearing tank is estimated to be 1.3 to 1.4 times more than that consumed by the larvae, since the live foods flow out of the tank with the circulating water.

4) Mass culture of Chlorella

At present Chlorella is still indispensible for mass culture of rotifers, although ω-yeast has also been developed as a food for rotifers[24]. The tank capacity used for Chlorella culture ranges from 10 to 800 m^3, and must be 1.5 to 3 times greater than that for rotifers. At the initiation of mass culture of Chlorella, agricultural fertilizers are added to the tank filled with seawater at a rate of 70–100 g ammonium sulfate/m^3, 10–50 g calcium superphosphate/m^3 and 5–50 g urea/m^3, and reagents for plankton culture are also added as needed. Then inocula of marine Chlorella are introduced into the tank at a concentration of about $300–600 \times 10^4$ cells/ml and the culture is aerated vigorously through an air stone and/or pipe. The cell number increases up to $2500–4000 \times 10^4$/ml. The maximum temperature for optimum growth of Chlorella is about 26°C. During the summer a reduced production is often observed due to the high temperature, and is improved by diluting the saline culture medium by 10–20% with fresh water.

5) Mass culture of rotifers

Rotifers have been used most extensively and are very important as the initial live food for rearing larval fish. At present, without the mass culture of rotifers, larval rearing of marine fishes would be virtually impossible.

Tanks less than 100 m^3 in capacity are usually used to culture rotifers. The density of rotifers at the start of culture is between 30 and 60 individuals/ml, and they are harvested for larval feeding when the density exceeds 100 individuals/ml. Recently, baker's yeast (*Saccharomyces cerevisiae*) and ω-yeast, prepared by adding fish oils to the culture medium for bakers' yeast, have been used as food organisms for rotifers, as well as Chlorella. These feedstuffs are useful when there is a shortage of marine Chlorella and also for saving tank capacity. The usual amount of yeast given to rotifers is about 1 g per million rotifers per day. Alcohol fermentation lees have been used experimentally to culture rotifers[13,21].

Problems in mariculture in Japan

In spite of the advanced state of mariculture in Japan various problems still remain, which should be solved as quickly as possible. These include the cultivation of broodstock to supply the needed quantities of high-quality eggs; the quality and mass propagation of live foods for the fish larvae and the development of good artificial diets for larval and juvenile fish; and the existence of defects such as lordosis and abnormal coloration in the cultured fish. Approaches to solving these problems will be discussed below.

1) Cultivation of broodstock

Recently, more than 300 different species of finfish have been cultivated throughout the world. Most fish seed comes from natural sources, which clearly are inadequate to supply the present and future needs of farmers in the world. In other words, economically productive aquaculture is heavily dependent upon an adequate supply of seed (fertile eggs or juvenile fish) with which to stock the ponds, enclosures, and other cultivation systems. Thus it is important to ensure an adequate and year-round, rather than seasonal, supply of high-quality fertile eggs that produce survival and growth rates as high as those occurring naturally.

In southern Japan, nearly 20 million seed of red sea bream are reared for release and for culture in net cages (Table 1). In cage culture farms, red sea bream has been gradually replacing yellowtail for several years. This may be attributed to sufficient supply of red sea bream seed by successful mass propagation, as opposed to the wild catch of yellowtail. Successful rearing of yellowtail and of other species from naturally spawned eggs should in the near future supply sufficient seed for net cage culture, and decrease the need to catch wild seed.

The cultivation of broodstock that can produce eggs of high quality is important. Nutrition has a profound effect upon gonadal growth and fecundity. Recent investigations[57,70,71,72] on the nutrition of broodstock have demonstrated that the nutritional quality of broodstock diets is very important for gonadal maturation, spawning and egg quality in both freshwater and marine fishes, as are environmental, physiological and genetic conditions. As one example, in female red sea bream broodstock fed on nutritionally low diets such as a phosphorus- or essential fatty acid (EFA)-deficient diet or a low-protein diet, the percentage of buoyant eggs, the total hatch, and the normality of hatched larvae were significantly low, leading to a low production of larvae suitable as seed[70]. Furthermore, cuttlefish meal was superior to white fish meal as a protein source in diets for red sea bream broodstock. In a later experiment by Watanabe *et al.*[72], spawning and egg quality were greatly affected by the quality of feed given to red sea bream shortly before spawning.

2) Nutritional quality of mass-cultured live foods

In mass propagation of marine fish, rotifers are indispensible as the initial diet for the larvae. However, some problems have been encountered regarding the dietary value of rotifers and the other Artemia. Rotifers were mass cultured on marine Chlorella as a feed organism until baker's yeast was also found to be suitable[43]. When baker's yeast was used as a food for rotifers, the density of the rotifer culture reached about 10 times that obtained by using marine Chlorella. Thus, recently, yeast has been increasingly used as a food for rotifers. However, when rotifers cultured with yeast were given to larval fish, sudden heavy losses of the larvae frequently occurred[15,18,29,30,45]. These high mortalities could be prevented by culturing the rotifers with both yeast and marine Chlorella, or by culturing the rotifers with yeast and then feeding them on marine Chlorella for a short period before feeding them to the fish[15,18,30,45]. This was one of the most important findings for the mass production of juvenile fish.

Our recent investigation of the relationship between the nutritional quality of living feeds and their food organisms has demonstrated that the content of essential fatty acids in the living feeds is the principle determinant of their dietary value.

Table 2 (ref.[60]) shows some of the ω3 fatty acids of rotifers cultured with baker's yeast, marine Chlorella or both organisms. The most striking difference was the content of essential fatty acids. The rotifers cultured with yeast were quite low in ω3 highly unsaturated fatty acids (ω3 HUFA) such as 20:5ω3, and high in monoenoic fatty acids such as 16:1 and 18:1. Those cultured with marine Chlorella contained a high amount of 20:5ω3, which is one of the EFA for marine fish[68]. The rotifers fed on both yeast and Chlorella showed intermediate values. These results may explain why rotifers cultured with yeast are always

Table 2. Some ω3 fatty acids in the total lipids from rotifer *Brachionus plicatilis* cultured with baker's yeast, *Saccharomyces cerevisiae,* and marine Chlorella

Fatty acid	Rotifer fed on: Yeast	Yeast + Chlorella	Chlorella
20:4ω3	0.4–0.5	0.4– 0.6	tr– 0.2
20:5ω3	1.0–1.9	8.1–12.0	22.8–30.5
22:5ω3	tr–0.3	1.7– 2.9	3.0– 3.8
22:6ω3	tr–0.5	tr– 0.9	tr– 0.5
Σω3 HUFA*	2.2–3.1	11.3–14.7	26.2–31.9
Lipid in rotifer (%)	1.4–2.3	2.2– 2.8	3.7– 4.2

* $C_{20:3}$ < ω3 fatty acids.

inferior to those cultured with marine Chlorella in their nutritional quality as a living feed.

Table 3 (ref.[69]) lists some fatty acids in the total lipids of the culture organisms. The baker's yeast used for mass culture of rotifers contains a fairly high amount (52–82%) of monoethylenic fatty acids, 16:1 and 18:1, and essentially no $\omega3$ HUFA. On the other hand, marine Chlorella contains a high level of $20:5\omega3$. These results show why rotifers cultured with yeast always contain little $\omega3$ HUFA. The fatty acid composition of freshwater Chlorella is quite different from that of marine Chlorella. The freshwater Chlorella contains high amounts of $18:2\omega6$ and $18:3\omega3$, but is low in $\omega3$ HUFA. Consequently, rotifers cultured with freshwater Chlorella had high levels of $18:2\omega6$ and $18:3\omega3$. Thus marine Chlorella should be used as the culture organism for rotifers intended as a food for larval marine fish.

Table 3. Fatty acid composition in total lipids of bakers' yeast and marine Chlorella

<table>
<tr><th></th><th>Baker's yeast
Saccharomyces cereviciae</th><th>Marine Chlorella
Chlorella minutissima</th><th>Freshwater Chlorella
Chlorella regularis</th></tr>
<tr><td>14:0</td><td>3.1</td><td>4.3</td><td>0.5</td></tr>
<tr><td>16:0</td><td>20.0</td><td>22.5</td><td>16.9</td></tr>
<tr><td>$16:1\omega7$</td><td>27.2</td><td>22.3</td><td>2.7</td></tr>
<tr><td>18:0</td><td>4.7</td><td>1.0</td><td>4.1</td></tr>
<tr><td>$18:1\omega9$</td><td>26.1</td><td>3.1</td><td>3.5</td></tr>
<tr><td>$18:2\omega6$</td><td>10.9</td><td>3.4</td><td>37.3</td></tr>
<tr><td>$18:3\omega3$</td><td>3.2</td><td>0.1</td><td>9.1</td></tr>
<tr><td>20:1</td><td>0.8</td><td>0.1</td><td>0.1</td></tr>
<tr><td>$20:3\omega3$</td><td>–</td><td rowspan="2">4.7</td><td rowspan="2">t</td></tr>
<tr><td>$20:4\omega6$</td><td></td></tr>
<tr><td>$20:4\omega3$</td><td>–</td><td>0.1</td><td>–</td></tr>
<tr><td>$20:5\omega3$</td><td>–</td><td>31.8</td><td>0.2</td></tr>
<tr><td>$22:5\omega3$</td><td>–</td><td>–</td><td>–</td></tr>
<tr><td>$22:6\omega3$</td><td>–</td><td>–</td><td>–</td></tr>
</table>

These results have shown that the fatty acid composition of rotifers is readily affected by the fatty acids of their food organism. Experiments were then conducted to verify the relationship between the dietary value of rotifers and their $\omega3$ HUFA content by feeding them marine Chlorella or freshwater Chlorella and feeding these rotifers to larval red sea bream[31,62]. When marine Chlorella was used as the culture organism, the low level of $\omega3$ HUFA in yeast-fed rotifers increased in proportion to the length of the culture period with Chlorella, due to incorporation of $20:5\omega3$ from the marine Chlorella. The dietary value of the rotifers for red sea bream was also found to be significantly improved by secondary culture with marine Chlorella for more than

6 h, but not with freshwater Chlorella. Thus, the content of ω3 HUFA in the rotifers correlates with their dietary value as feed, and the high mortalities frequently observed in various kinds of fish larvae receiving rotifers fed on yeast as their sole feed is due to an EFA deficiency in their food source.

Based on these results, a new kind of yeast has been developed as a culture organism for rotifers so as to improve the nutritional value of rotifers grown on them for fish larvae[24]. This new type of yeast (designated ω-yeast) was produced by adding fish oil or cuttlefish liver oil as a supplement to the culture medium for baker's yeast, which resulted in a high content of lipid and ω3 HUFA, the EFA for marine fish. The rotifers cultured with ω-yeast were higher in lipid content in general and in ω3 HUFA as a result of the oil added to the baker's yeast. Furthermore, the dietary value of the rotifers to fish larvae was significantly improved, compared to that of rotifers cultured on marine Chlorella[32,46,67]. These results clearly indicate that rotifers grown on the newly developed yeast are superior in food value to those grown on unmodified yeast. Arakawa *et al.*[1] also found that rotifers grown on ω-yeast had superior food value for larval puffer (*Fugu rubripes rubripes*). Recently, Fukusho *et al.*[17] mass cultured Tigriopus with ω-yeast and found that its nutritional quality for mud dab was much enhanced. This method may be called the indirect method for improving the dietary value of living feeds.

A direct method has also been developed[67]. Lipids containing ω3 HUFA were homogenized with a small amount of raw egg yolk and water, and the resulting emulsion was included in the culture medium for rotifers, together with baker's yeast. The rotifers took up the lipids readily, and the concentration of ω3 HUFA reached a maximum after 6 to 12 h of feeding, as also observed with the indirect method. The two methods, direct and indirect, are also very effective for improving the dietary value of other living feeds. With both methods, it was possible to further improve the dietary value of living feeds by allowing them to take up from the culture medium not only ω3 HUFA but also fat-soluble vitamins[67].

The nauplii of *Artemia salina* have been widely used as a food in the production of juvenile marine fish[8,22,36,38,52]. However, feeding with Artemia alone frequently resulted in high mortalities in various marine fish[8,19,61], although this phenomenon depended upon the fish species as well as the site of production of the Artemia. Some species of flounder, one species of mullet (*Liza haemotocheila*), one salmonid (*Plecoglossus altivelis*) and some gobiid fish are not easily affected, but the larvae of yellowtail (*Seriola quinqueradiata*) are very susceptible to this phenomenon[8]. In addition, many investigators have reported heavy losses of larval prawns, crabs, and marine fish fed on Artemia nauplii from Utah[2,11,36,53,72]. The dietary value of Artemia nauplii was improved when nauplii were fed to fish together with marine copepods such as Tigriopus and Acartia[14].

Watanabe *et al.*[65] analyzed the fatty acid composition of Artemia eggs and nauplii from different locations and found that Artemia could be classified into two types according to the fatty acid composition: one (the freshwater type) containing a high concentration of 18:3ω3, which is an EFA for freshwater fish, and the other (the marine type) high in 20:5ω3, which is an EFA for marine fish. Furthermore, Artemia of the marine type were a satisfactory food for juvenile red sea bream. The dietary value of the nauplii was improved by feeding them marine Chlorella and ω-yeast, both containing substantial amounts of the EFA required by marine fish. These results suggest that the class of EFA contained in Artemia is the principal factor in the variation in its food value for fish, as demonstrated in the case of rotifers. Wickens[73] reported that the food value of Artemia from Utah is improved by allowing them to feed on *Isochrysis galbana*, which generally contains high amounts of 20:5ω3 and 22:6ω3[59]. Moreover, the food value of Artemia is effectively improved when they are fed to fish together with Tigriopus or Acartia[14], both rich in 20:5ω3 and 22:6ω3[62].

Watanabe *et al.*[64,65] conducted further experiments to improve the dietary value of freshwater-type Artemia nauplii by the direct and indirect methods that were used for rotifers. The dietary value of the nauplii to fish larvae was improved by incorporating ω3 HUFA from emulsified lipids, and was proportional to the ω3 HUFA content of the nauplii. The dietary value of the nauplii fed on baker's yeast alone or on corn oil, which contain little ω3 HUFA, was very low for marine fish. Although any type of Artemia may be satisfactory for freshwater fish, judging from their EFA requirement[3,55,56,57,75], it is necessary to check the fatty acid composition of Artemia for use as a food for marine fish. If its fatty acid composition is not known, the Artemia should be fed to fish together with other marine copepods or should be fed on lipids containing ω3 HUFA to prevent heavy fish losses.

3) Development of formulated larval microdiet in place of live foods

Mass culture of rotifers together with Chlorella or other organisms not only requires excessive manpower and costly equipment, but also depends upon natural conditions such as the weather. For example, the mass propagation of one million fry of red sea bream requires 5×10^9 rotifers per day, and four 200-ton tanks are necessary for the production of this amount of rotifers. In addition, other large tanks are required to culture a high amount of marine Chlorella as a food for the rotifers, as described above. The success of mass culture of Chlorella also is greatly affected by the weather conditions.

Thus, the rearing of larval marine fish from the beginning, using artificial diets without rotifers, has long been desired by fish breeders. In Japan a formidable effort to meet this challenge has been made since 1979 by many workers[25,26].

Artificial microdiets prepared by various methods must:

1. float in the water for more than 30 min without leakage of any nutrients from the diets
2. suit the larva's taste
3. contain all the nutrients required by the larval fish
4. be digested and absorbed by the larval fish
5. be almost the same size as rotifers after absorption of water

From these viewpoints microdiets for larvae have been prepared by various methods, and techniques of diet preparation and quality of microdiets have been gradually improving, although it may take several years to scale up these laboratory techniques to the practical hatchery level.

4) Development of formulated feeds for juvenile marine fish

Formulated feeds are still used much less in mariculture of juvenile fish than in cultivation of freshwater fish. This may mainly be due to the markedly lower cost of trash fish such as anchovy, sand lance and mackeral available in the coastal region of Japan. For the main species in mariculture, yellowtail (more of which is produced than of any other species), dry artificial feeds are not acceptable and are being replaced by moist pellets, prepared by mixing minced fish meat with a formulated mash diet.

Feeding with trash fish, for the densely cultured fish, frequently results in deterioration of the environmental water due to loss of the diets and excretion from the fish. About 10% of the weight of frozen trash fish is generally lost as drippings during thawing before feeding them to fish[35]. Some of the drippings are clarified by sanitation facilities and some are discharged directly into the ambient water. Recently, frozen fish have been found to be directly acceptable to yellowtail. The total amount of nitrogen discharged into the ambient water after feeding with trash fish is estimated to be 35–50% of the total amount in the feed. About 80% of the water pollution in fish farming grounds is due to water-soluble substances from trash fish and the excreta from the cultured fish, and about 20% of the subsoil contamination is due to sediments from feces and uneaten food.

Besides nitrogen, a large amount of phosphorus, mainly derived from the bones of the trash fish, is also discharged, providing a synergistic action on water pollution together with nitrogen. Phosphorus contained in both raw fish and fish meal, mainly in the form of tricalcium phosphate, varies in its availability to fish from 0 to 65%, and is especially low in carp without stomachs[44].

Self-pollution due to use of trash fish also results in the occurrence of various fish diseases and of red tide on a large scale, leading to a high mortality of the cultured fish.

5) *Deformity in hatchery-reared juveniles*

Such deformities as short-tail, pughead, curvature of the skeleton, *etc.*, are observed to some extent in hatchery-reared juveniles of each species. In many cases they lead to a serious problem because the market price is reduced. Among these abnormalities in mass-produced juveniles, skeletal deformity in red sea bream and abnormal coloration in Japanese flounder have recently been significant.

Lordotic deformity in hatchery-reared red sea bream. Lordosis has been observed frequently in hatchery-reared red sea bream at almost all hatcheries since about 1973. Recent studies have demonstrated that lordosis occurred in fish with a deflated swim bladder, due to their failure to gulp air at the water surface at a body length of 4–4.5 mm, 10–15 days after hatching[34]. The fish with deflated swim bladders had difficulty in keeping themselves in the upper or middle layer of the water. They swam upward obliquely with rapid fin strokes. Thus, lordosis seemed to be induced by their compensating for the oblique direction of the body axis by a distortion in the spinal column.

It was also found that feeding rotifers with a low content of $\omega 3$ HUFA to larval red sea bream resulted in low swimming activity, lack of endurance and lack of reflex responses. Consequently, they had difficulty in gulping air after reaching the water surface while withstanding the water current set up by the aeration. Moreover, it was postulated that swift water currents resulting from too much aeration made it impossible for larvae to swim up toward the water surface and to gulp air there.

In addition to red sea bream, lordotic deformity is also observed in other species of hatchery-reared fish with deflated air bladders[48], and appears to be induced by the same factors as in red sea bream.

Abnormal coloration in Japanese flounder. Recently in Japan, mass propagation of juvenile flounder *Paralichthys olivaceus* has been markedly increasing; however, serious problems with abnormal coloration have been encountered. As a result of the advancement of hatchery techniques, millions of juvenile flounders are now produced as seed for both cultivation and release in Japanese coastal waters. Anomalous individuals are, in general, low in market value, and it is doubtful whether they can survive when released.

Abnormal coloration in wild Heterosomata has been reported by many authors[5,39,42]. The frequencies of albino and ambicolored wild plaice[6], *Pleuronectes platessa*, from the North Sea are at most 2.0% and 4.0%, respectively. On the other hand, those of hatchery-reared Heterosomata are very high. Suggested causes of the abnormal coloration are the lighting conditions during egg incubation[4], feed intake during the larval stage[49], and larval population density in the rearing tanks[52].

A recent study by Seikai[50] has demonstrated that there is a close relationship between the time of starting to feed with Artemia, a major feed during the larval stage, and abnormal coloration in hatchery-reared flounder, and that the abnormality was effectively reduced by feeding with natural plankton. Seikai and Watanabe[51] also reported that the frequency of occurrence of abnormal coloration was related to the kind of Artemia used and its origin. Differences in the concentrations of heavy metals and pesticides that may be responsible for the anomaly have not been detected in Artemia, although the studies are not yet complete.

6) Quality of maricultured products

As shown in Figure 1, the total production of marine fish is increasing every year; however, hatchery-reared and cultured fish are, in general, lower in quality, in terms of both taste and coloration, which leads to lower market prices than for wild fish. The poor taste of cultured fish is mainly due to a high concentration of body lipids induced by overfeeding, although it depends upon the fish species. Cultured yellowtail, red sea bream, and Ayu are typically poor in quality due to excess lipid accumulation in muscle and viscera. Marine fish in captivity are generally forced to accept the amounts of diet required for their optimum growth, to produce them in time for season, and also they are marketed not by quality but by weight.

The coloration of cultured fish such as yellowtail, red sea bream and Ayu constitutes another problem. The pigmentation or color tone in the skin of cultured fish is usually far inferior to that of wild fish, leading to a lower market price, about half that of the wild fish in the case of red sea bream. It is well known that cultured red sea bream turn dark when fed on carotenoid-free diets. Several attempts have successfully been made to enhance the pigmentation by feeding with krill, Mysis, shrimp and crab wastes, green alga *Haemotococcus* sp., *etc.*[12]. The Antarctic krill *Euphausia superba* and its meal are widely used, mainly because they are readily available commercially. The oil extracted from krill, composed almost exclusively of astaxanthin esters[12], is now also used for this purpose. It thus appears that the problem of color improvement in cultured red sea bream has been solved, although pigmentation artificially enhanced with carotenoids has a color tone somewhat different from that of wild fish.

The color of cultured yellowtail and salmonids was also effectively improved by adding krill oil to the formulated diets. Lutein was also effective in improving the coloration of Ayu fish.

Conclusions

During the last ten years aquaculture, especially mariculture, has developed remarkably to fill the gap between the supply and demand for fishery products, which is supported by the traditional preference of the Japanese for seafood. However, mariculture must now change its structure due to adverse circumstances such as increases in the costs of energy, feeds and other materials, low market prices, deterioration of environmental conditions, *etc.* For the resolution of these constraints, modern scientific knowledge and technology must be integrated to develop a new type of aquaculture. Expansion of mariculture grounds to more open-sea areas would be one solution. Technologies for development of formulated diets not only for young but also for adults and larvae, development of feeding standards, and prevention of epizootic diseases will enable the establishment of modern aquaculture.

Culture-based fishery, or marine-ranching, is becoming a target for further development of coastal fisheries in Japan. Many research studies have been conducted to this end with a great degree of governmental support. Establishment of networks of seed production farms has made it possible to carry out releasing experiments on a large scale, and a newly integrated restocking system should be developed by the analyses of the results of these experiments, even though the profitability of artificial restocking is not yet thoroughly verified. The commencement of a national scheme for culture-based fishery would be a manifestation of the foresight of the Japanese fish industry.

References

1 Arakawa T, Yogata T and Watanabe T 1979 Rearing of Larvae of puffer (*Fugu rubripes rubripes*) by rotifer (*Brachionus plicatilis*) cultured with various foods. Bull. Nagasake Pref. Inst. Fish. 5, 5–8.

2 Bookhout C T and Costlow J D 1970 Nutritional effect of *Artemia salina* from different locations on larval development of carp. Helgol. Wiss. Meeresunters. 20, 435–442.

3 Castell J D, Sinnhuber R D, Lee D J and Wales J H 1972 Essential fatty acids in the diets of rainbow trout (*Salmo gairdneri*): growth, feed conversion and some gross deficiency symptoms. J. Nutr. 102, 77–86.

4 Dannevig A and Hanson S 1952 Faktorerav betydaing for fiskeegenes og. fiskegnelens oppvekst. Fisk, Skr. Havundersok. 10, 1–36.

5 Dawsen C E 1962 Note on anomalous American heterosomata with description of five new records. Copeia 1, 171–174.

6 de Veen J F 1969 Abnormal pigmentation of the plaice (*Pleuronectes platessa* L.). J. Cons. Int. Explor. Mer., Copenhagen, 32, 344–384.

7 Fujita S 1962 Studies on the life history and culture of the puffers in Japan. Special Rep. Nagasaki Pref. Inst. Fish. 2, 121.

8 Fujita S 1972 Food value of *Acartia clausi*, *Tigriopus japonicus* and *Artemia salina* to juveniles of red sea bream. Report on selection of proper species and mass-culture of zooplanktons as food for marine fish larvae I. Rep. Nagasake Pref. Inst. Fish. 352, 27–28.

9 Fujita S 1973 Importance of zooplankton mass culture in producing marine fish seed for fish farming. Bull. Plankton Soc. Japan. 20, 49–53.

10 Fujita S 1979 Culture of red sea bream, *Pugrus major*, and its food. European Mariculture Society, Special Publication No. 4, EMS, Bredene, Belgium, 183–197.

11 Fujita S, Watanabe T and Kitajima C 1980 Nutritional quality of *Artemia* from different locations as a living feed for marine fish from the viewpoint of essential fatty acids. The Brine Shrimp *Artemia*, 1980, Vol. 3. Ecology, Culturing. Use in Aquaculture. Universal Press, Wetteren, Belgium.

12 Fujita T, Satake M, Watanabe T, Kitajima C, Miki W, Yamaguchi K and Konosu S 1983 Pigmentation of cultured red sea bream with astaxanthin diester purified from krill oil. Bull. Japan. Soc. Sci. Fish. 49, 1855–1861.

13 Fukuhara O, Yukawa H and Nishimura T 1982 Culture of rotifer, *Brachionus plicatilis* (Müller), in an air lift aquarium using alcohol fermentation slops. Bull. Nansei Reg. Fish. Res. Lab. 14, 1–11.

14 Fukusho K 1974 Artemia nauplii as a food for rock sea bream juveniles. Suisan Zoshoku, 21, 71–75.

15 Fukusho K 1977 Nutritional effects of the rotifers, *Brachionus plicatilis*, raised by baking yeast on larval fish of *Opelegnathus faciatus*, by enrichment with *Chlorella* sp. before feeding. Bull. Nagasaki Pref. Inst. Fish. 3, 152–157.

16 Fukusho K 1979 Studies on fry production of Japanese striped knifejaw *Oplegnathus fasciatus*, with special reference to feeding ecology and mass-culture of food organism. Special Rept. Nagasaki Pref. Inst. Fish. 6, 1–173.

17 Fukusho K, Arakawa T and Watanabe T 1980 Food value of copepod, *Tigriopus japonicus* cultured with ω-yeast for larvae and juveniles of mud dab, *Limanda yokochamae.* Bull. Japan. Soc. Sci. Fish. 46, 499–503.

18 Fushimi T 1968 Studies on the production of red sea bream II. Effects of live foods on growth and survival of larvae. J. Hiroshima Pref. Fish. Expt. Stn. 1, 50–51.

19 Fushimi T 1971 Studies on the seed production of red sea bream VI. Dietary value of *Tigriopus japonicus* together with *Artemia salina* to red sea bream juveniles. J. Hiroshima Pref. Fish. Exp. Stn. 3, 29–34.

20 Fushimi T 1972 Studies on egg-taking from cultured red sea bream, *Pagurus major* III. Tech. Rep. Farm. Fish. 1, 15–20.

21 Higashihara T, Fukuoka S, Abe T, Imada O and Hirano R 1983 Culture of the rotifer *Brachionus plicatilis* using a microbial flock produced from alcohol fermentation's slop. Bull. Japan. Soc. Sci. Fish. 49, 1001–1013.

22 Hirano R and Ohshima Y 1963 On the rearing of larvae of marine animals with special reference to their food organisms. Bull. Japan. Soc. Sci. Fish. 29, 282–297.

23 Hirata M, Fujita T, Hamada S and Nagata F 1975 Studies on mass production of red sea bream. Tech. Fish. 4, 61–68.

24 Imada O, Kageyama Y, Watanabe T, Kitajima C, Fujita S and Yone Y 1979 Development of a new yeast as a culture medium for living feeds used in the production of fish feed. Full. Japan. Soc. Sci. Fish. 45, 955–959.
25 Kanazawa A, Teshima S, Iwashita T and Nagao A 1981 Effect of phospholipids on growth, survival rate, and incidence of malformation in the larval ayu. Mem. Fac. Fish. Kagoshima Univ. 30, 301–309.
26 Kanazawa A, Teshima S, Sumida S and Iwashita T 1982 Rearing of larval red sea bream and Ayu with artificial diets. Mem. Fac. Fish. Kagoshima Univ. 31, 185–192.
27 Kitajima C 1976 Amount of the copepod, *Tgriopus japonicus*, consumed by red sea bream, *Pugrus major.* Bull. Nagasaki Pref. Inst. Fish. 2, 101–104.
28 Kitajima C, Fukusho K, Iwamoto H and Yamamoto H 1976 Amount of rotifer, *Brachionus plicatilis*, consumed by red sea bream larvae, *Pagurus major.* Bull. Nagasaki Pref. Inst. Fish. 2, 105–112.
29 Kitajima C and Koda T 1976 Lethal effects of the rotifer cultured with baking yeast on the larval red sea bream, *Pagrus major*, and the increase rate using the rotifer recultured with *Chlorella* sp. Bull. Nagasaki Pref. Inst. Fish. 2, 113–116.
30 Kitajima C 1978 Aquisition of fertilized eggs and mass-culture of juvenile of red sea bream, *Pugrus major.* Special Rept. Nagasaki Pref. Inst. Fish. 5, 1–92.
31 Kitajima C, Fujita S, Yone Y and Watanabe T 1979 Improvement of dietary value for red sea bream larvae of rotifers, *Brachionus plicatilis*, cultured with bakers' yeast, *Saccharomyces cerevisiae.* Bull. Japan. Soc. Sci. Fish. 45, 469–471.
32 Kitajima C, Arakawa T, Fujita S, Watanabe T and Yone Y 1980 Dietary value for red sea bream larvae of rotifers, *Brachionus plicatilis*, cultured with a new type of yeast. Bull. Japan. Soc. Sci. Fish. 46, 43–46.
33 Kitajima C, Yoshida M and Watanabe T 1980 Dietary value for Ayu fish, *Plecoglossus altivelis*, of rotifer, *Brachionus plicatilis*, cultured with baker's yeast supplemented with cuttlefish liver oil. Bull. Japan. Soc. Sci. Fish. 46, 47–50.
34 Kitajima C, Tsukashima Y, Fujita S, Watanabe T and Yone Y 1981 Relationship between uninflated swim bladder and lordotic deformity in hatchery-reared red sea bream *Pagrus major.* Bull. Japan. Soc. Sci. Fish. 47, 1289–1294.
35 Kubota T 1977 The present status of self-pollution. *In* Self-Pollution due to Mariculture. Japan. Soc. Sci. Fish. pp 9–18. Koseisha-Koseikaku, Tokyo.
36 Kurata H 1959 Preliminary report on the rearing of the herring larvae. Bull. Hokkaido Reg. Fish. Res. Lab. 2, 117–138.
37 Little G 1969 The larval development of the shrimp, *Plaemon macrodactylus* Rathfurn, reared in the laboratory and the effect of eyestalk extirpation on development. Crustacean 17, 69–87.
38 Morris R W 1956 Some aspect of the problem of rearing marine fish. Bull. Inst. Oceanogr. 108, 1–61.
39 Matsubara K 1955 Fish morphology and hierarchy. Ishizaki Shoten, Tokyo, Japan.
40 Nambu T 1977 Studies on the seed production of sea bass *Lateolabrax japonicus*-IX. The experiments on the feeding and digestive time of living food by larvae and juveniles. Annu. Rep. Kumamoto Pref. Exp. Stn. 224–229.
41 Noguchi T 1968 Natural spawning of red sea bream in an aquarium. Fish Culture 5, 8–85.
42 Norman J R 1934 A systematic monograph of the flat fish (Heterosomata). 1. Psettoidae, Bothidae, Pleuronectidae. Brit. Mus. Nat. Host., London.
43 Nozawa T, Ohara S, Kitamura S and Nakagawa K 1972 Studies on the mass-culture of rotifer. Annu. Meet. Japan. Soc. Sic. Fish. April, Tokyo, Abstr. 126.
44 Ogino C, Takeuchi L, Takeda H and Watanabe T 1979 Availability of dietary phosphorus in carp and rainbow trout. Bull. Japan. Soc. Sci. Fish. 45, 1527–1532.
45 Ohara S, Nozawa T, Kobayashi S and Kitamura S 1974 The mass-culture of rotifers with baker's yeast and their dietary value to Ayu fish. Annu. Meet. Japan. Soc. Sci. Fish. April, Tokyo, Abstr. 101.
46 Oka A, Suzuki N and Watanabe T 1980 Effects of fatty acids in rotifers on growth and fatty acid composition of larval ayu, *Plecoglossus altivelis.* Bull. Japan. Soc. Sci. Fish. 46, 1413–1418.

47 Okauchi M, Oshiro T, Kitamura S and Fukusho K 1980 Number of rotifers, *Brachionus plicatilis*, consumed daily by a larvae and juvenile of porgy, *Acanthopagrus schlegelis.* Bull. Natl. Res. Inst. Aquaculture 1, 39–45.

48 Paperna I 1978 Swimbladder and skeletal deformations in hatchery bred *Sparus aurata.* J. Fish. Biol. 12, 109–114.

49 Rilley J D 1966 Marine fish culture in Britain VII. Plaice (*Pleuronectas platessa* L.) post-larval feeding on *Artemia salina* L. nauplii and the effects of varying feeding levels. J. Cons. Perm. Inst. Explor. Mer. 30, 204–221.

50 Seikai T and Sinoda M 1981 Variation of frequency occurrence of color abnormalities in hatchery-reared flounder, *Paralichthys olivaceus*, due to duration of *Artemia salina* feed. Bull. Kyoto Ocean Fish. Sci. 5, 29–37.

51 Seikai T and Watanabe T 1983 Effect of feeding *Artemia* nauplii from different locations and natural planktons on the occurrence of abnormal coloration of hatchery-reared flounder. Annu. Meet. Japan. Soc. Sci. Fish. April, Tokyo, Abstr. 48.

52 Shelbourn J E 1964 The artificial propagation of marine fish. *In* Advances in Marine Biology, Vol. 2. Ed. F S Russel. pp 1–83. Academic Press, New York.

53 Shelbourn J E 1974 Population effects on the survival, growth and pigment of tank-reared plaice larvae. *In* Sea Fisheries Research. Ed. F R Harden Jones. pp 357–377. Eleck Science, London.

54 Slobodkin L B 1969 Aspects of the future of ecology. Biol. Sci. Tokyo 18, 16–21.

55 Takeuchi T and Watanabe T 1977 Requirement of carp for essential fatty acids. Bull. Japan. Soc. Sci. Fish. 43, 451–551.

56 Takeuchi T and Watanabe T 1977 Effect of eicosapentaenoic acid and docosahexaenoic acid in pollock liver oil on growth and fatty acid composition of rainbow trout. Bull. Japan. Soc. Sci. Fish. 43, 947–953.

57 Takeuchi T and Watanabe T 1979 Requirement for essential fatty acids of chum salmon (*Oncorhynchus keta*) in freshwater environment. Bull. Japan. Soc. Sci. Fish. 45, 1319–1323.

58 Takeuchi T, Watanabe T, Ogino C, Saito M, Nishimura K and Nose T 1981 Effects of low protein high calorie diets and deletion of trace elements from a fish meal diet on reproduction of rainbow trout. Bull. Japan. Soc. Sci. Fish. 47, 645–654.

59 Watanabe T and Ackman R G 1974 Lipids and fatty acids of American (*Crassostrea virginica*) and European flat (*Ostrea edulis*) oysters from a common habitat, and after one feeding with *Dierateria inomate* or *Isochaysis galkana.* J. Fish. Res. Board Can. 31, 403–409.

60 Watanabe T, Kitajima C, Arakawa T, Fukusho K and Fujita S 1978 Nutritional quality of rotifer, *Brachionus plicatilis*, as a living feed from the viewpoint of essential fatty acids for fish. Bull. Japan. Soc. Sci. Fish. 44, 1106–1114.

61 Watanabe T, Oowa F, Kitajima C and Fujita S 1978 Nutritional quality of Brine shrimp, *Artemia salina*, as a living feed from the viewpoint of essential fatty acids for fish. Bull. Japan. Soc. Sci. Fish. 44, 1115–1121.

62 Watanabe T, Arakawa T, Kitajima C and Fujita S 1978 Nutritional quality of living feed from the viewpoint of essential fatty acids for fish. Bull. Japan. Soc. Sci. Fish. 44, 1223–1227.

63 Watanabe T, Oowa F, Kitajima S, Fujita S and Yone Y 1979 Relationship between the dietary value of rotifers, *Brachionus plicatilis*, and their content of ω3 highly unsaturated fatty acids. Bull. Japan. Soc. Sci. Fish. 46, 35–41.

64 Watanabe T, Oowa F, Kitajima C and Fujita S 1980 Relationship between dietary value of brine shrimp *Artemia salina* and their content of ω3 highly unsaturated fatty acids. Bull. Japan. Soc. Sci. Fish. 45, 35–41.

65 Watanabe T, Ohta M, Kitajima C and Fujita S 1982 Improvement of dietary value of brine shrimp *Artemia salina* for fish larvae by feeding them on ω3 highly unsaturated fatty acids. Bull. Japan. Soc. Sci. Fish. 48, 1775–1782.

66 Watanabe T 1982 Nutritional quality of live foods used in the seed production of fish from the viewpoint of essential fatty acids. Yukagaku, 31, 77–90.

67 Watanabe T, Tamiya T, Oka A, Hirata M, Kitajima C and Fujita S 1983 Improvement of dietary value of live foods for fish larvae by feeding them on ω3 highly unsaturated fatty acids and fat soluble vitamins. Bull. Japan. Soc. Sci. Fish. 46, 499–503.
68 Watanabe T 1983 Dietary value of live foods. *In* The Rotifer *Brachionus plicatilis* – Biology and Mass Culture. Japan. Soc. Sci. Fish. 161 p. Koseisha-Kosekaku Tokyo.
69 Watanabe T, Kitajima C and Fujita S 1983 Nutritional values of live organisms used in Japan for mass propagation of fish: review. Aquaculture 34, 115–143.
70 Watanabe T, Arakawa T, Kitajima C and Fujita S 1984 Effect of nutritional quality of broodstock diets on reproduction of red sea bream. Bull. Japan. Soc. Sci. Fish. 50, 495–501.
71 Watanabe T, Itoh A, Kitajima C and Fujita S 1984 Effect of dietary protein levels on reproduction of red sea bream. Bull. Japan. Soc. Sci. Fish. 50, 1015–1022.
72 Watanabe T, Itoh A, Murakami A, Tsukashima Y, Kitajima C and Fujita S 1984 Effect of nutritional quality of diets given to broodstock on the verge of spawning on reproduction of red sea bream. Bull. Japan. Soc. Sci. Fish. 50, 1023–1028.
73 Wickins J F 1972 The food value of the brine shrimp, *Artemia salina* L., to larvae of the prawn, *Plaenon senatus* Pennant. J. Exp. Mar. Biol. Ecol. 10, 151–170.
74 Yasunaga C 1976 Amount on the feeding habit and growth of the plaice, *Parakuchthys olivaceus*, in the larval stage. Bull. Tokai Reg. Fish. Res. Lab. 68, 31–44.
75 Yone Y 1978 Essential fatty acids and lipid requirement of marine fish. *In* Dietary Lipids in Aquaculture. Japan. Soc. Sci. Fish. pp 43–59. Koseisha-Koseikaku, Tokyo.

GPSR Compliance

The European Union's (EU) General Product Safety Regulation (GPSR) is a set of rules that requires consumer products to be safe and our obligations to ensure this.

If you have any concerns about our products, you can contact us on ProductSafety@springernature.com

In case Publisher is established outside the EU, the EU authorized representative is:

Springer Nature Customer Service Center GmbH
Europaplatz 3
69115 Heidelberg, Germany

Batch number: 10816188

Printed by Printforce, the Netherlands